Oscar Zariski (24.4.1899–4.7.1986) was born in
Kobryn, Poland, and studied at the universities
of Kiev and Rome. He held positions at Rome University, John Hopkins University, the University of
Illinois and from 1947 at Harvard University.

Zariski's main fields of activity were in algebraic
geometry, algebra, algebraic function theory and
topology. His most influential results bear on
algebraic surfaces, the resolution of singularities
and the foundations of algebraic geometry over
arbitrary fields.

Classics in Mathematics

Oscar Zariski   Algebraic Surfaces

Oscar Zariski

Algebraic Surfaces

Reprint of the 1971 Edition

 Springer

Oscar Zariski

Originally published as Vol. 61 of the
Ergebnisse der Mathematik und ihrer Grenzgebiete, 2nd sequence

Mathematics Subject Classification (1991):
Primary 14JXX, 14CXX, 14EXX
Secondary 32–XX

ISBN-13: 978-3-540-58658-6   e-ISBN-13: 978-3-642-61991-5
DOI: 10.1007/978-3-642-61991-5

Photograph by kind permission of George Bergman

CIP data applied for

© Springer-Verlag Berlin Heidelberg 1995

SPIN 10485286   41/3140 - 543210–Printed on acid-free paper

Oscar Zariski

Algebraic Surfaces

Second Supplemented Edition

With Appendices by
S. S. Abhyankar, J. Lipman, and D. Mumford

Springer-Verlag Berlin Heidelberg New York
1971

The first edition was published 1935 as Band III, Heft 5
of the series Ergebnisse der Mathematik und ihrer Grenzgebiete

AMS Subject Classifications (1970):

Primary 14 J XX, 14 C XX, 14 E XX. Secondary 32-XX

ISBN-13: 978-3-540-58658-6

Preface to the First Edition

The aim of the present monograph is to give a systematic exposition of the theory of algebraic surfaces emphasizing the interrelations between the various aspects of the theory: algebro-geometric, topological and transcendental. To achieve this aim, and still remain inside the limits of the allotted space, it was necessary to confine the exposition to topics which are absolutely fundamental. The present work therefore makes no claim to completeness, but it does, however, cover most of the central points of the theory.

A presentation of the theory of surfaces, to be effective at all, must above all give the typical methods of proof used in the theory and their underlying ideas. It is especially true of algebraic geometry that in this domain the methods employed are at least as important as the results. The author has therefore avoided, as much as possible, purely formal accounts of results. The proofs given are of necessity condensed, for reasons of space, but no attempt has been made to condense them beyond the point of intelligibility. In many instances, due to exigencies of simplicity and rigor, the proofs given in the text differ, to a greater or less extent, from the proofs given in the original papers.

The author regrets that he has not been able, for the reasons outlined above, to include in his work two interesting and important developments of the theory: (I) the classification of surfaces by means of their invariants, due chiefly to ENRIQUES; (II) the theory of real algebraic surfaces, due to COMESSATTI. Fortunately, excellent and quite recent accounts of these two developments are available (I. GEPPERT, a; II. COMESSATTI, b; see "Bibliography").

Thanks are due to Dr. S. F. BARBER, National Research Fellow, and to Dr. R. J. WALKER of Princeton University, for careful reading of the manuscript and for many valuable suggestions.

Baltimore, June 12, 1934   O. ZARISKI

Preface to the Appendices

The many changes in mathematical taste and terminology and our limited knowledge of the literature have made all but impossible our task of satisfactorily updating ZARISKI's definitive account of the classical theory of algebraic surfaces. When, as the chief author of the appendices to the present edition, I sent the manuscripts to ZARISKI for his inspection, I felt acutely the deficiencies of our contributions. Is any potential reader skilled enough to be familiar with all the diverse foundations and abstract tools referred to in these appendices, patient enough to unwind the tangled relationships between old and new lines of argument, indulgent enough to forgive the gaps and gross oversimplifications caused by our parochial point of view and interested enough to want to read our hodge-podge that jumps back and forth between references and brief allusions?

The original edition of this book came at a very opportune moment. The Italian school, judged by its own standard, had completed a mature theory of algebraic surfaces. ZARISKI brought together the techniques of topology, analysis and algebraic geometry proper and put together a coherent and essentially complete account of this theory. It seems to us that the original text of the book is an excellent place for a student to learn the methods of classical geometry and to find the old results — some of them familiar results in the modern theory but still stated here clearly without the trappings of any abstract machinery. To help such a student one of the aims of our appendices is to clarify the connection between the modern and the Italian terminology and between essentially equivalent modern and Italian theorems. Chapters 2 and 3 particularly are hard for a modern reader to follow. The reason for this is that to the Italians, a surface was essentially a birational equivalence class and the models used were almost always (non-normal) surfaces in $\mathbb{P}^3$; whereas today 2 surfaces are thought to be "the same" only if they are biregularly equivalent (i.e. isomorphic as schemes), and since the models used must usually lie in at least $\mathbb{P}^5$, the basic definitions are made by appealing to methods familiar in differential and analytic geometry (charts, tangent vectors and differentials, line bundles) rather than by projective methods. Thus for instance the text uses extensively the concept of a linear system *with assigned base points*. This is a complicated notion which is forced on you if you want to set up a birationally invariant theory of linear systems.

However it seems to me that it has relatively few applications — the main one in this book is ENRIQUES' proof in Chapter 4 of the birational invariance of the arithmetic genus — and the reader should be advised that for most of Chapter 4 and Chapter 5 through 8, the concept of an ordinary linear system will suffice. On the other hand, as the reader of LIPMAN's appendix to Chapter 2 will see, it can be used together with the "ZARISKI RIEMANN surface" of the function field to give a very clear idea of the birationally-invariant geometry of the surface.

The largest change in the scope of the modern theory (as opposed to changes in foundation or in technique or generalizations of results on surfaces to varieties of arbitrary dimension) is probably the extension to characteristic p and the consequent tie-up with arithmetic problems over finite fields. In almost every appendix, we have come back repeatedly to the question: which results are still true in char. p and where and how are they proven. For instance the extension to characteristic p of the theory of PICARD varieties with its application to the theory of correspondences lead WEIL to initiate the abstract theory of abelian varieties and to seek methods in char. p of constructing auxiliary spaces which could be defined in char. 0 by analytic methods. In the case of the topology of a surface, the extension to char. p has gone hand-in-hand with a whole new technique for defining cohomology groups — the theory of étale cohomology — and we have sketched a few of its parallels with the simplicial homology of a surface in the appendix to Chapter 6. Another major advance since this book first appeared is HODGE's theory of KÄHLER manifolds and the (p, q)-decomposition of their cohomology. This is briefly outlined in the appendix to Chapter 7. Finally the theory of the resolution and structure of singularities, in the hands of ZARISKI, ABHYANKAR, HIRONAKA and others, has grown tremendously. In fact, it has grown way beyond the scope of this book and it seemed impossible in a short appendix to do justice to the results as well as the many new geometric ideas that have been introduced here. Therefore, with regret, we abandoned the project of adding an Appendix to Chapter 1.

Like the original text, we have tried to point out clearly the main gaps in the classical theory — e.g. the problem of the completeness of the characteristic systems of various algebraic families (cf. Chapter 5 and its appendix), and the "strong LEFSCHETZ theorem" (cf. the appendix to Chapter 6). The first of these is now known to be sometimes true, sometimes false, but much work remains to be done before we have a clear idea which of these is the exception and which the rule. The second has been proven in char. 0 by trancendental means but is still unknown in char. p. In the case of char. p, I would like to mention what seem to me to be the 2 most important outstanding problems in the theory of algebraic surfaces: a) to find a theory of "integrals" in char. p, leading to

DeRham cohomology groups over a suitable p-adic coefficient field; b) to prove or disprove Tate's conjecture on the existence of divisors on an algebraic surface in terms of the eigenvalues of the Frobenius acting on H^2.

Warwick, December 1970   David Mumford

Table of Contents

Chapter I

Theory and Reduction of Singularities

1. Algebraic varieties and birational transformations (KRON-
ECKER, 1; MACAULAY, a; VAN DER WAERDEN, a₂, 4, 5; BERTINI, a;
SEVERI, c, 26). Let $x_1, x_2, \ldots, x_{r+1}$ be homogeneous point coördinates
in a complex projective r-dimensional space S_r. An *algebraic variety*
V in S_r is the locus of points (x) satisfying a system of algebraic
equations,

$$(1) \qquad f_1(x_1, \ldots, x_{r+1}) = 0, \ldots, f_n(x_1, \ldots, x_{r+1}) = 0,$$

where $f_1, f_2, \ldots, f_n$ are homogeneous polynomials. If φ is a homogeneous
polynomial in the x's which vanishes at all the common zeros of $f_1, \ldots, f_n$,
i. e. at every point of V, we say briefly that φ *vanishes* ($\varphi = 0$) *on* V.
The variety V is *irreducible*, if from $\varphi\psi = 0$ on V it follows necessarily
that one at least of the polynomials φ, ψ vanishes on V. In the language
of the theory of ideals this definition can be formulated as follows:
V is irreducible if the homogeneous polynomial ideal $(f_1, \ldots, f_n)$ *(H-ideal)
is a primary ideal*[1] (MACAULAY, a, p. 33; VAN DER WAERDEN, a₂, p. 54).
From the theorem of HILBERT-NETTO (MACAULAY, a, p. 48; VAN DER
WAERDEN, a₂, p. 11) it follows then that either φ^ϱ or ψ^ϱ is a member of
$(f_1, \ldots, f_n)$, where ϱ is a convenient integer.

The set of all homogeneous polynomials which vanish on V constitutes
an ideal, and, if V is irreducible, this ideal is *prime*, i. e. if $\varphi\psi$ is a member
of the ideal, then at least one of the polynomials φ and ψ is a member
of the ideal. There is only one prime polynomial ideal associated with
a given irreducible algebraic variety (VAN DER WAERDEN, a₂, p. 53).

The application of KRONECKER's method of elimination (KRON-
ECKER, 1) toward the actual determination of the solutions of the
system of equations (1) leads to important conclusions concerning the
parametric representation of an irreducible variety V and allows us to
give a rigorous definition of *the dimension* of V (MACAULAY, a, p. 27).

---

[1] If we were dealing with non-homogeneous coördinates $x_1, x_2, \ldots, x_r$ and
with a system of non-homogeneous equations $\psi_1 = 0, \ldots, \psi_n = 0$, then from
the hypothesis that $(\psi_1, \ldots, \psi_n)$ is a primary ideal it also would follow that
V is irreducible, provided that we include in V only such points at infinity
which are limit points of points at finite distance on V.

It is then found that for a generic choice of the coördinate system in S_r an irreducible variety V admits a representation of the type:

$$(2) \quad \begin{cases} \varrho y_i = t_i \varphi', & i = 1, 2, \ldots, k+2, k \le r, \\ \varrho y_{k+2+j} = \psi_j(t_1, \ldots, t_{k+2}), & j = 1, 2, \ldots, r-k-1, \\ \varphi(t_1, t_2, \ldots, t_{k+2}) = 0, \end{cases}$$

where φ and the ψ_j's are homogeneous polynomials of like degree, ϱ is a factor of proportionality, φ is irreducible and $\varphi' = \dfrac{\partial \varphi}{\partial t_1}$. This representation gives *all* the points of V provided that we include all the limit points, which arise when φ' and the polynomials ψ_j vanish simultaneously at a zero of φ. This statement follows from the following lemma, due to RITT: *If a polynomial g does not vanish on an irreducible variety V, then any point of V, at which $g = 0$, is a limit point of such points of V, at which $g \ne 0$.* In the present case the polynomial g is φ'. A simple proof of the above lemma was given by VAN DER WAERDEN (4). The integer k which occurs in (2) is called the *dimension* of V. To indicate that V is of dimension k we denote the variety by V_k. The form of the equations (2) permits us to prove without difficulties the following theorem: *An arbitrary linear subspace S_{r-k} of S_r has always points in common with V_k (and for a generic S_{r-k} the number of common points is finite), while there exist subspaces S_{r-k-1} which have no points in common with V_k.* This property of an irreducible V_k can be used for the definition of k.

The dimension k can also be defined as follows: $k+1$ is the maximum number of variables x_i, say $x_1, x_2, \ldots, x_{k+1}$, such that no homogeneous polynomial in these $k+1$ variables is a member of the prime H-ideal associated with V.

A *neighborhood* on V of a point $P_0(x_1^0, \ldots, x_{r+1}^0)$ of V is the set of all points $P(x_1, \ldots, x_{r+1})$ on V such that $|x_i - x_i^0| < \varepsilon$, where ε is a small real positive number. An irreducible V_k is also characterized by the property that if P_0 is *any* point of V_k, any algebraic variety which contains a neighborhood of P_0 contains the whole V_k (MACAULAY, a, p. 28).

In agreement with the above definitions it can be proved that: (1) the locus of points common to two or more algebraic varieties in S_r, or, more generally, satisfying a system of algebraic equations involving the point coördinates x_i and possibly certain parameters, consists of a finite number of irreducible algebraic varieties, possibly of different dimensions; (2) a V_{r-1} in S_r, i. e., an *hypersurface* in S_r, can be represented by *one* equation between the point coördinates x_i.

The parametric representation of an irreducible V_k is frequently used as the point of departure for the definition of a V_k (see, for instance,

SEVERI, b, Introduction; d, p. 14). An irreducible V_k in S_r is then defined as the locus of points admitting a parametric representation

$$(3) \qquad \varrho x_i = \varphi_i(t_1, t_2, \ldots, t_{k+2}), \qquad i = 1, 2, \ldots, r+1,$$

where the $k + 2$ parameters t_j satisfy an *irreducible* homogeneous algebraic equation

$$(4) \qquad f(t_1, t_2, \ldots, t_{k+2}) = 0,$$

and where the φ_i's are forms of like degree such that the Jacobian matrix of the $r + 2$ polynomials φ_i and f is of rank $k + 2$ on (4). This last condition implies that there exists a $(k + 2)$-rowed determinant, which is not divisible by the irreducible polynomial f, and is necessary in order to insure that the variety be exactly of dimension k and not less. The points (x) represented by (3) and (4), together with their limit points, which arise when the polynomials φ_i vanish simultaneously at a zero of f, constitute an irreducible V_k also in the sense of the first definition. The associated prime polynomial ideal is the set of all homogeneous polynomials $\psi(x_1, \ldots, x_{r+1})$ such that $\psi(\varphi_1, \ldots, \varphi_{r+1})$ is divisible by f.

Another definition of the dimension of an irreducible variety, independent of the elimination theory, has been given by VAN DER WAERDEN (a_2, pp. 61—64).

Let V_k and V'_l be two irreducible algebraic varieties in $S_r(x_1, x_2, \ldots, x_{r+1})$ and $S'_\varrho(y_1, y_2, \ldots, y_{\varrho+1})$ respectively. An *algebraic correspondence* T between the two varieties is a correspondence between their points such that the coördinates of corresponding points (x) and (y) $(= T(x))$ satisfy a system of algebraic equations

$$(5) \qquad \begin{cases} \psi_1(x_1, \ldots, x_{r+1}; y_1, \ldots, y_{\varrho+1}) = 0, \\ \psi_2(x_1, \ldots, x_{r+1}; y_1, \ldots, y_{\varrho+1}) = 0, \ldots, \end{cases}$$

where $\psi_1, \psi_2, \ldots$ are polynomials homogeneous in each set of variables: $x_1, \ldots, x_{r+1}; y_1, \ldots, y_{\varrho+1}$. The possibility of the correspondence not being defined for all points of V_k is not excluded, in consequence of the fact that for generic points (x) on V_k the equations (5) may be inconsistent on V'_l. In this case the points (x) of V_k for which homologous points $(y) = T(x)$ on V'_l exist, lie on a finite number of algebraic varieties on V_k of dimension $< k$. Similar considerations apply to V'_l and to the inverse correspondence T^{-1}.

In various questions concerning algebraic correspondences between two varieties V_k and V'_l it is found useful to consider the *variety of pairs of points* of V_k and V'_l. We put

$$(5') \qquad \sigma X_{ij} = x_i y_j, \qquad i = 1, 2, \ldots, r+1; \, j = 1, 2, \ldots, \varrho+1,$$

where σ is a factor of proportionality, and we interpret the X_{ij}'s as homogeneous point coördinates in a complex projective space S_N, of dimension $N = (r + 1)(\varrho + 1) - 1$. If the x's and y's are considered as arbitrary independent parameters, the equations (5') give the parametric representation of an algebraic variety $\overline{V}$ immersed in S_N, referred to as a *variety of* SEGRE (see SEVERI, d, p. 69). This variety is irreducible, is of dimension $r + \varrho$, and its points are in $(1, 1)$ correspondence with the pairs of points (x) and (y) of S_r and S_ϱ. If we now restrict the point (x) to vary on V_k and the point (y) to vary on V'_l, then the equations (5') together with the parametric equations of V_k and V'_l define an irreducible algebraic variety V_{k+l} immersed in $\overline{V}$, called the variety of pairs of points of V_k and V'_l, since its points are in $(1, 1)$ correspondence with the pairs of points (x) and (y) of V_k and V'_l.

The pairs of homologous points (x) and (y) of the above correspondence T are mapped upon an algebraic variety V immersed in V_{k+l}. Conversely, any such variety V on V_{k+l} defines an algebraic correspondence between V_k and V'_l. The correspondence T is said to be *irreducible*, if V is irreducible.

The notion of an algebraic correspondence is of fundamental importance in algebraic geometry. Some general aspects of the theory of correspondences will be dealt with in VI, 13 and in Appendix B. For detailed reports see LEFSCHETZ, c and BERZOLARI, a. The algebraic foundations of the theory of correspondences have been investigated recently by VAN DER WAERDEN (5), who gave complete algebraic proofs of most of the statements made below.

Let us assume that the following conditions are satisfied: (I) T is irreducible; (II) T is defined for generic points of V_k and T^{-1} is defined for generic points of V_l; (III) to a *generic* point (x) of V_k there corresponds one and only one point of V'_l. The elimination method of KRONECKER applied to the system of equations (5) and the equations of V_k and V'_l always enables us to decide whether these conditions are or are not satisfied. In view of the irreducibility of T the condition (III) has a sense, and is satisfied if there exists at least *one* point on V_k to which there corresponds a unique point of V'_l. If the above conditions are satisfied, then the elimination method of KRONECKER permits us, by using the equations of V_k and V'_l, to solve the equations (5) with respect to $y_1, \ldots, y_{\varrho+1}$ and to express the y's rationally in terms of the x's:

$$(6) \qquad \sigma y_j = \varphi_j(x_1, x_2, \ldots, x_{r+1}), \qquad j = 1, 2, \ldots, \varrho + 1,$$

where $\varphi_1, \ldots, \varphi_{\varrho+1}$ are homogeneous polynomials of like degree. The correspondence T is then called a *rational transformation*, and V'_l is a *rational transform* of V_k. It follows from (6) that necessarily $l \leq k$. If $l = k$, then to a generic point (y) of V'_k there corresponds on V_k a finite number μ of points (x) [1]. The correspondence between V_k and V'_k is $(1, \mu)$.

We observe that the definition of an irreducible V_k as given above, by means of its parametric representation (3), simply implies that V_k is a rational transform (with μ finite) of the algebraic hypersurface (4) in S_{k+1}.

If $\mu = 1$, then the transformation is *birational*. It will then be possible to invert the equations (6) and to express the x_i's rationally in terms of $y_1, \ldots, y_{\varrho+1}$ (as long as the point (y) remains on V'_k!):

$$(6') \qquad \sigma' x_i = \psi_i(y_1, y_2, \ldots, y_{\varrho+1}), \qquad i = 1, 2, \ldots, r+1.$$

If at a given point (x) of V_k all the polynomials φ_j vanish, then the corresponding point (y) of V'_k is *a priori* undetermined. From continuity considerations it is then seen that to the point (x) there may correspond on V'_k infinitely many points constituting one or more algebraic varieties of dimensions $\leq k - 1$. The point (x) is called a *fundamental point* of the transformation, and the corresponding varieties on V'_k are also called *fundamental varieties*. If $k > 2$, the fundamental points on V_k may fill up algebraic varieties, and the situation may be somewhat complicated. However, the following general theorem holds: If T *has fundamental points on* V_k, *then the totality of all point pairs* (x, y) *of* T, *in which* x *is any fundamental point on* V_k, *consists of a finite number of algebraic subvarieties of* V, *all of dimension* $k - 1$ (for the proof see, for instance, VAN DER WAERDEN, 5). For $k = 2$ it follows that a birational transformation between two algebraic surfaces possesses on either surface at most a finite number of fundamental points. Further properties of the fundamental elements of a birational transformation between two surfaces will be given in II, 6.

A CREMONA transformation is a birational transformation between two linear (projective) spaces.

In the present work we shall be concerned mainly with properties of algebraic surfaces which are invariant under birational transformations. The study of these properties constitutes what is referred to as *the geometry on an algebraic surface*. The birational theory of algebraic curves will be occasionally recalled but in the main will be supposed known.

2. Singularities of plane algebraic curves. In this section we shall report briefly on the theory of singularities of algebraic curves as developed in ENRIQUES-CHISINI, a_2. The conclusions of this theory,

---

[1] More generally, if to a generic point of V_k (or of V'_l) there corresponds by T (or by T^{-1}) an algebraic variety of dimension $a \geq 0$ ($b \geq 0$) on V'_l (or on V_k), then it can. be proved with complete rigor (see, for instance, VAN DER WAERDEN, 5) that $k + a = l + b = q$, where q is the dimension of the *irreducible* variety V (*the principle of counting constants*).

especially the rigorous development of the notion of infinitely near multiple points, are necessary as an introduction to the resolution of singularities of surfaces and also provide a rigorous foundation for the general theory of linear systems on an algebraic surface, given in Chapter II.

If a singular point of a plane algebraic curve is reduced by quadratic transformations into a certain number of ordinary multiple points, the intermediate steps of the reduction yield a basis for the analysis of the given singularity and for the definition of infinitely near multiple points. This is essentially NOETHER's method. The method of EN-RIQUES is more direct and is based on the computation of the multiplicity of intersection of two algebraic (or algebroid) branches given by their PUISEUX expansions.

Let $y = a x + a_1 x^{\frac{\nu + \nu'}{\nu}} + a_2 x^{\frac{\nu + \nu' + \nu''}{\nu}} + \cdots$ $(a \neq 0, \nu, \nu', \nu'', \ldots$ —positive integers) be a given branch γ of order ν in the (x, y) plane, possessing a ν-fold point at the origin $O (x = y = 0)$, and let us cut it by another arbitrary branch $\bar{\gamma}$ of origin O:

$$\bar{y} = b x + b_1 x^{\frac{\mu + \mu'}{\mu}} + b_2 x^{\frac{\mu + \mu' + \mu''}{\mu}} + \cdots.$$

The multiplicity of intersection, λ, of the two branches at O is the exponent of the lowest term in x in the power series $\prod\limits_{i=1}^{\nu} \prod\limits_{j=1}^{\mu} (y_i - \bar{y}_j)$, where $y_1, \ldots, y_\nu$ are the ν determinations of y and $\bar{y}_1, \ldots, \bar{y}_\mu$ are the μ determinations of $\bar{y}$. We find that if $a \neq b$, then $\lambda = \nu \mu =$ product of the multiplicities of the common multiple point O. If $a = b$, the two branches have the same principal tangent $y = a x$, and we find that $\lambda \geq \nu \mu + \nu_1 \mu_1$, where $\nu_1 = \nu$ or $\nu_1 = \nu'$ according as $\nu' \geq \nu$ or $\nu' < \nu$, and μ_1 is defined in a similar manner for the branch $\bar{\gamma}$. Moreover, there exist branches $\bar{\gamma}$ for which $\lambda = \nu \mu + \nu_1 \mu_1$, for instance the linear branch $(\mu = 1) y = a x + b_1 x^2 + \cdots$, with the condition: $b_1 \neq 0$, if $\nu' > \nu$, $b_1 \neq a_1$, if $\nu' = \nu$ (b_1-arbitrary, if $\nu' < \nu$). We express this fact by saying that *the branch γ possesses infinitely near O a ν_1-fold point O' in the direction of the line $y = a x$*. The branch $\bar{\gamma}$, possessing the same principal tangent as γ, passes through O' with the multiplicity μ_1, and $\nu \mu + \nu_1 \mu_1$ is the amount contributed to λ by the common multiple points O and O'. We see that this amount is the same as if we dealt with two multiple points at a finite distance from each other. If $\lambda > \nu \mu + \nu_1 \mu_1$, the two branches have another point O'' in common, infinitely near to O' and in the second order neighborhood of O. This point O'' and, more generally, a succession of infinitely near points $O, O', O'', \ldots, O^{(s)}, O^{(s+1)}$ on the branch γ, and their multiplicities for γ, may now be defined by induction as follows:

Let us suppose that we have already introduced the succession of points $O, O', O'', \ldots, O^{(s)}$, that their multiplicities $\nu, \nu_1, \nu_2, \ldots, \nu_s$ for γ

have been determined, and that it is known under what conditions and with what multiplicities any other branch $\bar{\gamma}$ passes through these points. We also suppose that these incidence conditions imply that if $\bar{\gamma}$ passes through $O, O', O'', \ldots, O^{(s)}$ with the multiplicities $\mu, \mu_1, \ldots, \mu_s$, then the two branches $\gamma, \bar{\gamma}$ have at O an intersection multiplicity $\lambda \geqq \nu\mu + \nu_1\mu_1 + \cdots + \nu_s\mu_s$, and that there exist branches $\bar{\gamma}$ for which the equality holds. We say *that $\bar{\gamma}$ passes through the point $O^{(s+1)}$, which follows $O^{(s)}$ on γ, if $\lambda > \nu\mu + \nu_1\mu_1 + \cdots + \nu_s\mu_s$. If ν_{s+1} is the minimum value of the positive difference $\lambda - (\nu\mu + \nu_1\mu_1 + \cdots + \nu_s\mu_s)$, we say that $O^{(s+1)}$ is a ν_{s+1}-fold point of γ* (this minimum corresponds to branches $\bar{\gamma}$ passing simply through $O^{(s+1)}$). It is then not difficult to prove the above mentioned implication of the incidence conditions for any two branches $\bar{\gamma}_1$ and $\bar{\gamma}_2$ passing through the points $O, O', \ldots, O^{(s)}$, $O^{(s+1)}$.

The actual carrying out of the steps required by the above definitions reveals some interesting features of the composition of the given singularity, which are closely associated with the existence of certain terms of the Puiseux expansion, called *characteristic terms*. The importance of these terms has been pointed out by Smith and Halphen. Let ϱ_i be the g.c.d. of $\nu, \nu', \ldots, \nu^{(i)}$ and let m be the smallest value of i for which $\varrho_i = 1$. A characteristic term is a term $a_i x^{\frac{\nu + \nu' + \cdots + \nu^{(i)}}{\nu}}$ such that $\varrho_i < \varrho_{i-1}$ and $i \leqq m$. It is easy to verify that the exponents of the characteristic terms are invariant under analytic transformations on the two variables x and y, regular at O. We rewrite the Puiseux expansion of the branch γ so as to display the characteristic terms:

$$(7) \quad \begin{cases} y = \sum_{i=1}^{k_1} a_{1,i} x^i + b_1 x^{\frac{m_1}{n_1}} + \sum_{i=1}^{k_2} a_{2,i} x^{\frac{m_1+i}{n_1}} + b_2 x^{\frac{m_2}{n_1 n_2}} + \cdots \\ \quad + \sum_{i=1}^{k_g} a_{g,i} x^{\frac{m_{g-1}+i}{n_1 n_2 \ldots n_{g-1}}} + b_g x^{\frac{m_g}{n_1 n_2 \ldots n_g}} + \sum_{i=1}^{+\infty} c_i x^{\frac{m_g+i}{\nu}}, \end{cases}$$

with the following significance of the symbols: $n_j > 1$, $k_j = \left[\dfrac{m_j - m_{j-1} n_j}{n_j}\right]$ $(m_0 = 0)$, $b_j \neq 0$, m_j and n_j are relatively prime $(j = 1, 2, \ldots, g)$, and $n_1 n_2 \ldots n_g = \nu$. The pairs of numbers (m_j, n_j) are called the *characteristic pairs*. The g terms with the coefficients b_j are the characteristic terms. Some of the coefficients $a_{j,i}$ or c_i may also vanish. Let $\nu_j = n_j n_{j+1} \ldots n_g$ $(\nu_1 = \nu)$ and let us develop $\dfrac{(m_j - m_{j-1} n_j)\nu_{j+1}}{\nu_j} = \dfrac{m_j - m_{j-1} n_j}{n_j}$ $(\nu_{g+1} = 1)$ into a continued fraction $(k_j, k_{j1}, \ldots, k_{j,q_j})$:

$$(m_j - m_{j-1} n_j)\nu_{j+1} = k_j \nu_j + \nu_{j1},$$
$$\nu_j = k_{j1}\nu_{j1} + \nu_{j2},$$
$$\cdots \cdots \cdots \cdots$$
$$\nu_{j,q_j-1} = k_{j,q_j}\nu_{j,q_j},$$

where $\nu_{j,q_j} = \nu_{j+1}$.

With these notations the composition of the given singularity, as to the multiplicities and the positions of the successive, infinitely near, multiple points in the order in which they follow each other on the branch γ, can now be described. The branch γ possesses infinitely near the ν-fold point O:

(a_1). $k_1 - 1$ other ν_1-fold points ($\nu_1 = \nu$) and a ν_{11}-fold point on the linear branch (the position of a succession of infinitely near points will always be indicated by a branch of the lowest order to which they belong):

$$y = \sum_{i=1}^{k_1} a_{1i} x^i.$$

(b_1). $k_{11} - 1$ other ν_{11}-fold points and $k_{1j} \nu_{1j}$-fold points ($j = 2, \ldots, q_1$). The first $k_{11} + k_{12} + \cdots + k_{1j} + t$ of these points, beginning with the last (ν_{11}-fold) point of the set (a_1), ($0 \le j < q_1$; $2 \le t \le k_{1,j+1} + 1$, if $j < q_1 - 1$; $2 \le t \le k_{1,\,q_1}$, if $j = q_1 - 1$) lie on the branch

$$y = \sum_{i=1}^{k_1} a_{1i} x^i + d x^{\frac{m}{n}},$$

where m/n is the continued fraction $(k_1, k_{11}, \ldots, k_{1j}, t)$ *and d is arbitrary*. It should be noted that this branch passes simply through the last t points.

(a_2). $k_2 \nu_2$-fold points ($\nu_2 = \nu_{1q_1}$) and one ν_{21}-fold point belonging to the branch:

$$y = \sum_{i=1}^{k_1} a_{1i} x^i + b_1 x^{\frac{m_1}{n_1}} + \sum_{i=1}^{k_1} a_{2i} x^{\frac{m_1+i}{n_1}}.$$

(b_2). $k_{21} - 1$ other ν_{21}-fold points and $k_{2j} \nu_{2j}$-fold points ($j = 2, 3, \ldots, q_2$). The first $k_{21} + \cdots + k_{2j} + t$ of these points, beginning with the last (ν_{21}-fold) point of the set (a_2) ($0 \le j < q_2$; $2 \le t \le k_{2,j+1} + 1$, if $j < q_2 - 1$; $2 \le t \le k_{2,\,q_2}$, if $j = q_2 - 1$) lie on the branch

$$y = \sum_{i=1}^{k_1} a_{1i} x^i + b_1 x^{\frac{m_1}{n_1}} + \sum_{i=1}^{k_1} a_{2i} x^{\frac{m_1+i}{n_1}} + d x^{\frac{m}{n_1 n}},$$

where d is arbitrary, and $\dfrac{m - n m_1}{n} = (k_2, k_{21}, \ldots, k_{2j}, t)$. Etc. The set ($a_g$) will consist of k_g points of multiplicity $\nu_g = n_g$ and of one ν_{g1}-fold point, while the set (b_g) will end up with *simple* points. All the successive points will also be simple and their positions are dependent on the coefficients c_i of the expansion (7).

The positions of the $k_j + 1$ points of the set (a_j) vary as the coefficients a_{ji} vary. On the contrary the points of a set (b_j) do not depend, as to position, on any variable parameter, but only on the existence of the corresponding characteristic pair (m_j, n_j). For this reason the points of a set (a_j) are called *free points*, while the points of a set (b_j) are called *satellites*. The points of γ which follow the last point of the set (b_q) are all simple and free. From the preceding considerations it is easy to deduce the following characteristic property of a satellite $O^{(s)}$ belonging to a succession $O, O', \ldots, O^{(s-1)}, O^{(s)}$ of infinitely near points of a branch of order ν: if $O^{(s-1)}$ is a free point, then any branch of order $\geq \nu$ which passes through $O, O', \ldots, O^{(s-1)}$ necessarily passes through $O^{(s)}$; if $O^{(s-1)}$ is also a satellite, then the branches of order $\geq \nu$ passing through $O, O', \ldots, O^{(s-1)}$ divide into two classes, the branches of one class passing through $O^{(s)}$, and the branches of the second class passing through another fixed point $O^{(s)}$, itself a satellite following $O^{(s-1)}$. For instance, if we suppose that $O^{(s-1)}$ belongs to the set (b_1), we may assume that is the last point of a set (b_1) relative to a conveniently chosen

branch: $y = \sum_{i=1}^{k_1} a_{1i} x^i + b_1 x^{\frac{m_1}{n_1}} + \cdots$, where $\frac{m_1}{n_1} = (k_1, k_{11}, \ldots, k_{1q}), k_{1q} \geq 2$.

Then the point $O^{(s)}$ is the last point of the set (b_1) relative to the branch

$y = \sum_{i=1}^{k_1} a_{1i} x^i + c_1 x^{\frac{m_1}{\bar{n}_1}} + \cdots$, where $\frac{\bar{m}_1}{\bar{n}_1} = (k_1, k_{11}, \ldots, k_{1q} + 1)$, and $O^{(s)}_1$

is the last point of the set (b_1) relative to the analogous branch, where

now $\frac{\bar{m}_1}{\bar{n}_1} = (k_1, k_{11}, \ldots, k_{1q} - 1, 2)$.

The points of the set (a_1) lie on a linear branch, but no linear branch exists which passes through points following the set (a_1). Hence a necessary and sufficient condition that a given succession of infinitely near points lie on a linear branch is that all the points of the succession be free. The existence of infinitely near points not lying on linear branches was pointed out by C. SEGRE (3).

The passage from singularities composed of one branch to singularities composed of several branches of origin O is immediate. The points infinitely near and in the neighborhoods of orders $1, 2, \ldots$ of O are the points following O on the different branches. If a point $O^{(s)}$, in the ith neighborhood of O, is common to several branches, then the multiplicity of $O^{(s)}$ for the given singularity is the sum of its multiplicities for the branches passing through it. Instead of a simple succession of points, one in each neighborhood of a given order i, as was the case of one branch, we now have a constellation of points, of which more than one may belong to a neighborhood of a given order i. A point $O^{(s)}$ in the ith neighborhood may be followed in the $(i+1)^{\text{th}}$ neighborhood (1) by any number of free points; (2) by not more than one satellite, if $O^{(s)}$ is a free point; (3) by not more than two satellites, if $O^{(s)}$ is itself a satellite.

An important notion, especially useful in a general theory of linear systems of curves having arbitrary base singularities (see II, 2), is that of the set of *proximate points (punti prossimi) of a given point* $O^{(s)}$ in a succession or constellation of infinitely near points. We consider first the case of a single branch γ. Let ν_s be the multiplicity of $O^{(s)}$ on γ and let v_{s+1} be the multiplicity of the point $O^{(s+1)}$ which immediately follows $O^{(s)}$. Necessarily then $v_{s+1} \leqq v_s$. If $v_{s+1} = v_s$ the set of proximate points of $O^{(s)}$ will consist, by definition, of the single point $O^{(s+1)}$. If $v_{s+1} < v_s$ let $v_s = h v_{s+1} + v'$ $(0 \leqq v' < v_{s+1})$. From the preceding analysis of the composition of the singularity of a given branch, it follows immediately that $O^{(s)}$ is followed on γ by h successive v_{s+1}-fold points $O^{(s+1)}$, $O^{(s+2)}$, ..., $O^{(s+h)}$ and by one v'-fold point $O^{(s+h+1)}$ if $v' > 0$. These $h + 1$ points are the proximate points of $O^{(s)}$. If $v' = 0$, the point $O^{(s+h+1)}$ naturally disappears, and the set of proximate points of $O^{(s)}$ consists of h points. In the case of a singularity composed of several branches the set of proximate points of a given point $O^{(s)}$ will consist, by definition, of all the proximate points of $O^{(s)}$ on the different branches passing through it. In either case the following fundamental property of proximate points holds: *the multiplicity of $O^{(s)}$ is equal to the sum of the multiplicities of its proximate points.*

It follows easily from the preceding considerations that the above definition of proximate points is independent of the particular branch γ used in the definition. In exact terms, *if $O_{i+\alpha}$ is in the set of proximate points of O_i on a given branch γ, then it belongs also to the set of proximate points of O_i on any other branch passing through $O_{i+\alpha}$* (and hence also through O_i).

We add that the scheme of a given singularity, as analyzed in this section, can be graphically represented by a very convenient diagram. See the quoted treatise of ENRIQUES-CHISINI.

The multiplicity λ of intersection of two curves at a common multiple point O, of multiplicities s and σ respectively, is given by NOETHER's formula: $\lambda = s\sigma + \sum s_i \sigma_i + \sum s_{ij} \sigma_{ij} + \cdots$, where the summations are extended to all the common multiple points of the two curves, infinitely near and in the neighborhoods of the 1st, the 2nd, ... order of O.

The application of this formula to the polar curves of a given curve f gives the diminution of the class and of the genus of f due to its singularity at O. It is found that the diminution of the genus is equal to $\frac{s(s-1)}{2} + \sum \frac{s_i(s_i-1)}{2} + \sum \frac{s_{ij}(s_{ij}-1)}{2} + \cdots$, as in the case of ordinary multiple points at a finite distance from each other.

3. Singularities of space algebraic curves (C. SEGRE, 3; ENRIQUES-CHISINI, a_2, pp. 545—577). The parametric representation by means of power series of an algebraic (or algebroid) branch γ of order v in S_3 takes its canonical form, if the origin O of the branch, its principal tangent and its osculating plane are assumed as the origin of the coördinates x, y, z, the y-axis and the plane $z = 0$ respectively. The parametric representation of γ is then the following: $x = t^v$, $y = a t^{v+\mu} + \cdots$, $z = b t^{v+\mu+\lambda} + \cdots$ $(a \neq 0, b \neq 0, \mu \geqq 1, \lambda \geqq 1)$. The integers μ and λ are projective characters of γ, called *first* and *second class* (ENRIQUES; also *first* and *second rank* [BERTINI], or *class* and *rank* [HALPHEN]).

The space transformations which are commonly used to reduce the singularity at O and hence also to define its composition are the *quadratic transformations of the 1st kind*, given by equations of the following type:

$$x' = \frac{\varphi_1}{\varphi_4},\, y' = \frac{\varphi_2}{\varphi_4},\, z' = \frac{\varphi_3}{\varphi_4}, \quad \text{where} \quad \varphi_i = \varphi_i(x, y, z) = 0 \quad (i = 1, 2, 3, 4)$$

are 4 independent quadrics passing through O and through a conic K. Instead of quadratic transformations it is possible to use, with the same effect, any CREMONA transformation with O as a simple isolated base point, or more generally, any analytic transformation (φ_i — an analytic function, regular at O), such that $\varphi_i(0, 0, 0) = 0$ and that the Jacobian matrix of the 4 functions φ_i be of rank 3 at O (transformations *locally quadratic*). In either case to a direction $x : y : z = a : b : c$ about the fundamental point O corresponds in the space S_3' the point

$$x' = \frac{a_1 a + b_1 b + c_1 c}{a_4 a + b_4 b + c_4 c},\; y' = \cdots,\; z' = \cdots, \quad \text{where} \quad \varphi_i(x, y, z) = a_i x + b_i y$$

$+ c_i z + \cdots$. This point obviously varies in a plane π'.

The transform of γ by a transformation T locally quadratic at O will be an algebroid branch γ', whose origin A' is the point of the fundamental plane π' corresponding to the direction of the principal tangent of γ and whose order $v_1 \leq v$ is equal to the smaller of the two numbers v and μ. We say that γ *possesses a v_1-fold point O_1 infinitely near O* (in the direction of the principal tangent).

If $v_1 > 1$, we transform γ' by a transformation locally quadratic at A' and we obtain an algebroid branch γ'' in the space S_3'', whose origin A'' is in the fundamental plane π'' and whose order is $v_2 \leq v_1$. *We then say that O_1 is followed on γ by a v_2-fold point O_2.*

Continuing in this manner, we define the (non-increasing) multiplicities $v, v_1, v_2, \ldots, v_s, \ldots$ of the successive points $O, O_1, O_2, \ldots, O_s$ on γ. It is a simple matter to show that the numbers v_i are independent of the particular successive transformations used and that after a finite number of steps only simple points are obtained.

Two branches γ and δ of origin O have the successive points $O, O_1, \ldots, O_s$ in common, if, for $i = 1, 2, \ldots, s$, their transforms $\gamma^{(i)}, \delta^{(i)}$ by one and the same sequence of transformations have the same origin $A^{(i)}$. This incidence condition completes the definition of the infinitely near points O_i as regards their position in space.

The notions of free points, satellites, proximate points etc., can be generalized to space singularities (ENRIQUES-CHISINI, a_2), but give rise to new and more complicated relations, completely analyzed only in the simplest cases.

In concluding let us observe that *the composition of a generic plane projection of a singular branch γ in space is in general not identical with* *the composition of γ itself, as defined by the above method of successive quadratic transformations.* Most authors seem to have missed the point in the belief that the compositions of the branch γ and of its projection are identical, as a matter of intuition. In ENRIQUES-CHISINI, a_2, p. 559, the following proof is given. It is permissible to use special space quadratic transformations with a degenerate fundamental conic K. Let T be such a transformation, and let Q, a generic point of S_3, be the double point of K. Let similarly Q' be the double point of the fundamental conic K' in the space S_3'. It is shown then that the branch γ_1 obtained by projecting γ from Q into a plane in S_3 and the branch γ_1' obtained by projecting $\gamma' = T(\gamma)$ from Q' into a plane in S_3' can be transformed into each other by a quadratic transformation between the two planes (completely defined by T), possessing O, the origin of γ, as a fundamental point. The objection to this proof is that either for T or for some of the successive transformations, the point Q' may (and in general does) assume some special position with respect to γ' (for instance may lie in the osculating plane of γ'), so that γ_1' is not a *generic* projection of γ'.

If γ is given in its canonical representation: $x = t^\nu$, $y = f(t) = a t^{\nu+\mu} + \cdots$, $z = \varphi(t) = b t^{\nu+\mu+\lambda} + \cdots$, the multiplicities of the successive multiple points on γ depend in general on the exponents of *both* series $f(t)$ and $\varphi(t)$. On the other hand if we project γ into the (x, y) plane from the point at infinity on the z-axis (which we may suppose to be a generic point of the space), we obtain the branch $x = t^\nu$, $y = f(t)$, whose composition depends only on the exponents of the series $f(t)$. The following example will serve as an illustration: $x = t^4$, $y = t^6 + t^7$, $z = t^7$. The multiplicities $\nu, \nu_1, \nu_2, \ldots$ of the successive points of γ are: $\nu = 4$, $\nu_1 = 2$, $\nu_2 = \nu_3 = \cdots = 1$. The projection γ_1 of γ into the plane (x, y) from a generic point (a, b, c) of the space is the following:
$x = t^4 - \dfrac{a}{c} t^7 + \cdots$, $y = t^6 - \dfrac{b-c}{c} t^7 + \cdots$, and the composition of γ_1 is: $\nu = 4$, $\nu_1 = \nu_2 = 2$, $\nu_3 = \nu_4 = \cdots = 1$.

The composition indices are in general increased by projection. If γ can be imbedded in a surface element having at O a *simple* point, then it can be easily shown that the composition is not altered by projection. Perhaps the question is related to the minimum multiplicity and the type of singularity which an algebroid surface element passing through γ must possess at O. This multiplicity is clearly an analytical invariant of the branch.

4. Topological classification of singularities. BRAUNER (1) gave a topological classification of algebroid singularities in the plane which completes from an entirely new point of view the classical algebro-geometric theory. This classification establishes an interesting connection between the theory of singularities and the theory of knots in the real three-space.

The neighborhood of an algebroid singularity in the plane, of origin $O(x = y = 0)$, is a two-dimensional complex K_2 immersed in the space of 4 real dimensions of the complex variables x and y. The intersection of K_2 with the boundary of a small 4-cell, $|x| <$ const., $|y| <$ const., is a system of one-dimensional circuits, equal in number to the number of branches constituting the singularity. By a stereographic projection of the boundary of the 4-cell (3-sphere) on to the ordinary 3-space, we obtain in this space a system of knots. While the linking coefficients of this system depend on the orders of contact between the different branches, the individual knots of the system can be described as follows. If Γ is any knot, a knot Γ' drawn on the surface of a torus of axis Γ shall be called a *tubular knot (Schlauchknot) of axis* Γ. If Γ' makes μ turns along the meridians and ν turns along the parallels of the torus (meridians being the circuits on the torus which are ~ 0 in the inner space of the torus, while the parallels are those whose linking number with Γ is zero), the numbers μ and ν will be referred to as the *indices* of Γ'. If Γ is unknotted, then Γ' is a *torusknot*. We define the symbol $\Gamma_i^{\mu_1, \nu_1; \ldots; \mu_{i-1}, \nu_{i-1}; \mu_i, \nu_i}$ by induction, as follows:

1. $\Gamma_1^{\mu_1, \nu_1}$ is a torusknot of indices μ_1, ν_1;
2. $\Gamma_i^{\mu_1, \nu_1; \ldots; \mu_{i-1}, \nu_{i-1}; \mu_i, \nu_i}$ is a tubular knot of indices μ_i, ν_i, whose axis is $\Gamma_{i-1}^{\mu_1, \nu_1; \ldots; \mu_{i-1}, \nu_{i-1}}$.

Brauner proves that *the knot which corresponds to the branch given by the expansion* (7) *of section* 2 *is a* $\Gamma_g^{m_1, n_1; \ldots; m_g, n_g}$.

The *fundamental group* of this knot is generated by $g + 1$ elements $P_1, Q_1, \ldots, Q_g$, satisfying the following generating relations [Brauner, l. c., also Kähler (1), Zariski (6)]: $Q_i^{n_i} = P_i^{\overline{m}_i} Q_{i-1}^{n_i} (i = 1, 2, \ldots, g, Q_0 = 1)$ where $P_{i+1} P_i^{y_i} Q_{i-1}^{n_i} {}^{z_i} = Q_i^{z_i}$ $(i = 1, 2, \ldots, g - 1)$. Here x_i and y_i are integers such that $x_i \overline{m}_i + y_i n_i = 1$, and $\overline{m}_i = m_i - m_{i-1} n_i$.

Burau (1) and Zariski (6) have proved independently that the characteristic pairs (m_i, n_i) are topological invariants of the above knot. It follows that two algebroid singularities which are distinct with respect to analytical transformations are also topologically distinct.

5. Singularities of algebraic surfaces. The foundations of a geometric theory of singularities of algebraic surfaces have been laid down by C. Segre in his important paper (3). Using the method of successive quadratic transformations of the 1st kind he arrives at a theoretically satisfactory definition of infinitely near multiple points on a surface in the following manner.

Let
$$(8) \qquad f(x, y, z) = f_s(x, y, z) + f_{s+1}(x, y, z) + \cdots = 0$$
be the equation of a surface f having an s-fold point at the origin $O(x = y = z = 0)$. Here f_i denotes a homogeneous polynomial of degree i in x, y and z and $f_s = 0$ is the equation of the tangent cone at O. We apply the special quadratic transformation: $x = x'z', y = y'z'$,

$z = z'$, whose fundamental elements in S_3 are the point O, the line at infinity in the plane $z = 0$ (counted twice) and the plane at infinity; and in S_3' the point at infinity of the z'-axis, the line at infinity (counted twice) of the plane $z' = 0$ and the plane π', $z' = 0$. The equation of the transformed surface f' is the following:

$$f'(x', y', z') = f_s(x', y', 1) + z' f_{s+1}(x', y', 1) + \cdots = 0.$$

To the points infinitely near O on f corresponds the curve of intersection of f' with the fundamental plane $z' = 0$, outside the fundamental line at infinity, i. e. the curve of order s, $f_s(x', y', 1) = 0$. The points of this curve correspond to the directions about O of the elements of the tangent cone at O. If a point O' of $f_s = 0$, $z' = 0$, which corresponds to a tangential direction t about O, is s'-fold for f', then we say that *the surface f possesses an s'-fold point O' infinitely near O in the direction t*. Evidently, the multiplicity of the point O' for f' cannot be greater than its multiplicity for the curve $f_s(x', y', 1) = 0$, *hence $s' \leq s$*. For the equality to hold it is necessary (but not sufficient) that this curve should consist of s distinct or coincident lines through O'.

   If $f_s = 0$ is reducible, and if one of its components of order μ consists of s'-fold points of the surface f', then we say that *f possesses infinitely near the s-fold point O an (infinitesimal) s'-fold curve of order μ*. A second special quadratic transformation with O' as fundamental point (but otherwise generic with respect to the tangent cone at O') will determine the multiplicity s'' of any point O'' (and virtually of any curve) which follows on f the point O'. The multiplicities $s, s', s'', \ldots, s^{(k)}$ of a succession of points $O, O', O'', \ldots, O^{(k)}$ form a sequence of non-increasing integers. The position of the points $O^{(i)}$ can be assigned by a branch γ of a space algebraic curve, such that after k quadratic transformation the transformed branch $\gamma^{(k)}$ will pass simply through $O^{(k)}$ and will not touch the fundamental plane $\pi^{(k)}$. It is a simple matter to show that, if any branch γ passes through $O, O', \ldots, O^{(k)}$ with the multiplicities $\mu, \mu', \ldots, \mu^{(k)}$ and does not pass through points of f which follow $O^{(k)}$, then the multiplicity of intersection of f with γ at O is $s\mu + s'\mu' + \cdots + s^{(k)}\mu^{(k)}$.

   An important question arises: is it possible to assign a finite upper limit for the number of points in a succession $O, O', \ldots, O^{(k)}$, for which $s^{(k)} > 1$? This question is treated in DEL-PEZZO (1), KOBB (1), and especially B. LEVI (1, 3), whose proof is entirely rigorous and who arrives at the final correct result. The question is evidently equivalent to the following one: *under what conditions, if any, is a succession of infinitely near points, all of the same multiplicity s, necessarily finite?* LEVI proves that the condition is that *no two or more (necessarily consecutive) points of the succession should lie together on an s-fold curve through O or on an s-fold line following O (immediately or not)*. His method of proof consists in showing that if the number of points of a succession satisfying the above condition is sufficiently large, then the intersection of f and of its pth polar surface with respect to a generic pole (where p is a conveniently chosen integer >0 and $\leqq s - 1$) passes through O with one or more branches, *one of which passes through all the points of the succession with multiplicities* >1.

A trivial case of an infinite succession of s-fold points is the one in which all the points $O^{(i+1)}, O^{(i+2)}, \ldots$, following a point $O^{(i)}$, lie on an s-fold line infinitely near $O^{(i)}$. In (3) LEVI shows that this is not the only case and that the above stated condition for the finiteness of a succession is not only sufficient but also necessary. For instance, if O, O', O'' are 3 successive points of a cuspidal curve, then infinitely near O'' there always exists a multiple curve which is also cuspidal.

There is an obvious connection between the finiteness of the composition process, as discussed above, and the following theorem first stated by KOBB (1): *It is possible to represent the complete neighborhood of a singular point O of an algebraic surface by a finite number of integral power series of two parameters.* The complete neighborhood of the s-fold point O is in $(1, 1)$ correspondence with the neighborhood of the curve $f_s(x', y', 1) = 0$. If this curve does not contain s-fold points of the transformed surface f', the theorem may be considered as proved by induction on s, since it is true for $s = 1$, and since from the hypothesis that it is true for the neighborhood of every point of the curve $f_s = 0$, it follows by analytical prolongation that it is true for the neighborhood of the whole curve, and hence of O. If the curve $f_s = 0$ possesses s-fold points (possibly all its points are s-fold), the above reasoning may be repeated for each of its s-fold points. In the belief of having established *in every case* the finiteness of a sequence of s-fold points, KOBB in this manner deduces the above theorem. In view of LEVI's results this deduction is, however, not permissible.

Important contributions toward the proof of the theorem of KOBB are due to HENSEL (1), but the only complete and rigorous proofs of this theorem were given by BLACK (1) and JUNG (1). The methods of HENSEL and JUNG are function-theoretic in nature and deal with the expansions of the algebraic function z of x and y in the neighborhood of a point $x = a, y = b$.

Let us assume that the point O of F is at the origin $x = y = z = 0$, and let $z_1, \ldots, z_m$ be the branches of z which approach zero as $x \to 0$ and $y \to 0$. For any preassigned algebroid branch of origin $x = y = 0$,

$$(9) \qquad\qquad y = y_0(x),$$

HENSEL derives expansions for the roots z_i *which converge in the neighborhood of the preassigned branch.* These expansions are fractional or integral power series of $y - y_0$, according as the branch (9) does or does not belong to the *branch curve* of the function z. The coefficients of these power series are algebroid functions of x in the neighborhood of $x = 0$, *which may become infinite at* $x = 0$. This happens necessarily only in the following cases: (I) there passes through $x = y = 0$ a branch of the branch curve of z, distinct from the branch (9); (II) the branch (9) belongs to the branch curve of z and is of order > 1, or is of order 1 and touches the line $x = 0$. In either one of these cases HENSEL proves that the region of convergence of any of the resulting power series is given by inequalities of the type:

$$(10) \qquad\qquad |x| < r, \qquad |y - y_0| < s |x|^\varepsilon,$$

where r, s and ε are positive numbers. It is thus seen that the radius of convergence of the power series, for $x = \text{const.}$, approaches zero as $x \to 0$. HENSEL does not investigate the question, essential in the point of proof of KOBB's theorem, of the possibility of assigning a finite number of branches (9) (including all the branches of the branch curve) in such a manner that the (possibly overlapping) regions of convergence (10) to which they give rise should cover the entire neighborhood of the point $x = y = 0$.

JUNG (1), whose proof we now proceed to outline, applies successive plane quadratic transformations in order to reduce the singularity of the branch curve of the function z at the point $x = a$, $y = b$ under consideration. This eminently correct method of approach simplifies considerably the entire procedure and reduces to a minimum the analytical apparatus employed by HENSEL. We first make two preliminary remarks.

(I) Let the point $O(x = y = 0)$ be a *simple* point of the branch curve D of the function z and let $y = y_0(x) = a_1 x + a_2 x^2 + \cdots$ be the expansion of the (linear) branch of D of origin O. *The various roots* z_i, *near* $z = 0$, *can be expanded in fractional power series of* $y - y_0$ *whose coefficients are functions of* x, *regular at* $x = 0$, *and which converge in a whole neighborhood* $|x| < \varrho$, $|y| < \sigma$ of O. This is an early result by HALPHEN (see ENRIQUES-CHISINI, a_2, p. 648) and is also a special case of HENSEL's result quoted above.

(II) Let $O(x = y = 0)$ be an ordinary double point of the branch curve D and let $y = y_1(x) = a_{11} x + a_{12} x^2 + \cdots$, $y = y_2(x) = a_{21} x + a_{22} x^2 + \cdots$ $(a_{11} \neq a_{22})$, be the expansions of the two branches of D passing through O. *The roots* z_i, *near* $z = 0$, *can be expanded in fractional power series of* $y - y_1$ *and* $y - y_2$ *whose coefficients are functions of* x, *regular at* $x = 0$, *and which converge in a whole neighborhood* $|x| < \varrho$, $|y| < \sigma$ *of* O (JUNG, 1). The situation at an ordinary double point of the branch curve is thus relatively simple. If this situation is analyzed from the point of view of the fundamental group of the residual space of the singularity (see section 4), it is seen that the simplifying factor is the *permutability* of the two substitutions on the branches of z relative to the two branches of D at O. In fact, it is easily shown that for an ordinary double point the fundamental group is abelian (ENRIQUES, 18; ZARISKI, 3).

If a root z is a power series of $(y - y_1)^{1/k}$ and $(y - y_2)^{1/l}$, then putting $y - y_1 = t^k$, $y - y_2 = \tau^l$, we find x, y and z expressed as integral power series of the parameters t and τ. These power series and their relationship to the monodromy group of the function z in the neighborhood of O are thoroughly studied by JUNG.

After these preliminary remarks we pass to the general case in which the branch curve D has an arbitrary singularity at O. We assume that the principal tangents of D at O are distinct from the axes x and y and we apply the quadratic transformation T:

$$x' = x; \qquad y' = \frac{y}{x}.$$

Let D' be the transform of D in the plane (x', y'). A neighborhood g of O, $|x| < \varrho$, $|y| < \sigma$, is transformed by T into an infinite region G along the axis $x' = 0$, given by the following inequalities: $|x'| < \varrho$, if $|y'| < \dfrac{\sigma}{\varrho}$ and $|x'| < \dfrac{\sigma}{|y'|}$, if $|y'| \geqq \dfrac{\sigma}{\varrho}$. These inequalities clearly indicate a division of G into two regions, which we shall denote by G_1 and G_2 respectively. The branches of D which are in g are transformed into branches of D' which lie in G and whose origins are points $O_1, O_2, \ldots$ on the axis $x' = 0$. We may assume, by taking ϱ sufficiently small, that all these branches lie entirely in the region G_1. Since the branch locus in G of the function z consists of the branches of D' and possibly of the line $x' = 0$, it easily follows that in G_2 the various roots of z can be given by HENSEL expansions. In fact, the portion of g which corresponds to G_2 appears as a neighborhood of the line $x = 0$; this line now plays the rôle of the branch (9).

The only multiple points of the branch locus of z in G_1 are the points $O_1, O_2, \ldots$, and hence, in virtue of the remark (I), we need only study the expansions of z in the neighborhoods of these points. Now, if D has originally an *ordinary multiple point* at O, then $O_1, O_2, \ldots$ are ordinary double points and the remark (II) can be applied. The theorem of KOBB now follows from the fact that by successive quadratic transformations it is always possible to reduce the singularity of D at O into a number of ordinary multiple points.

6. The reduction of singularities of an algebraic surface. We recall that the problem of the reduction of singularities of an algebraic curve gives rise to the following two theorems:

1. *An algebraic curve can always be birationally transformed into a curve in S_3 free from singularities and then, by projection, into a plane curve with ordinary double points only.*

2. *A plane algebraic curve can always be transformed by a* Cremona *transformation into a plane curve with ordinary multiple points only.*

The numerous existing proofs of the theorem 1 can be roughly classified into two groups: geometric proofs, originated by Noether, Halphen and Bertini, and algebraic proofs based on Kronecker's elimination theory and following up Hensel-Landsberg's arithmetic theory of algebraic functions. An account of these various proofs can be found in Bliss (a). At the present time both theorems 1. and 2. can be considered as rigorously acquired results.

The reduction theorem for algebraic surfaces, analogous to the above theorem 1, is the following: *Every algebraic surface f can be birationally transformed into a surface in S_5, free from singularities, and then, by projection, into a surface in S_3 with ordinary singularities only,* i. e. a nodal curve, on which there is a finite number of ordinary cuspidal points and of triple points for the curve, which are also triple and triplanar points for the surface. Proofs of this theorem have been proposed by Del-Pezzo (1), B. Levi (2), Severi (18) and Albanese (3). In explanation we add that at a cuspidal point P of the nodal curve (i. e. a point at which the two tangent planes coincide, a uniplanar point) the polar curves of the surface (i. e. the curves cut out on f by its first polar surfaces *outside the multiple curves of f*) have a base point with a fixed tangent t, and that the cuspidal point is *ordinary* if t is distinct from the tangent to the nodal curve at P and if the polar curves have at P a *simple* contact with each other. These two conditions imply that P is followed by an *ordinary* double point in the direction t.

A reduction theorem for surfaces analogous to the above theorem 2 has been obtained by Chisini (1) and is the following: *Every algebraic surface f in S_3 can be transformed by a* Cremona *transformation into a surface having only ordinary multiple curves* (i. e. with distinct tangent planes) *free from singularities and possessing a finite number of ordinary cuspidal points; in addition any two distinct multiple curves have at any of their common points P distinct tangents and P is not a base point of the polar curves of f.*

The proofs of these theorems are very elaborate and involve a mass of details which it would be impossible to reproduce in a condensed form. It is important, however, to bear in mind that in the theory of singularities the details of the proofs acquire a special importance and make all the difference between theorems which are rigorously proved and those which are only rendered highly plausible.

For Del-Pezzo's proof we refer the reader to C. Segre (3), where this proof is critically examined.

Levi first simplifies the singularities of the given surface f by transformations alternately quadratic of the first kind and birational, and obtains a surface f' whose only singularities are multiple curves, free from singularities and carrying no point of multiplicity higher than the multiplicity of the generic point of the curve. The possibility of this simplification is a consequence of the theorem proved by Levi in (1) and quoted in section 5, showing that a sequence of infinitely near multiple points of the same multiplicity s, each lying on a multiple curve of multiplicity $<s$, is necessarily finite. Each quadratic transformation is followed by a birational transformation eliminating the "accidental" singularities introduced by the quadratic transformation.

To this new surface f' Levi applies transformations alternately monoidal and birational. A monoidal transformation for which a given s-fold curve of f' is fundamental operates on the generic plane section of f' through the vertex of the monoids as a de Jonquières transformation. It is therefore clear that the s-fold curve will be ultimately replaced by curves of multiplicity $<s$. Here also after every monoidal transformation it is necessary to apply a birational transformation in order to eliminate the "accidental" singularities introduced by the monoidal transformation. The transformed curves of multiplicity $<s$ may however contain isolated s-fold points. In this case it is again necessary to apply quadratic transformations with such an isolated point P as fundamental. As a result of the resolution of the singularity at P, new s-fold curves may appear on the transformed surface f'', in which case these curves must again be resolved by monoidal transformations. The last part of the proof tends to show that this process terminates after a finite number of steps, so that ultimately the transformed surface will possess only singularities of multiplicities $<s$.

In regard to the "accidental" singularities introduced by the quadratic and the monoidal transformations and in regard to the manner in which they should be eliminated by birational transformations, Levi's proof is not sufficiently explicit. As we understand it, the situation is as follows. Let the surface $f(x, y, z) = 0$ be transformed by a Cremona transformation $[x' = x'(x, y, z), y' = y'(x, y, z), z' = z'(x,y,z)]$ into a surface $f'(x', y', z') = 0$. The equation $f = 0$ together with the equations of the transformation defines a surface F in the space $S_6(x, y, z, x', y', z')$, of which both surfaces f and f' are projections. It should be proved that the surface F possesses only singularities which are common to both f and f' (in a sense which should be clearly defined). A generic projection of F into S_3 will give a surface on which the "accidental" singularities of f' have disappeared and only ordinary singularities have been created by the projection.

Severi's proof starts with an entirely different conception of the problem. According to this conception, to which C. Segre (3) has already called attention, the problem of transforming birationally a given surface f into a surface free from singularities in $S_r (r \geq 3)$ is equivalent to the construction of a *simple* (II, 6) linear system Σ of curves on f, of dimension $r \geqq 3$, enjoying the following property: if n is the degree of Σ (i. e. the number of *variable* intersections of two curves of Σ, see II, 1), *any* linear subsystem of Σ, of dimension $r - 1$, is of degree *not less* than $n - 1$. If the curves of Σ are projectively referred to the hyperplanes of an S_r (see II, 6), f is birationally transformed into a surface free from singularities.

A preliminary but essential step in SEVERI's proof is the elimination of the so-called *proper* multiple points of f. A point P of a surface F in S_r is a proper multiple point, if the genus of the section of F by a generic hyperplane through P is less than the genus of a generic hyperplane section of the surface. The problem of transforming birationally the given surface f into a surface F in S_r free from proper multiple points depends on the construction on f of a simple linear system Σ, ∞^r, of effective genus p (i. e., $p =$ effective genus of the generic curve of Σ) not containing linear subsystems of dimension $r - 1$ and of effective genus $<p$. SEVERI shows that such a system is given by the system of curves cut out on f by the surfaces of a sufficiently high order, passing with convenient multiplicities through *all the base points*, including the infinitely near base points, *of the polar curves of f*.

Having obtained a birational transform F of f, free from proper multiple points, SEVERI considers the linear system Σ_1 cut out on F by the hypersurfaces of a sufficiently high order which are adjoints of the generic hyperplane section of F. It follows by relatively simple considerations of the geometry on a curve, that by means of the system Σ_1 the surface F is transformed into a surface free from singularities.

There are a few details in SEVERI's proof which should be further elaborated, preferably by analytical methods. For instance, in the first part of the proof (the elimination of proper multiple points) SEVERI uses the following lemma: *The genus of a continuously varying space algebraic curve cannot diminish without the curve acquiring a new double point.* In the proof of this lemma the genus of a space curve is evaluated by the formula, used universally:

$$p = \frac{(n - 1)\,(n - 2)}{2} - \tau - h,$$

where n is the order of the curve, h the order of the congruence of its bisecants and τ is the number of the double points of the curve. If the curve has more complicated singularities than double points, each s-fold point of the curve (including the infinitely near multiple points) is considered as the equivalent of $s(s - 1)/2$ double points. What matters, however, and is essential for the application which SEVERI makes of this lemma, is that the composition of the singularities of the curve should be defined by the method of space quadratic transformations and that τ should be evaluated accordingly. We have seen, however, in section 3 that if intended in this sense the composition of a singularity of a space curve may differ from the composition of its plane projection. *Hence the above formula is not correct.* Since the composition indices are *not diminished* by projection, we can only write $p \leqq \dfrac{(n-1)(n-2)}{2} - \tau - h$.

ABLANESE's proof is an extension of the method used by the same author in (2) to prove the analogous theorem for curves. His proof of this last theorem is of striking simplicity and is as follows: Let f be a plane algebraic curve of order n and let g_m^r, $m = \lambda n$, $r = \dfrac{(\lambda+1)(\lambda+2)}{2} - \dfrac{(\lambda-n+1)(\lambda-n+2)}{2} - 1$, be the linear series, free from fixed points, cut out on f by the curves of a sufficiently high order λ. Let g_m^r, $g_{m_1}^{r-1}, \ldots, g_{m_s}^{r-s}$ be a sequence of linear series, free from fixed points, such that: (1) $g_{m_i}^{r-i}$ is partially contained in $g_{m_{i-1}}^{r-i+1}$; (2) $m_i \leqq m_{i-1} - 2$. We have $0 \leqq m_s \leqq m - 2s$ and consequently $r - s \geqq r - \dfrac{m}{2} = \dfrac{n(\lambda - n + 3)}{2} - 1$. It follows, for λ sufficiently large, that *there exists a series* $g_{m_s}^{r-s}$, *of dimension* $\geqq 2$, *which does not contain partially series of dimension* $r - s - 1$ *and of order* $< m_s - 1$. If the sets of this series are referred to the hyperplanes of an S_{r-s}, the curve is birationally transformed into a curve in that space free from singularities.

In extending this method of proof to algebraic surfaces, ALBANESE proceeds as follows. Let Σ_m^r be the linear system of curves, of degree m and of dimension r, cut out on the given algebraic surface f in S_3 by the surfaces of a sufficiently high order λ, and let $\Sigma_m^r, \Sigma_{m_1}^{r_1}, \ldots, \Sigma_{m_s}^{r_s}$ be a sequence of linear systems of curves on f, such that: (1) $r_{i-1} > r_i \geqq r_{i-1} - 4$; (2) each $\Sigma_{m_i}^{r_i}$ is partially or totally contained in the preceding system $\Sigma_{m_{i-1}}^{r_{i-1}}$ (see II, 4); (3) if $\Sigma_{m_i}^{r_i}$ is totally contained in $\Sigma_{m_{i-1}}^{r_{i-1}}$, then $m_i \leqq m_{i-1} - 2(r_{i-1} - r_i) - 1$. ALBANESE proves that $r_s > \lambda n + 4$, and that consequently *there exists a linear system* $\Sigma_{m_s}^{r_s}$, *of dimension* r_s *arbitrarily large, such that every linear subsystem* $\Sigma_{m_s-j}^{r_s-i}$ *of* $\Sigma_{m_s}^{r_s}$, $0 < i \leqq 4$, *is irreducible, is totally contained in* $\Sigma_{m_s}^{r_s}$ *and is of degree* $m_s - j$, *where* $j < 2i + 1$. Adding (see II, 4), if necessary, to $\Sigma_{m_s}^{r_s}$ the system cut out on f by surfaces of a given order, a linear system Σ_k^h is obtained, which satisfies the same conditions as $\Sigma_{m_s}^{r_s}$ and which in addition is simple (II, 6). By means of this system the surface f is birationally transformed into a surface F or order k in an S_h. From the properties of the system Σ_k^h it is then deduced that F possesses only double points (and not points of higher multiplicity), to wit: a nodal curve with a finite number of ordinary cuspidal points, and a finite number of isolated double points, *followed in their neighborhoods of different orders by at most a finite number of double points.* These isolated double points can therefore be reduced by a finite number of quadratic transformations.

In Albanese's proof extensive use is made of properties of linear systems and especially of intersection properties of curves on a surface. These notions are well defined on surfaces without singularities. It may be readily granted that some of these notions can be immediately extended to surfaces with any singularities, but some others, as, for instance, the definition of the multiplicity of intersection of two curves, present new difficulties. With this distinction in mind Albanese's proof should be closely scrutinized.

We now turn to Chisini's method of reducing the singularities of an algebraic surface by Cremona transformations only. In Chisini's proof the first step consists in the reduction of the *isolated* singularities of f, i. e. of the multiple points of f which are base points of the polar curves. Such are all the multiple points which do not lie on multiple curves of f, and also certain special points on the multiple curves. We may also consider as isolated singularities the multiple points of a multiple curve (in particular, the common points of distinct multiple curves), although these points are not necessarily base points of the polar curves.

For the reduction of a given isolated singularity P Chisini uses *space quadratic transformation of the 2nd kind* (quadrato-cubic transformations). The homaloidal net in S_3 of such a transformation T is given by a web of quadrics *through a generic line g through P* and through 3 generic points A_1, A_2, A_3 of the space. The homaloidal net in S_3' is given by a web of cubic surfaces on 4 lines d_1', d_2', d_3', g', where g' is a trisecant of the triple of lines d_i'. It should be noted that the transformation T^{-1} does not possess (in S_3') *isolated* fundamental points. This is the reason why these transformations must be used in preference to quadratic transformations of the 1st kind.

The fundamental lines d_i' and g' correspond to the planes $(g A_i)$ and (A_1, A_2, A_3) respectively. The transformed surface f' in S_3' will have the same singularities as f, except for the multiple lines d_i' and g', created by the transformation. Moreover, to the multiple point P there will correspond a certain number of points $P_1', P_2', \ldots$, on the line g'. Chisini shows that: (1) *the lines d_i' and g' are ordinary multiple lines (i. e., with distinct tangent planes)*; (2) *the only isolated singularities on these lines, outside the points $P_1', P_2', \ldots$, are ordinary cuspidal points and normal crossings*. By a *normal crossing* we mean a common point of two distinct multiple curves through which both curves pass simply with distinct tangents and which is not a base point of the polar curves.

For the analysis of the points P_i', to which the singularity P has been reduced by the transformation, Chisini defines an important controlling character of a singularity P: *its rank* $\varrho = \varrho(P)$. This character is defined as follows:

$$\varrho = 4v - 5c - r(r - 1),$$

where r is the multiplicity of P on f, c is the number of distinct branches at P of the multiple curves of f (if P is not on a multiple curve, then $c = 0$), and $v = v(P)$, the *valence* of P, is another character of P, which we proceed to define. We consider the cone which projects from a *generic* point O of the space the curve composed of the polar curve of O and of all the multiple curves of f. *The valence v of P is the diminution of the class of this cone due to the line OP*. This signifies in particular that if OP is not an element of the considered cone or is a simple element of the cone, then $v(P) = 0$.

Comparing the rank of P with the rank of any of the points P_i', CHISINI shows that always $\varrho(P) > \varrho(P')$, except when P is (1) a normal crossing, or (2) a simple point of a multiple curve, which is a base point of the polar curves, and through which the generic polar curves pass simply with a tangent distinct from the tangent of the multiple curve. The values of $\varrho(P)$ in these two cases are -4 and 1 respectively. It is easily seen that in all other cases $\varrho(P) \geqq 3$.

It follows immediately that *by a finite number of quadratic transformations of the 2nd kind f can be transformed into a surface whose multiple curves are free from singularities and whose only isolated singularities are of the two types just described.*

For the resolution of the multiple curves CHISINI employs, as DEL PEZZO and B. LEVI do also, monoidal transformations. This part of the proof is rather complicated, the source of complication being the new isolated singularities created by the monoidal transformations. Each of these singularities must be carefully investigated and resolved again by quadratic transformations.

We could not comprehend CHISINI's geometric reasoning by which he proves the invariance of the valence $v(P)$ under analytic transformation regular at P (CHISINI, 1, p. 13). Otherwise his proof is outstanding for the standard of rigor achieved and for the carefulness with which the most minute details are treated, leaving little, if anything, to intuition.[1]

---

[1] *Note added during the reading of the proofs.* It has come to my knowledge that R. J. WALKER in his Princeton dissertation in course of publication in the Annals of Mathematics gives a complete function-theoretic proof of the reduction theorem for algebraic surfaces. Having read the thesis by the courtesy of the author we believe that WALKER's proof stands the most critical examination and settles the validity of the theorem beyond any doubt.

Chapter II
Linear Systems of Curves

1. Definitions and general properties. In the sequel we shall consider only algebraic surfaces without singularities in S_r, or with ordinary singularities in S_3, obtained from the former surfaces by a generic projection. The points of the nodal curve, with the exception of the cuspidal points, should be considered as pairs of distinct points of the surface. It clearly appears from the projection that the cuspidal points are the branch points of the doubly covered nodal curve.

An *algebraic system*[1] Σ of curves C on a surface f in $S_k(x_0, x_1, \ldots, x_k)$ is a system of curves cut out on f by a system of hypersurfaces

$$(1) \qquad \varphi(x_0, x_1, \ldots, x_k; \lambda_0, \lambda_1, \ldots, \lambda_r) = 0,$$

depending rationally on $r + 1$ homogeneous parameters λ_i connected by an algebraic relation $\psi(\lambda_0, \lambda_1, \ldots, \lambda_r) = 0$. Let φ be of degree m in the variables x_i. If we interpret the literal coefficients of the general polynomial of degree m in the x's as homogeneous point coördinates in an S_N, $N = \binom{m+k}{k} - 1$, then the system (1) is represented in S_N by an algebraic variety V_l $(l \leqq r)$, which is a rational transform (I, 1) of the hypersurface $\psi(\lambda) = 0$. The system Σ is a rational transform of V_l, the fundamental elements on V_l being those hypersurfaces (1) which pass through the surface f. If ψ is irreducible, then also V_l and Σ are irreducible. If s is the dimension of Σ and if a generic curve of Σ is cut out on f by ∞^a hypersurfaces (1), then, by the principle of counting constants (I, 1), we have $l = s + a$.

If φ depends linearly on *independent* parameters, i. e. if (1) is of the form

$$(2) \qquad \begin{cases} \lambda_0 \varphi_0(x_0, x_1, \ldots, x_k) + \lambda_1 \varphi_1(x_0, x_1, \ldots, x_k) + \cdots \\ \qquad + \lambda_r \varphi_r(x_0, x_1, \ldots, x_k) = 0, \end{cases}$$

Σ is called a *linear system*. We suppose that the number of arbitrary parameters λ_i in (2) cannot be reduced, i. e. that the polynomials φ_i are linearly independent. The system Σ is *of dimension* r, i. e. the curves C of Σ are in $(1 - 1)$ correspondence with the points of the linear space

---

[1] In Chapter V, 1 a more general definition of an algebraic system will be given.

$S_r(\lambda_0, \lambda_1, \ldots, \lambda_r)$, if and only if the system (2) does not contain hypersurfaces passing through the surface f. In the contrary case, the hypersurfaces of the system (2) which contain f form a linear subsystem of (2), of dimension $h - 1\ (h \geqq 1)$, represented in S_r by an S_{h-1}, and the dimension of Σ is $r - h$. In this case it is sufficient to consider in S_r a subspace S_{r-h} skew to S_{h-1} in order to have a linear system of hypersurfaces which cuts out on f the system Σ and is of the same dimension as Σ. We shall suppose that the original system (2) already satisfies this condition.

The hypersurfaces of the system (2) may all pass through fixed isolated points of f—the *base points* of Σ—and through fixed curves of f. These may be either included as *fixed components* of the total curves C, or may be left out entirely.

If $r = 1, 2, 3, \Sigma$ is called a *linear pencil*, a *net*, a *web* respectively.

If f is birationally transformed into a surface f', any linear system on f of dimension r is transformed into a linear system on f' of the same dimension.

The system Σ enjoys the property that through r generic points of f there passes one and only one curve of Σ. This property is characteristic to the following extent: *If an irreducible algebraic system Σ of dimension $r > 1$ is such that through r generic points of f there passes only one curve of Σ, then Σ is a linear system or is made up of curves of a linear system, each counted a certain number of times* (ENRIQUES, 1, C. SEGRE, 1). The second alternative simply means that the equation (1) of the defining system of hypersurfaces is of the form: $(\lambda_0 f_0 + \cdots + \lambda_r f_r)^n = 0$.

The case $r = 1$ of a pencil is a real exception. It is possible to have on a surface f an *irrational* pencil Σ, i. e. a pencil which, considered as a one-dimensional variety whose elements are its curves, is of genus $p > 0$. The pencil Σ is then certainly not linear. An elementary consideration shows that a rational pencil ($p = 0$) is always linear. The generators of a cone projecting a plane curve of genus p (*cone of genus p*) or the generators of a ruled surface of genus p (birationally equivalent to a cone of genus p) constitute the simplest examples of an irrational pencil (in this case, of rational curves). Another important example is given by the surfaces representing the pairs of points of two algebraic curves, of which one at least is not rational (see I, 1 and also VI, 13). The class of surfaces possessing irrational pencils is birationally special (V, 1, 3).

A pencil Σ possessing a base point at a *simple* point O of f is necessarily rational, and consequently linear. This is obvious in the simplest case, in which the curves of Σ have at O a simple point with a variable tangent line. The curves of Σ are then in $(1 - 1)$ correspondence with the tangent lines to f at O, these lines forming a flat pencil. The most general case of a multiple base point or of infinitely near base points can be reduced to the simple case by operating on the neighborhood of O on

f by transformations locally quadratic. As a consequence, *an irrational pencil on a surface without singularities or with ordinary singularities in S_3 does not possess base points*.

A linear system Σ is *irreducible* or *reducible* according as its generic curve is irreducible or reducible. Reducible linear systems without fixed components can be constructed in the following manner. Let Σ_0 be a linear pencil, without fixed components, cut out on f by $\lambda_0 \varphi_0 + \lambda_1 \varphi_1 = 0$, and let $\psi_0(\lambda_0, \lambda_1), \ldots, \psi_r(\lambda_0, \lambda_1)$ be $r+1$ linearly independent and relatively prime binary forms of degree n. If in

$$(3) \qquad t_0 \psi_0(\lambda_0, \lambda_1) + t_1 \psi_1(\lambda_0, \lambda_1) + \cdots + t_r \psi_r(\lambda_0, \lambda_1) = 0,$$

we replace λ_0 and λ_1 by φ_1 and $-\varphi_0$, we obtain a linear system of curves, Σ, of dimension r, in which each curve is composed of n curves of Σ_0. We say that Σ *is composed of the curves of the pencil Σ_0*.

In a similar manner it is possible to construct reducible systems composed of the curves of an irrational pencil Σ_0 existing on f by considering in Σ_0 a linear series g_n^r of ∞^r sets of n curves of Σ_0. That the system Σ obtained in this manner is linear follows, for $r > 1$, from the preceding theorem on the characteristic property of linear systems and from the well known fact that a g_n^r on a curve does not possess variable multiple points; for $r = 1$ from the fact that Σ is rational.

The extended theorems of BERTINI:

1. *The generic curve of an irreducible linear system cannot have multiple points outside the base points of the system.*

2. *A reducible linear system, without fixed components, is necessarily composed of the curves of a pencil* (ENRIQUES, 5). The proofs are the same as for plane algebraic curves.

Remark. The first theorem of BERTINI can be extended to infinitely near multiple points. If O is a multiple base point of a linear system Σ, at which the generic curve C of Σ has an arbitrarily complicated singularity, then all the multiple points of C infinitely near and in the successive neighborhoods of O must be in fixed positions (in the sense of I, 2). If, in fact, O_i were a variable multiple point of C in the neighborhood of order i of O, then operating locally by successive quadratic transformations which resolve the singularity at O up to its ith neighborhood, we would arrive at a linear system Σ' whose curves have a variable multiple point on a straight line, which is impossible. But we can say even more: if O_i is a simple free point followed by satellites, it also must be fixed (if O_i is itself a satellite, the statement is obvious). In fact, it is easily seen that the existence of a variable, free, simple point $O^{(i)}$ followed by satellites implies a variable contact of the curves of Σ' with a straight line, and this is impossible. For instance, if the curves of a linear system have a cuspidal base point, then the cusp tangent must be fixed. The preceding considerations can be summarized in the case of a single branch at O, as follows: *if x and y are local coördinates on f in the neighborhood of O and if $\varphi(x, y) + t\psi(x, y) = 0$ is the equation of a pencil having O as a base point, then in the expansion of y in terms of x the coefficients of all the terms preceding the last characteristic term are independent of t.*

If Σ is irreducible and of dimension $r \geq 1$, its *effective degree* n is the number of *variable* intersections of two curves of Σ and its *effective genus* p is the genus of a generic curve of Σ. The linear series g_n^{r-1} cut out by the curves of Σ on a generic curve C of Σ is called the *characteristic series of Σ on C*. We have obviously $n \geq r - 1$. If the g_n^{r-1} is contained in a complete series g_n^{n-p+i}, where i is the index of speciality of the series, then we have the following relation: $r = n - p + i - \delta + 1$, where $\delta \geq 0$ is called the *deficiency* of the g_n^{r-1}.

2. On the conditions imposed by infinitely near base points. It is well known that if we impose on the curves of a linear system on a surface the condition of passing with given multiplicities through given points of the surface we obtain a *linear* subsystem of the given system. This statement still holds true if some of the imposed base points become infinitely near, following each other on an algebroid branch of origin O, in the sense defined in I, 2. For the proof (which is the same for surfaces, having at O a simple point, as for the plane) it is sufficient to observe that if $O, O_1, O_2, \ldots, O_k$ is a succession of infinitely near points on the above branch, the condition that a curve pass through these points with assigned multiplicities $s, s_1, s_2, \ldots, s_k$ is equivalent to the condition that the curve have contacts of certain orders with branches passing through $O; O, O_1; O, O_1, O_2; \ldots; O, O_1, \ldots, O_k$. To express this condition we may use, for each i, a branch γ_i of the lowest order passing through $O, O_1, \ldots, O_i$. Let $\nu^{(i)}, \nu_1^{(i)}, \ldots, \nu_i^{(i)} (= 1)$ be the multiplicities of $O, O_1 \ldots, O_i$ for γ_i. Then, if λ_i is the order of contact of our curve with γ_i, we must have

$$(4) \qquad \lambda_i \geq \nu^{(i)} s + \nu_1^{(i)} s_1 + \cdots + \nu_i^{(i)} s_i, \qquad i = 0, 1, \ldots, k.$$

The conditions (4) are obviously expressed analytically by relations which are *linear* in the parameters of the linear system.

In order that there should exist curves (at least of sufficiently high order) which pass through the infinitely near points $O, O_1, \ldots, O_k$ with effective multiplicities equal to the assigned multiplicities $s, s_1, \ldots, s_k$, it is necessary (and sufficient) *that any multiplicity s_i be not greater than the sum of the multiplicities of the assigned base points which are proximate to O_i* (see I, 2; ENRIQUES-CHISINI, a_2, p. 390 and p. 427). If these conditions, referred to in the sequel as the *proximity inequalities*, are not satisfied, the effective multiplicities at $O, O_1, \ldots, O_k$ will be different from (and not necessarily greater than) the assigned multiplicities. A rule for determining in this case the effective multiplicities has been given by ENRIQUES (a_2, pp. 427—433) and has been referred to by this author as the *"principle of discharge"* (*principio di scaricamento*). We quote, as an illustration of this principle, the following two cases:

1. Let $O, O_1, \ldots, O_k$ be free points on a linear branch of origin O, and let $s \geq s_1 \geq s_2 \geq \cdots \geq s_{k-1}, s_{k-1} < s_k$. Let h be the smallest value of i for which $s_i < \frac{s_{i+1} + \cdots + s_k}{k-i}$. The effective multiplicities s'_i are the following:

$$s'_i = s_i (i < h), \quad s'_{h+i} = \left[\frac{s_h + s_{h+1} + \cdots + s_k + k - h - i}{k - h + 1}\right], \quad (0 \leq i \leq k - h).$$

2. Let $O_{i+1}, O_{i+2}, \ldots, O_{i+h}$ be proximate points of O_i and let $s_i < s_{i+1} + \cdots + s_{i+h}$. Then $s'_i = s_i + x$, $s'_j = s_j - x$, $(j = i+1, \ldots, i+h)$, where x is the smallest positive number satisfying the inequality $s_i + x(h+1) \geq s_{i+1} + \cdots + s_{i+h}$.

In all cases the effective multiplicities s'_i must satisfy the inequalities

$$(5) \quad v^{(i)}(s' - s) + v_1^{(i)}(s'_1 - s_1) + \cdots + v_i^{(i)}(s'_i - s_i) \geq 0, \quad i = 0, 1, \ldots, k,$$

derived from (4), and are completely determined by these inequalities and by the condition that they be exhibited by a branch of the lowest possible order.

In the general theory of linear systems it is impossible to confine the range of assigned multiplicities to numbers satisfying the proximity inequalities. Thus, given two linear systems, whose base points are assigned with multiplicities satisfying the proximity inequalities, the assigned multiplicities of the base points of the system difference (section 4) do not necessarily satisfy these inequalities. It may be, for instance, necessary to regard a double base point O of a linear system as a simple base point infinitely near which there are r (r—arbitrary) assigned simple base points $O_1, O_2, \ldots, O_r$ in distinct directions. Here $s = s_1 = s_2 = \cdots = s_r = 1$, but by case 2. of the principle of discharge we have necessarily $s' = 2$, $s'_1 = s'_2 = \cdots = s'_r = 0$. By regarding in this manner the double base point of the system, we must assume that the number of fixed intersections of two curves of the system absorbed at O is $r + 1$ and correspondingly define the virtual degree (see section 5) of the system. It should be noted that the above number of fixed intersections may be *greater* than the effective intersection multiplicity at O.

For the determination of the effective multiplicities s'_i we have used branches of the lowest order and we have applied the inequalities (4), or (5). However, if the effective multiplicities s'_i are different from the assigned multiplicities s_i, then it is not at all obvious that the order of contact at O of our curve with an *arbitrary* branch, passing through $O, O_1, \ldots, O_k$ with effective multiplicities $r, r_1, \ldots, r_k$, is not less that $rs + r_1 s_1 + \cdots + r_k s_k$. We therefore prove the following theorem:

Let a given branch pass through $O, O_1, \ldots, O_k$ *with effective multiplicities* $s', s'_1, \ldots, s'_k$ *and let* $s, s_1, \ldots, s_k$ *be arbitrary (positive or nega-*

tive) integers. If the inequalities (5) *hold and if γ ia any branch passing through the points $O, O_1, \ldots, O_k$ with effective multiplicities $r, r_1, \ldots, r_k$, then $r(s'-s) + r_1(s_1'-s_1) + \cdots + r_k(s_k'-s_k) \geqq 0$.*

Proof. Let $O_{i+1}, O_{i+2}, \ldots, O_{i+\mu_i}$ $(i + \mu_i \leqq k)$ belong to the set of proximate points of O_i and let us consider the following linear form in the variables $x, x_1, \ldots, x_k$:

$$\varphi(x, x_1, \ldots, x_k) = \sum_{i=0}^{k} (r_i - r_{i+1} - \cdots - r_{i+\mu_i}) (\nu^{(i)} x + \nu_1^{(i)} x_1 + \cdots + \nu_i^{(i)} x_i).$$

We have $r_i \geqq r_{i+1} + \cdots + r_{i+\mu_i}$, and hence we deduce from (5) that $\varphi(s'-s, s_1'-s_1, \ldots, s_k'-s_k) \geqq 0$. We now prove the identity:

$$\varphi(x, x_1, \ldots, x_k) = r x + r_1 x_1 + \cdots + r_k x_k,$$

and from this our theorem follows. Let O_σ be the first point in the sequence $O, O_1, \ldots, O_k$ such that O_i is in the set of proximate points of O_σ. From the analysis of plane singularities given in (I, 2) it follows that O_i is a proximate point of O_{i-1} but is not a proximate point of $O_{\sigma+1}, \ldots, O_{i-2}$. Hence the coefficient of r_i in $\varphi(x, x_1, \ldots, x_k)$ is

$$\sum_{j=0}^{\sigma} (\nu_j^{(i)} - \nu_j^{(\sigma)} - \nu_j^{(i-1)}) x_j + \sum_{j=\sigma+1}^{i-1} (\nu_j^{(i)} - \nu_j^{(i-1)}) x_j + x_i,$$

if $\sigma < i - 1$, and is

$$\sum_{j=0}^{i-1} (\nu_j^{(i)} - \nu_j^{(i-1)}) x_j + x_i,$$

if $\sigma = i - 1$. By using the fundamental property of proximate points we may calculate successively the multiplicities $\nu_{i-1}^{(\alpha)}, \nu_{i-2}^{(\alpha)}, \ldots, \nu^{(\alpha)}$ $(\alpha = \sigma, i - 1, i)$, and we find in the first case: $\nu_j^{(i-1)} + \nu_j^{(\sigma)} = \nu_j^{(i)}$ $(j = 0, 1, \ldots, \sigma)$, $\nu_j^{(i-1)} = \nu_j^{(i)}$ $(j = \sigma + 1, \ldots, i - 1)$, and in the second case: $\nu_j^{(i-1)} = \nu_j^{(i)}$, $j = 0, 1, \ldots, i - 1$, q. e. d.

3. Complete linear systems (ENRIQUES, 2, 5, 9). Let Σ_1 and Σ_2 be two linear systems of curves of the same order on f, cut out by the systems of hypersurfaces, $\sum_{i=0}^{r_1} \lambda_i \varphi_i = 0$ and $\sum_{j=0}^{r_2} \mu_j \psi_j = 0$, respectively. If the two systems have a curve in common, say the curve cut out on f by $\varphi_0 = 0$ and $\psi_0 = 0$ respectively, then the consideration of the system of hypersurfaces

$$\lambda_0 \varphi_0 \psi_0 + \psi_0 (\lambda_1 \varphi_1 + \cdots + \lambda_{r_1} \varphi_{r_1}) + \varphi_0 (\mu_1 \psi_1 + \cdots + \mu_{r_2} \psi_{r_2}) = 0$$

shows that the curves of both Σ_1 and Σ_2 all belong as total curves to a more inclusive linear system Σ. We say that Σ_1 and Σ_2 are *totally contained* in Σ. A linear system is *complete* if it is not totally contained in a linear system of higher dimension. It follows that *any given linear system is either complete or is totally contained in a complete system which is uniquely determined.*

Two curves C and D are said to be *linearly equivalent*, in symbols, $C \equiv D$, if they are total curves of some linear system of curves. If $C \equiv D$ and if the two curves have no common components, there exists on f a rational function φ/ψ, which vanishes on C and becomes infinite on D, in such a manner that if either C or D is reducible and possesses multiple components, then φ/ψ has zeros or poles of order k along any k-fold component of C or D respectively. The complete system containing a given curve C and uniquely determined is denoted by $|C|$ and is made up of all curves equivalent to C. If $C \equiv D$, then $|C|$ and $|D|$ coincide. The equivalence relation is transitive: if $C \equiv D$ and $D \equiv E$, then $C \equiv E$.

The above definition of complete systems can be generalized by the imposition of a set of base points. Let $O_1, O_2, \ldots, O_k$ be a set of (distinct or infinitely near) points on f, and let $s_1, s_2, \ldots, s_k$ be a set of arbitrary non-negative integers[1]. If there exist curves equivalent to C and passing through each point O_i with a multiplicity $\geqq s_i$, their totality is a linear system, called the *complete system relative to H*, where $H = O_1^{s_1} O_2^{s_2} \ldots O_k^{s_k}$ is the set of the assigned base points with their assigned multiplicities. This system shall be denoted by $|C|_H$.

It may well happen that in virtue of the imposed conditions the curves of $|C|_H$ necessarily pass through other base points not in H, or that they all have at O_i a multiplicity $s_i' \neq s_i$.[2] To keep track of this possibility, in itself significant, it is convenient to distinguish: (1) between the assigned base points (in H) and the possible base points of $|C|_H$ not in H, which are referred to as *accidental* base points; (2) between the *assigned* or *virtual* multiplicity s_i of O_i and its *effective multiplicity s_i'* (for plane curves, see CASTLENUOVO, 1). An accidental base point can be considered as a base point of virtual multiplicity 0. These conventions have naturally only a relative character. We may, if we wish, convert all the effective multiplicities of the effective base point of $|C|_H$ into virtual multiplicities, without altering the system. On the other hand it is not excluded that $|C|_H$ can be equally defined by a set H' different from H, and that consequently, according to the new definition of $|C|_H = |C|_{H'}$, some of the points O_i become accidental base points. If the assigned multiplicities s_i satisfy the proximity conditions, then neces-

---

[1] In the next section the condition that the integers s_i be non-negative will be removed.

[2] It is not necessary that $s_i' > s_i$, even if the integers s_i satisfy the proximity inequalities (section 2). For instance, let $|C|_H$ be a system of plane curves and let $k = 3$. We assume that O_2 is infinitely near to O_1 and that O_3 is on the line $O_1 O_2$ and at finite distance from O_1. It is then seen that if the C's are quartics and if $s_1 = s_2 = 2$ and $s_3 = 1$, then $s_1' = 3$, $s_2' = s_3' = 1$.

sarily $s'_i = s_i$, provided the curves C are of a sufficiently high order. In the contrary case this may not be true (see section 2).

We have defined in the preceding section the effective degree and the effective genus of an irreducible linear system of dimension ≥ 1, having regard to all the effective base points of the system taken with their effective multiplicities. Taking into account that an r-fold base point diminishes the degree by r^2 and the genus by $\dfrac{r(r-1)}{2}$, we are led to define the *virtual degree* and the *virtual genus* of $|C|_H$, supposed irreducible and of dimension ≥ 1, as follows:

$$\text{virtual degree of } C = n + \sum_i (s'^2_i - s^2_i),$$

$$\text{virtual genus of } C = p + \tfrac{1}{2}\sum_i [s'_i(s'_i - 1) - s_i(s_i - 1)],$$

where n and p are the effective degree and the effective genus of $|C|_H$ and where the summation is extended to all the effective base points of $|C|_H$ of effective multiplicities s'_i and virtual multiplicities s_i.

We observe that in addition to accidental base points, $|C|_H$ may possess accidental fixed components. On the other hand, if we add to the curves of any complete system a fixed component A, the resulting linear system may always be considered as a complete system with A as an *assigned* fixed component.

In the sequel we shall at times omit the subscript H and we shall use the symbol $|C|$ for any complete system, with or without assigned base points.

4. Addition and subtraction of linear systems (ENRIQUES, 2, 5, 9). If $C_1 \equiv D_1$ and $C_2 \equiv D_2$, then from the consideration of the product of the two rational functions on f, having C_i and D_i $(i = 1, 2)$ as zero and polar curves respectively, it follows that $C_1 + C_2 \equiv D_1 + D_2$. If then $|C|$ and $|D|$ are two complete systems, there exists a complete system $|C + D|$, containing, *among other curves*, *also* every reducible curve $C + D$, where C is any curve in $|C|$ and D is any curve in $|D|$. This system is called the *sum* of the two systems $|C|$ and $|D|$:

$$|C + D| = |C| + |D|.$$

As for the base points of $|C + D|$, each base point of virtual multiplicity s for $[C|$ and σ for $|D|$ shall be considered as a base point of $|C + D|$, of virtual multiplicity $s + \sigma$.

Any accidental base point or base point of accidental hypermultiplicity (effective multiplicity > virtual multiplicity) of $|C + D|$ must be necessarily such for either $|C|$ or $|D|$, but not conversely: if $|C|$ or $|D|$ possess accidental base points, these will, in general, not be base points of $|C + D|$. $|C + D|$ is reducible if $|C|$ and $|D|$ are coincident pencils, or are composed of the curves of one and the same pencil, or if either of them has an assigned fixed component. It *may* be reducible, if either $|C|$ or $|D|$ has an accidental fixed component. $|C + D|$ is certainly irreducible in any other case.

If $|C|$ and $|D|$ coincide, we obtain, by repeated addition of $|C|$, the *multiples* of $|C|:|2C|$, $|3C|$, etc.

If $|C| = |A| + |B|$, then the system $|B| = |C| - |A| = |C - A|$ is called the *difference* of the two systems $|C|$ and $|A|$. The operation of subtraction is not always possible. Given $|C|$ and $|A|$, in order that $|C - A|$ exist it is necessary (but not sufficient) that: (1) a generic[1] curve A_0 of $|A|$ should be *partially* contained in $|C|$, i. e. that there should exist curves C containing A_0 as a component; (2) the virtual multiplicity of any base point of $|A|$ should not be greater than its virtual multiplicity for $|C|$.

The condition (1) will then hold true for *any* other curve A of $|A|$, and the *residual curves* of $|C|$ *with respect to* A_0, i. e. the curves $\overline{B}$ such that there exist curves $C = A_0 + \overline{B}$, form a complete linear system $|\overline{B}|$, independent of the particular curve A_0 of $|A|$, *as long as A_0 is generic in* $|A|$. This is the *residue theorem* in its invariant form.

The system $|\overline{B}|$ is not yet necessarily the difference of $|C|$ and $|A|$. In order that $|C| - |A|$ exist it is necessary that there should exist curves B in $|\overline{B}|$ having at each base point of virtual multiplicity s for $|C|$ and σ for $|A|$ a base point of *virtual* multiplicity $\geq s - \sigma$. If such curves B exist, they form the complete system $|B| = |C| - |A|$. If the virtual multiplicities of the base points of $|C|$ and $|A|$ are equal to their effective multiplicities, then the conditions (1) and (2) are sufficient, and $|\overline{B}| = |B|$.

The condition (2) is somewhat artificial, and the only purpose of introducing it is to insure that the virtual multiplicities of the base points of $|C - A|$ be non-negative. Moreover, it leads to untenable conclusions. In fact, if the condition (2) is not satisfied, it may well happen that by raising the virtual multiplicities of $|C|$ we can obtain a system $|\overline{C}|$, for which the condition (2) holds and such that $|\overline{C} - A|$ exists. Hence, while the system $|C - A|$ is not defined, there exists the system $|\overline{C} - A|$, although $|C|$ *contains* $|\overline{C}|$. We eliminate the condition (2) by allowing *negative virtual multiplicities* in the definition of complete linear systems relative to a set H of base points. This simply means that in the inequalities (4), which give a meaning to virtual multiplicities, we allow some or all of the integers s_i to be negative (see in this connection the theorem proved at the end of section 2). We keep the definitions of the virtual multiplicities of the base points

---

[1] It is sufficient to say: a curve A_0, whose effective multiplicities at the base points of $|C|$ are equal to the effective multiplicities of the generic curve of $|A|$.

of the sum and of the difference of two linear systems, given above and in the preceding section. It is then seen that virtual multiplicities, in particular negative virtual multiplicities, have an *operational significance*. For instance, let $|B|$ be a complete linear system, having at a point O a base point of virtual multiplicity = effective multiplicity $= s$, and let $|A|$ be another complete linear system, for which O is not an effective base point. The point O can be considered as a base point of $|A|$, of an arbitrary *non-positive* virtual multiplicity r. If $r = -s$, then the virtual multiplicity of O for the system $|A + B|$ is zero, i. e. the generic curve of this system is not required to pass through O. If $r = 0$, we obtain for $|A + B|$ a system, possibly less inclusive, whose generic curve passes through O with virtual multiplicity s.

Let $|B|$ have at O a base point of virtual multiplicity s, and let $|\overline{B}|$ be the subsystem of $|B|$ obtained by raising the virtual multiplicity of O to $s + r$, $r > 0$. According to our preceding definitions the system $|B - \overline{B}|$ is not defined, because the total curves of $|\overline{B}|$ are also total curves of $|B|$. However, for any system $|A|$, the system $|A| + (|B| - |\overline{B}|) = |A + B| - |\overline{B}|$ is defined and can be obtained directly from $|A|$ by assigning O as a base point of virtual multiplicity $\sigma - r$, where σ is the virtual multiplicity of O for $|A|$. Similarly the system $|A| - (|B| - |\overline{B}|)$ has at O a base point of virtual multiplicity $\sigma + r$. It is convenient to regard the system $|B - \overline{B}|$ as *the point O counted r times*: $|B - \overline{B}| = O^r$, and accordingly to define *the addition and subtraction of points to a linear system*. To add or to subtract O^r to $|A|$ means·to diminish or to increase, respectively, by r the virtual multiplicity of O for $|A|$.

Later on in this chapter (section 7) we shall have occasion to illustrate the above definitions and conventions from the point of view of birational transformations.

Given two irreducible complete systems $|C_1|_{H_1}$ and $|C_2|_{H_2}$, of dimensions $\geqq 1$, of virtual degrees n_1, n_2 and of virtual genera π_1, π_2 respectively, the virtual characters n and π of the system $|C_1 + C_2|_{H_1 + H_2}$ are given by the formulas (NOETHER, 1, ENRIQUES, 5):

(6) $$n = n_1 + n_2 + 2(C_1 \cdot C_2)_{H_1, H_2},$$

(6') $$\pi = \pi_1 + \pi_2 + (C_1 \cdot C_2)_{H_1, H_2} - 1,$$

where $(C_1 \cdot C_2)_{H_1, H_2}$ denotes the number of intersections of a generic C_1 with a generic C_2 *outside the assigned base points taken with their virtual multiplicities*. If, namely, $O_1, O_2, \ldots$ are the base points of $|C_1|$ and $|C_2|$, of virtual and effective multiplicities $s_1^{(j)}, s_2^{(j)}, \ldots$ and $\sigma_1^{(j)}, \sigma_2^{(j)}, \ldots$ respectively for $|C_j|$ $(j = 1, 2)$, and if we denote by i the number of *variable* intersections of a C_1 with a C_2, then

$$(C_1 \cdot C_2)_{H_1, H_2} = i + \sum_k (\sigma_k^{(1)} \sigma_k^{(2)} - s_k^{(1)} s_k^{(2)}).$$

In the sequel we shall often write $(C_1 \cdot C_2)$ instead of $(C_1 \cdot C_2)_{H_1, H_2}$, it being understood that when the indices H_1, H_2 are omitted they are fixed throughout the discussion or appear clearly from it.

The formulas (6) and (6') will follow immediately, if they are first proved for the effective characters of $|C_1|, |C_2|$ and $|C_1 + C_2|$. For these, (6) is immediate: the number of variable intersections of two reducible curves $C_1 + C_2$ and $C_1' + C_2'$ is $(C_1 \cdot C_1') + (C_2 \cdot C_2') + (C_1 \cdot C_2')$ $+ (C_2 \cdot C_1') = n_1 + n_2 + 2i$. Supposing, for simplicity, that f is in S_3, it is possible to give a very simple algebro-geometric proof of (6'), by projecting C_1, C_2 and a given curve C of $|C_1 + C_2|$ on to a plane (Picard and Simart, a_2, p. 106). It is possible to prove (6'), even more rapidly, by means of topological considerations (Enriques). As a variable curve C of $|C_1 + C_2|$ degenerates into a curve $C_1 + C_2$, i new double points $O_1, O_2, \ldots, O_i$ are acquired, namely the intersections of C_1 and C_2. Conversely, C is topologically given by identifying the two points O_j', O_j'' on the two branches through each of the new double points O_j and by replacing the singular point $O_j' = O_j''$ by a 1-sphere σ_j so as to have a manifold. Consequently, the 2π 1-cycles of C, i. e. of its Riemannian manifold, are given by the $2\pi_1$ cycles of C_1, the $2\pi_2$ cycles of C_2, the $i - 1$ cycles of the type $O_1 O_j + \overline{O_j O_1}$ ($j = 2, \ldots, i$), where $O_1 O_j$ is a path in C_1 and $\overline{O_j O_1}$ is a path in C_2, and finally by the i cycles σ_j, of which, however, only $i - 1$ are independent of the previous cycles. Hence $2\pi = 2\pi_1 + 2\pi_2 + 2(i - 1)$.

5. The virtual characters of an arbitrary linear system. In order to extend the definitions of the virtual degree and virtual genus of a complete system $|C|$ to reducible systems without *assigned* fixed components or to systems of dimension zero (i. e. consisting of only one curve), we use an auxiliary system $|K|$, which we shall suppose, for simplicity, to satisfy the following conditions: 1) $|K|$ is irreducible, of dimension $\geqq 1$, and without base points; 2) the system $|K + C|$ is irreducible, free from accidental base points or base points of accidental hypermultiplicity. We may assume as $|K + C|$, for instance, the system, completed if necessary, cut out on f by the hypersurfaces of a sufficiently high order and on which all the base points of $|C|$ are imposed. Let n and π be the virtual ($=$ effective) degree and genus of $|K|$, N and Π the analogous characters of $|K + C|$, and $i = (K \cdot C)$. We define the virtual degree x and the virtual genus y of C by using *formally* the relations (6) and (6'), i. e. as follows:

(7) $\qquad \pi + y + i - 1 = \Pi,$ $\qquad$ (7') $\quad n + x + 2i = N.$

The values of x and y derived from (7) and (7') *are independent of* $|K|$. To see this it is sufficient, given another system K_1 satisfying the conditions 1) and 2), to apply (6) and (6') to the system $|K + K_1 + C|$ $= |(K + C) + K_1| = |(K_1 + C) + K|$.[1]

We are now in position to speak of the virtual characters of any linear system whatsoever. The definitions given are such that the formulas (6) and (6') have now general validity. In the sequel we shall speak of the virtual degree and of the virtual genus of a curve C, meaning by it the corresponding characters of $|C|_H$, and we shall denote these characters by $(C^2)_H$ and by $[C]_H$ respectively, or simply by (C^2) and $[C]$ (SEVERI, 13).

We observe that the numerical function of C, $\varphi(C) = 2[C] - (C^2) - 2$, is an additive function of curves. One could use this property of $\varphi(C)$ for the definition of $[C]$ (LEFSCHETZ, d.).

If C is an *isolated* curve, i. e. if $|C|$ consists of C only, then (C^2) may be negative. Conversely, if $(C^2) < 0$, *and if the virtual multiplicities of the base points of* $|C|$ *satisfy the proximity inequalities*, then C is either an isolated curve, or one of its components must be an isolated curve and an accidental fixed component of $|C|$. *If the proximity inequalities are not satisfied, then* $|C|$ *may be of negative virtual degree even if it is of dimension* >0 *and irreducible* (see section 2).

The symbol $(C \cdot D)$, denoting the number of intersections of two curves C and D, has a definite meaning even if C and D have common components. Let

$$C = m_1 \Gamma_1 + m_2 \Gamma_2 + \cdots + m_k \Gamma_k,$$
$$D = \mu_1 \Gamma_1 + \mu_2 \Gamma_2 + \cdots + \mu_k \Gamma_k,$$

where the m_i's and the μ_i's are non-negative integers, and each Γ_i is irreducible. If C and D are virtually free from base points, then

$$(C \cdot D) = \sum_{i,j=1}^{k} m_i \mu_j (\Gamma_i \cdot \Gamma_j).$$

The value of $(C \cdot D)$ being known, the evaluation of $(C \cdot D)_{H_1, H_2}$ is immediate. In particular, if for each Γ_i two sets of base points, $H_1^{(i)}$ and $H_2^{(i)}$, can be found, such that $H_1 = \sum m_i H_1^{(i)}$, $H_2 = \sum \mu_i H_2^{(i)}$, then $(C \cdot D)_{H_1, H_2}$

$$= \sum_{i,j=1}^{k} m_i \mu_j (\Gamma_i \cdot \Gamma_j)_{H_1^{(i)}, H_2^{(j)}}.$$

In this case, if $(C \cdot D) < 0$, and if the proximity inequalities are satisfied, we must have for some value of i, $(\Gamma_i^2)_{H_1^{(i)}, H_2^{(i)}} < 0$, $m_i \neq 0$, $\mu_i \neq 0$,

---

[1] In the proof we make the tacit assumption that the virtual multiplicities of the base points of $|C|$ satisfy the proximity inequalities, because otherwise the auxiliary system $|K|$ would not exist. To define in this case the virtual characters of $|C|$ we may use the same method as in section 3, after having defined the virtual character of $|C|$, *supposed virtually free of base points*.

since $(\Gamma_i, \Gamma_j)_{H_1^{(i)}, H_2^{(j)}} \geqq 0$, for $i \neq j$. In particular, if C and D are virtually free from base points and if $(C \cdot D) < 0$, *the two curves must have in common at least one irreducible component of virtual negative degree.*

From the topological point of view, the virtual degree of a curve C is simply the intersection number of two 2-cycles homologous to C (VI, 3).

If C is irreducible, of effective genus p', and if the proximity inequalities are satisfied, then $[C]_H \geqq p'$, and $[C]_H = p'$ if and only if all the multiple points of C are assigned with virtual multiplicities equal to their effective multiplicities for C.

Let $|C|_H = m_1 |C_1|_{H_1} + \cdots + m_k |C_k|_{H_k}$, where the m_i's are non-negative integers and $|C_i|_{H_i}$ is irreducible, of virtual degree n_i and virtual genus p_i. By repeated application of (6) and (6') we find:

$$p = [C]_H = \sum m_i p_i + \sum \frac{m_i(m_i-1)}{2} n_i$$
$$+ \sum_{i<j} m_i m_j (C_i \cdot C_j)_{H_i, H_j} - \sum m_i + 1 .$$

C is said to be *disconnected* or *connected* according as it is or it is not the sum of two complementary components having no points in common *outside the assigned base points* (NOETHER, 1; CASTELNUOVO, 1). If C has no multiple components, then $[C]_H = \sum p_i + \sum_{i<j}(C_i \cdot C_j)_{H_i, H_j} - k + 1$, and if it is connected, then $\sum_{i<j} (C_i \cdot C_j)_{H_i, H_j} \geqq k - 1$. Hence, assuming always that the proximity inequalities are satisfied, we have that *the virtual genus of a connected curve free from multiple components is $\geqq 0$.* If the proximity inequalities are not satisfied, the genus may become negative even in the simplest case of an irreducible curve. For instance, if C is a plane curve of order 5 and if $H = O^2 O_1^4$, where O_1 is infinitely near O, then $[C]_H = -1$. Effectively O and O_1 are triple points and C is rational.

A very fruitful remark, due to ENRIQUES (10), is that *if a reducible curve C_0 is the limit of an irreducible curve, in particular, if C_0 is the total curve of an irreducible linear system, then C_0 is connected (Principle of degeneration).* If, in fact, C_0 is a total curve of an irreducible system $|C|_H$ and if $C_0 = C_0' + C_0''$, where C_0' and C_0'' have no common components, then it must be possible to join *any* two points of C_0, in particular a point of C_0' to a point of C_0'', by means of a continuous path on C_0 without crossing the base points in H, since this is possible to do on the generic *irreducible* curve of $|C|_H$. For this, however, it is necessary that C_0' and C_0'' should have intersections outside of H, i. e. that $(C_0' \cdot C_0'') > 0$.

6. Exceptional curves (NOETHER, 2; ENRIQUES, 5; CASTELNUOVO-ENRIQUES, 2; JUNG, 4). Given on the surface f an irreducible linear system Σ of dimension $r \geqq 3$, cut out by a system of hypersurfaces (2), the formulas

$$\varrho y_i = \varphi_i(x_0, x_1, \ldots, x_k), \qquad i = 0, 1, \ldots, r,$$

define a rational transformation T of f into a surface f' in $S_r(y_0, y_1, \ldots, y_r)$, on which to the curves C of Σ there correspond the hyperplane sections of f'. In order that T be *birational*, it is necessary and sufficient that Σ *be not composed of an involution of sets of points of order $\nu > 1$*, i. e. that the ∞^{r-1} curves of Σ constrained to pass through a generic point P of f do not pass automatically through other $\nu - 1$ points $P_1, P_2, \ldots, P_{\nu-1}$, which vary as P varies. If this condition is satisfied, Σ is said to be *simple*, and the order of the surface f' is equal to the effective degree n of Σ. If, however, Σ is composed of an involution of order ν, then the points $P, P_1, \ldots, P_{\nu-1}$ of a set of the involution correspond to one and the same point of f', and f is represented upon the ν-fold covered surface f', which is in this case of order n/ν.

The invariant properties of the linear system Σ are reflected in the projective properties of f'. For instance, if Σ is complete (all its base points being assigned with their effective multiplicities), then it is not difficult to see that in that case (and only in that case) f' is a *normal* surface, i. e. it is not the projection of a surface of the same order as f' and belonging to a space $S_{r'}, r' > r$ (for other examples, see I, 6).

The base points of Σ are *fundamental points* of T, i. e. they are transformed into curves of f'. On the other hand T may possess on f *fundamental curves*, which are transformed into (simple or singular) points of f'. If L is such a curve, then it cannot have variable intersections with the curves of Σ, so that the curves of Σ which contain a generic point of L form a linear subsystem ∞^{r-1}, having L as a fixed component. L is also said to be a fundamental curve of Σ. These exceptions to the $(1 - 1)$ character of a birational transformation between two surfaces present new features, which are not encountered in the theory of curves. It is well known, in fact, that in a birational transformation between two curves the fundamental points on either curve can be only among the multiple points. In view of this, the exceptions to the $(1 - 1)$ correspondence can be formally removed, if instead of the points we consider the branches of a curve. At any rate, a birational transformation between two curves free from singularities is a $(1 - 1)$ correspondence between their points, absolutely without exceptions. However, in the theory of surfaces, even if we make allowances for the singularities (for instance, by refusing to consider a singularity of a surface as a *point* of the surface), or if we consider only surfaces without singularities, there still remains the essentially new fact, that to certain curves of one of the surfaces may correspond *simple* points of the other surface. Such curves on f are called *exceptional curves*. We can give a definition of an exceptional curve, which is independent of the supposedly existing transformation. Let us say, for brevity, that L is a *total* fundamental curve of the linear system Σ, if the ∞^{r-1} curves of Σ containing L do not have other fixed components outside L. *A curve L (irreducible or not) is an exceptional curve, if there exists on f a linear system Σ, such that:*

1. *Σ is irreducible, simple, of dimension $r \geqq 3$;*

2. *L is a total fundamental curve of Σ;*

3. *if n is the effective degree of Σ, then the residual curves of L with respect to Σ, i. e. the curves C_1 such that $C_1 + L$ is a total curve of Σ, form a linear system $\Sigma_1 (\infty^{r-1})$ of effective degree $n - 1$[1].*

If it is possible to find the above system Σ in such a manner that, in addition, it should have no base points on L, then L is called an exceptional curve *of the 1st kind*. If that is not possible, L is called an exceptional curve *of the 2nd kind*. Hence, an exceptional curve L is of the 1st or of the 2nd kind, according as it is or it is not possible to transform L into a simple point in such a manner that no point of L is at the same time transformed into a curve.

An exceptional curve may be reducible (CASTELNUOVO-ENRIQUES, 2). A systematic treatment of reducible exceptional curves is not available. The following treatment is a contribution to the theory of exceptional curves of the 1st kind.

If L is an exceptional curve of the 1st kind, then $(L^2) = 1$ and $[L] = 0$[2].
Proof. Let H denote the set of effective base points of Σ. We have $(C \cdot L) = ((C_1 + L) \cdot L) = 0$, since no base point of Σ is on L. Consequently $(C_1 \cdot L) = -(L^2)$, and since $C_1 + L$ is connected, it follows $(L^2) < 0$. We also have $n = (C^2)_H = (C_1^2)_H + 2 (C_1 \cdot L) + (L^2) = (C_1^2)_H - (L^2)$. From the hypothesis that the effective degree of Σ_1 is $n - 1$, it follows that $(C_1^2)_H \geqq n - 1$, and that consequently $(L^2) \geqq -1$. Hence $(L^2) = -1$, and incidentally

$$(8) \qquad\qquad (C_1^2)_H = n - 1,$$

$$(8') \qquad\qquad (C_1 \cdot L) = 1.$$

Furthermore, $[C]_H = [C_1]_H + [L] + (L \cdot C_1) - 1 = [C_1]_H + [L]$. The curves C_1 correspond to the hyperplane sections of f' through a simple point O and have therefore the same effective genus as the curves C. By (8), the effective genus of C_1 is equal to $[C_1]_H$. Hence $[C]_H = [C_1]_H$, and $[L] = 0$, q. e. d.

Let $L = k_1 L_1 + k_2 L_2 + \cdots + k_s L_s$, where each L_i is irreducible and $k_i \geqq 1$. Choosing properly our notations we deduce from (8'):

---

[1] Σ_1 is necessarily irreducible, since by 2. it is free from fixed components and since by 1. ($r \geqq 3$) it is of positive effective degree and therefore cannot be composed of the curves of a pencil.

[2] L is considered as virtually free from base points.

$k_1 = 1$, $(L_1 \cdot C_1) = 1$, $(L_i \cdot C_1) = 0$ $(i = 2, 3, \ldots, s)$. Since $(C \cdot L_i) = 0$, it follows that $(L \cdot L_1) = -1$ and $(L \cdot L_i) = 0$, if $i > 1$. We have then the following set of relations:

(9)
$$\begin{cases} (L_1^2) + \sum_{j=2}^{s} k_j (L_1 \cdot L_j) = -1, \\ (L_1 \cdot L_i) + \sum_{j=2}^{s} k_j (L_i \cdot L_j) = 0, \qquad i = 2, 3, \ldots, s. \end{cases}$$

Furthermore $[L] = 0$ furnishes, by making use of the relation $(L^2) = -1$, the following relation:

(10)
$$\sum_{j=1}^{s} k_j (n_j - 2p_j + 2) = 1,$$

where $p_j = [L_j] \geq 0$ and $n_j = (L_j^2)$.

An obvious consequence of (9) is that $(L_i^2) < 0$ for any i, since for a given value of i one at least of the intersection numbers $(L_i \cdot L_j)$ must be positive (L is connected).

We shall now assume that all the components of L are simple[1]. In this case $k_1 = k_2 = \cdots = 1$ and the relations (9) and (10) become

(9′)
$$\begin{cases} \sum_{j=1}^{s} (L_1 \cdot L_j) = -1, \\ \sum_{j=1}^{s} (L_i \cdot L_j) = 0, \qquad\qquad i > 1, \end{cases}$$

and

(10′)
$$2\sum_{j=1}^{s} p_j - \sum_{j=1}^{s} (L_j^2) = 2s - 1.$$

In view of (9′), the relation (10′) can be written as follows:

$$2\sum_{j=1}^{s} p_j + \sum_{i \neq j} (L_i \cdot L_j) = 2s - 2.$$

Since L is connected, we must have

$$\sum_{i < j} (L_i \cdot L_j) \geq s - 1.$$

Consequently

(11) $p_j = 0$,

(12) $\sum_{i < j} (L_i \cdot L_j) = s - 1$.

From (12) it follows that if $\bar{L}$ and $\bar{\bar{L}}$ are any two connected components of L, then $(\bar{L} \cdot \bar{\bar{L}}) = 0$ or 1 [in particular all the intersection numbers $(L_i \cdot L_j)$, $i \neq j$, are 0 or 1]. Let us suppose that $L_2 + L_3 + \cdots + L_s$

---

[1] The general case will he studied in a joint paper by Dr. S. F. Barber and the present author to appear shortly in the American Journal of Mathematics.

consists of connected components L', L'', ..., $L^{(\sigma)}$ having two by two no points in common. Then $(L_1 \cdot L^{(\alpha)}) = 1$ and, by (9'), $(L_1^2) = -(\sigma + 1)$. *Each curve $L^{(i)}$ satisfies relations analogous to (9') and (10'), in particular* $(L^{(i)})^2 = -1$, $[L^{(i)}] = 0$. In fact, let, for instance, $L' = L_2 + L_3 + \cdots + L_{s'}$, and let $(L_2 \cdot L_1) = 1$, $(L_j \cdot L_1) = 0$ $(j = 2, \ldots, s')$. If we write the relations (9') for $i = 2, 3, \ldots, s'$, we obtain:

$$\sum_{j=2}^{s'} (L_2 \cdot L_j) = -1,$$

$$\sum_{j=2}^{s'} (L_i \cdot L_j) = 0, \qquad\qquad i = 3, 4, \ldots, s'.$$

We have moreover $[L] = p_1 + [L'] + [L''] + \cdots + [L^{(\sigma)}] = 0$. Since $p_1 = 0$ and since the virtual genus $[L^{(i)}]$ of the connected curve $L^{(i)}$ is $\geqq 0$, it follows that $[L^{(i)}] = 0$.

The character of the correspondence between the curve L and the neighborhood on f' of the simple point $O = T(L)$ can now be described. The curves C_1 meet L_1 in one variable point, but do not meet any of the remaining components of L. Hence the neighborhood proper of O (of the 1st order) is represented by the curve L_1, whose points are in $(1 - 1)$ correspondence with the directions about O. But there are σ singular directions. The corresponding points O_1', O_2', ..., O_σ', infinitely near O, are fundamental for T^{-1}, and to O_i' corresponds the curve $L^{(i)}$, which by itself is an exceptional curve of the 1st kind. That component of L', which we have denoted by L_2 and which is such that $(L_2 \cdot L_1) = 1$, has the same privileged rôle in L' as L_1 has in L. The 1st neighborhood proper of O_1' is represented by the curve L_2, whose points are in $(1 - 1)$ correspondence with the second order differential elements through O, O_1'. If L' is reducible, there will exist fundamental points of T^{-1} infinitely near O_1' (in the second order neighborhood of O), to which will correspond the connected components of $L' - L_2$, themselves exceptional curves of the 1st kind, etc. The virtual degree of any *irreducible* component of L, corresponding to a fundamental point $O_j^{(i)}$ in the ith neighborhood of O, is equal to $-(\sigma_{ij} + 1)$, where σ_{ij} is the number of the fundamental points on f' which immediately follow $O_j^{(i)}$. It can be shown that the fundamental points all lie on *linear branches* and that, conversely, this property of the fundamental points characterizes the case of exceptional curves free from multiple components[1].

---

[1] The proof is given in the joint paper by Dr. BARBER and myself referred to in the preceding footnote. We quote the following results established in this paper: 1) *The virtual degree of an irreducible component L_i of $L = \sum k_j L_j$, corresponding to a fundamental point O_{i-1} infinitely near O, equals $-(\sigma_i + 1)$, where σ_i is the number of fundamental points which follow O_{i-1} and are proximate points of O_{i-1}. 2) If the points $O, O_1, \ldots, O_{k-1}$ lie on a single branch, then L_i meets in one point $L_{i+\sigma_i}$, where $O_{i+\sigma_i-1}$ is the last proximate point of O_{i-1}, does not meet any other component L_j, $j > i$, and the branch of the lowest order passing through $O, O_1, \ldots, O_{i-1}$ is of order k_i.*

The importance of the distinction between exceptional curves of the 1st and of the 2nd kind is due to the following fact. If f possesses only exceptional curves of the 1st kind, then it can be proved that it is possible to transform f into a surface f' free altogether from exceptional curves (IV, 4). The difficulties inherent to the existence of exceptional curves are in this manner eliminated. On the other hand, *if f possesses exceptional curves of the 2nd kind, this will still be true of any birational transform of f*. The proof is immediate. Let L be an exceptional curve of the 2nd kind on f, and let us suppose, for simplicity, that L is irreducible. Then if L is transformed into a simple point O of a surface f', some point of L is transformed into a curve L', which is easily seen to be an exceptional curve of the 2nd kind. If L is transformed into a curve L' and a simple point of L is transformed into an exceptional curve L'' of the 1st kind, then both L' and L'' are exceptional curves of the 1st kind, but $L' + L''$ is of the 2nd kind. For further properties of exceptional curves of the second kind see section 7.

The class of surfaces possessing exceptional curves of the 2nd kind is, however, well known: they are the surfaces birationally equivalent to ruled surfaces (IV, 4).

7. Invariance of the virtual characters. In a birational transformation T between two surfaces f and f', every linear system Σ on f is transformed into a linear system Σ' on f'. The requirement that the virtual characters of Σ and Σ' be the same leads to certain conventions concerning the definition of the total curves C' of Σ' (ENRIQUES, 9). We may confine ourselves, for simplicity, to birational transformations which do not possess infinitely near fundamental points, since any birational transformation is the product of birational transformations of this type. If the fundamental points of T are outside the base points of Σ, no special conventions are necessary; C' is simply $T(C)$, the transform of the total curve C of Σ. If T has a fundamental point O, which is a base point of Σ, of virtual multiplicity i and effective multiplicity j, then necessarily $j \geqq i$, since by hypothesis O is a proper point of f, and we agree that the exceptional curve L' on f' which corresponds to O is to be counted $j - i$ times as an *accidental* fixed component of C'. As regards the virtual multiplicities of the base points of Σ', they are defined by the above conventions, if applied to T^{-1} and $\Sigma = T^{-1}(\Sigma')$, and by the condition that homologous base points of Σ and Σ', not fundamental for T and T^{-1}, should be assigned with the same virtual multiplicities. It is then easy to prove that the virtual characters of Σ and Σ' are the same, by using the auxiliary system K of section 5. To elucidate the necessary character of the preceding convention, we consider an example. Let O be an assigned simple base point of a complete system $|C|$. If we consider O as an accidental base point, then we may possibly obtain a larger system $|\ddot{C}|$, containing $|C|$, for which O is not a base point. It is then clear that as $\bar{C}$ varies and approaches a curve C, $T(\bar{C})$ approaches $T(C) + L'$.

It is seen immediately that according to the above definitions the sum or the difference of two systems on f goes into the sum or the difference of the two corresponding systems on f'.

As an illustration of the case in which the difference between the effective and the virtual multiplicities is brought about by the principle of discharge (section 2), let us suppose that Σ possesses, infinitely near the simple base point O, r simple base points O_i in distinct directions $(r > 1)$. O is an effective double base point and L' is an accidental fixed component of Σ'. Moreover, the r points O_i' on L', which correspond to the directions OO_i, should be considered as assigned base points of Σ'.

It is easily seen that the above rule holds also for birational transformations possessing infinitely near fundamental points. In this case, however, the possibility of having $j < i$ calls for an explanation. Let $O, O_1, \ldots, O_k$ be a sequence of infinitely near base points of Σ, which are fundamental for T, and let their virtual and effective multiplicities for Σ be $i, i_1, \ldots, i_k$; $j, j_1, \ldots, j_k$ respectively. To each point O_h there will correspond on f' an exceptional curve $L^{(h)}$, and every irreducible component of $L^{(h)}$ is also a component of $L^{(q)}$, if $q < h$. Our rule says that the curve $\Gamma = (j - i)L + (j_1 - i_1)L^{(1)} + \cdots + (j_k - i_k)L^{(k)}$ must be considered as an accidental fixed component of C'. Some of the coefficients $j_h - i_h$ may be negative, but for each negative coefficient $j_h - i_h$ there must be at least one positive coefficient $j_q - i_q$, such that $q < h$. In the final result each irreducible component of the curves $L^{(h)}$ will have to occur in Γ to a non-negative multiplicity.

The preceding considerations lead quite naturally to the definition of *addition of points to a complete linear system*, given in section 4. That definition was to the effect that if O is a base point of $|C|$, of virtual multiplicity i, the system $|C + O^k|$ is simply obtained by assigning O as a base point of virtual multiplicity $i - k$, instead of i. If O is a fundamental point of T, the corresponding effect on the transformed system $|C'|$ consists in the addition to $|C'|$ of the exceptional curve L' counted k times.

An application to exceptional curves of the second kind. We go back to the definition of an exceptional curve L given in the preceding section, and we suppose that L passes through some of the base points of Σ, so that L is an exceptional curve of the second kind. We take as sets of base points for Σ and Σ_1 the sets of their effective base points taken with their effective multiplicities. This defines a set G of base points for L, in virtue of the relation $L + C_1 \equiv C$. We consider the linear system of curves cut out on f by hypersurfaces of a sufficiently high order, passing through all the base points of Σ. Referring the curves of this system to the hyperplanes of a linear space (section 6) we obtain a surface $f' = T(f)$, birationally equivalent to f, on which the irreducible transforms $C' = T(C)$ of the curves C form a system Σ' effectively free from base points. Since Σ and Σ_1 are irreducible and since the virtual and effective multiplicities of their base points coincide, it follows that the same holds for the transformed systems Σ' and Σ_1', and that $C' = T(C)$ and $C_1' = T(C_1)$. Moreover Σ_1' is of dimension and degree one less than Σ'. Hence Σ' possesses a total fundamental curve L', which is an exceptional curve of the first kind (since Σ' is effectively free from base points). Since $C' \equiv C_1' + L'$, *it follows that L' is the transform of L, and we conclude that* $(L^2)_G = -1$, $[L]_G = 0$. These relations are analogous to those proved for exceptional curves of the first kind, the only difference being that now the virtual characters of L are evaluated with respect to a non-vacuous set of base points G.

8. Virtual characteristic series. Virtual curves (SEVERI, 4, 6). We have defined in section 1 the characteristic series g_n^{r-1} of an irreducible system of *effective* degree n and of dimension $r > 0$. It is obvious how this definition should to be modified for an irreducible complete system $|C|_H$ of *virtual* degree n, defined with respect to a set H of base points. The definition fails, however, if we deal with an irreducible *isolated* curve C. In this case we may proceed as follows. It is always possible to find (in many ways) an irreducible system $|E|$ such that the system $|D| = |E - C|$ exists and does not contain C as a fixed component. The *virtual characteristic series* of C is defined as *the difference of the two series cut out on C by $|E|$ and $|D|$* (in the sense of the theory of linear series of sets of points on an algebraic curve). It is easily proved that the series thus defined is independent of the auxiliary system $|E|$ and that its order is equal to the virtual degree of C. If $(C^2) \leqq 0$, the characteristic series of C is certainly not an effective series on C. But also if $(C^2) > 0$, the virtual characteristic series of C may not be effective. If the difference of the above two series effectively exists on C and if C is contained in a system $|C|$ of dimension > 0, the virtual (complete) characteristic series of C coincides with or totally contains the characteristic series of $|C|$ on C. If C is isolated (from the point of view of linear equivalence) but belongs to a *continuous system* of dimension > 0, the virtual characteristic series of C coincides with or totally contains the characteristic series of the continuous system (V, 1).

If $|A|$ and $|B|$ are two linear systems and if $|A - B|$ does not exist, there may nevertheless exist the system $|E + A - B| = |E + A| - |B|$, where $|E|$ is a system of a sufficiently high dimension. The symbol $A - B$ does not correspond to any effectively existent curve on f, but has a definite significance as an operation on linear systems. As a further step in the direction of, and in conformity with, the preceding definitions of virtual entities on a surface, we shall refer to the symbol

$A - B$ as a *virtual curve*. The relation of linear equivalence and the notions of virtual degree and genus can be immediately extended to virtual curves. Thus, $A - B \equiv A' - B'$ simply means $A + B' \equiv A' + B$. The virtual characters of a virtual curve are formally defined, in the same manner as for reducible curves, by means of the equations (6) and (6'). From the topological point of view, a virtual curve is simply an algebraic cycle which does not correspond to an effective algebraic curve on f (VI, 3). The notion of a virtual curve has been introduced and systematically developed by Severi. It is interesting, however, to observe that earlier investigations by Castelnuovo and Enriques on the *a priori* virtual canonical and pluricanonical systems (III, 9) on surfaces of genus $p_g = 0$ already clearly contain this notion without mentioning it explicitly.

Remark. The introduction of virtual curves allows us to assign on a given effective curve C a completely arbitrary set H of base points with arbitrary virtual multiplicities s_i, whereas until now these virtual multiplicities s_i had to satisfy the inequalities (5) of section 2, which involve the effective multiplicities s_i' of C at the points of H. We see this immediately, if we transform the points of the set H into exceptional curves. It is easily seen that the inequalities (5) give a necessary and sufficient condition that the transform C' of C contain any irreducible component of these exceptional curves to a non-negative multiplicity, i.e. that C' be an effective curve. We shall say in the sequel that C_H, i.e. the curve C with the assigned base set H on it, is an *effective curve*, if C is an effective curve and if the inequalities (5) hold.

Appendix to Chapter II

By Joseph Lipman

In this appendix we will reconsider, from a different point of view, the main ideas of the foregoing Chapter II (referred to as "the text").

1. Let f be a non-singular irreducible projective surface over an algebraically closed field k, with function field K/k. The (closed) points of f are in one-one correspondence with their (two-dimensional) local rings on f; accordingly we will simply call these local rings "points on f". In fact we will refer to *any* two-dimensional regular local ring with fraction field K as a "point". A point O is said to be "infinitely near" to f if O contains some point on f. This terminology is justified by the following fact (ABHYANKAR, 2, theorem 3, p. 343): *if O is infinitely near to f then there exists a (unique) sequence of points*

$$O_1 < O_2 < \cdots < O_r = O$$

such that O_1 is a point on f, and for each $i = 1, 2, \ldots, r-1$, O_{i+1} is a quadratic transform of O_i (i. e. O_{i+1} is a point on the surface obtained from f by blowing up successively $O_1, O_2, \ldots, O_i$).

By associating to each point O, with maximal ideal $m(O)$, the "order valuation" ord_O (determined by the condition that for non-zero $x \in O$, $\mathrm{ord}_O(x) = \max\{t \mid x \in m(O)^t\}$) we obtain a one-one correspondence between infinitely near points and prime divisors of the second kind with respect to f (i. e. valuations of K/k centered at a point on f, and with residue field transcendental over k) (cf. ABHYANKAR, 2, proposition 3, p. 336).

The text deals, in essence, with the free abelian group Δ on the set of *all* prime divisors, of first and second kind, on f. After identifying prime divisors of the first kind with integral curves on f (namely, their respective centers) and prime divisors of second kind with infinitely near points, we can represent any element W of Δ uniquely in the form $W = C + H$, where C is a divisor on f (= finite formal sum of integral curves) and

$$H = \sum_{i=1}^{r} s_i O_i$$ with infinitely near points O_i and integers s_i; s_i is called the "virtual multiplicity" of W at O_i. To conform with the text, we denote such a W by C_H and think of it as the divisor C together with the "base divisor" H.

Now let f' be a non-singular projective surface birationally equivalent to f, and let Δ' be the free abelian group generated by the prime divisors on f'. We will define an isomorphism $\Theta_{f',f}: \Delta \to \Delta'$; $\Theta_{f',f}(C_H) = C'_{H'}$ will be called the "transform" of C_H on f'. Suppose first that f' is obtained from f by a quadratic transformation, i. e. by blowing up a closed point O on f. Let $T: f' \to f$ be the domination map, and let $L' = T^{-1}(O)$ be the exceptional curve. Then, if i is the virtual multiplicity of C_H at O, we define $C'_{H'}$ by $C' = T^{-1}(C) - i L'$, $H' = H - i O$ where $T^{-1}(C)$ is the so-called total transform of C. Next assume only that the birational map from f' to f is defined everywhere on f' (i. e. f' dominates f). Then the Factorization Theorem of ZARISKI (cf. ZARISKI, 24, § II.1) asserts that *f' is obtained from f by a sequence of quadratic transformations.* Hence by repeated application of the preceding transformation process we obtain the transform of C_H on f'. (One checks that this definition does not depend on the choice of the sequence of quadratic transformations leading from f to f'.) Finally, in the most general case, we can always find a non-singular surface f'' dominating both f and f', and set $\Theta_{f',f} = (\Theta_{f'',f'})^{-1} \circ (\Theta_{f'',f})$. (This is easily seen to be independent of the choice of f''.)

One studies then those properties of C_H which are *invariant* under $\Theta_{f',f}$ for all f'. It is evident that in order to check that a property is invariant, it is enough to consider only the case when f' is obtained from f by a quadratic transformation. What is more, it can be shown that for given C_H there is an f' on which the transform $C'_{H'}$ of C_H is an ordinary divisor on f', i. e. $H' = 0$. Thus, *with each invariant property of C_H there is associated a property of ordinary divisors which is invariant under quadratic transformations, and conversely.* For example, if C and D are divisors on a surface f, if f' is obtained from f by a quadratic transformation, and if C' and D' are the total inverse images of C and D on f', then $(C' \cdot D') = (C \cdot D)$ and $\pi(C') = \pi(C)$ (where π is the virtual arithmetic genus). Therefore it is possible to define *invariantly* (and in just one way) the virtual arithmetic genus of a C_H, and the intersection number of two such objects. Explicitly, if $H = \Sigma \mu_i O_i$, $G = \Sigma \nu_i O_i$, then

$$\pi(C_H) = \pi(C) - \tfrac{1}{2} \Sigma \mu_i(\mu_i - 1),$$
$$(C_H \cdot D_G) = (C \cdot D) - \Sigma \mu_i \nu_i.$$

2. There is another interpretation of the theory which may make it seem more natural. For this purpose, one makes use of the ZARISKI-RIEMANN space Z of K/k, which is the set of valuation rings of K/k topologized by taking as basic open sets those of the form U_S, where S is a finite subset of K and

$$U_S = \{z \in Z \mid S \subseteq z\}$$

(cf. ZARISKI-SAMUEL, 1, Ch. VI, § 17). Z is also a ringed space with the structure sheaf $\mathcal{O}_Z$ whose ring of sections over any open subset U of Z is the intersection of all members of U. Thus it makes sense to speak of (locally principal) divisors on Z. For each surface f' as above, and each f'' dominating f', the domination map $f'' \to f'$ (resp. $Z \to f'$) induces the "inverse image" map from the group of divisors $\text{Div}(f')$ to the group of divisors $\text{Div}(f'')$ (resp. $\text{Div}(Z)$). It is easily shown that

(*)
$$\text{Div}(Z) = \varinjlim_{f'} \text{Div}(f') \,.$$

A key fact is that *there is a family of isomorphisms*

$$\Theta_{f'} : \Delta' \to \text{Div}(Z)$$

(f', Δ', as before) *such that, for any projective non-singular surface f'' birationally equivalent to f, we have*

$$\Theta_{f',f''} = (\Theta_{f'})^{-1} \circ \Theta_{f''} \,.$$

Thus a member of Δ' may be viewed as the representative on f' of a divisor on Z, and then its transform on f'' is the representative on f'' of the same divisor. In other words, *the "invariant" theory outlined above is nothing but the theory of divisors on Z.*

The existence of the isomorphisms $\Theta_{f'}$ follows at once from (*) and the fact that any $C'_{H'}$ on f' transforms into an ordinary divisor C'' on a suitable dominating surface: $\Theta_{f'}(C'_{H'})$ is then the inverse image of C'' on Z. We can obtain more explicit descriptions of Θ_f and Θ_f^{-1} in the following way:

For any divisor D on Z and any valuation v of K/k, there exists $x \in K$ such that D is equal to the divisor of x in some neighborhood on Z of the valuation ring of v (D is locally principal). If x' is another such element then $v(x) = v(x')$, and hence we can define $v(D)$ to be $v(x)$. For example, if H and $\Theta_f(H)$ are as above and O is infinitely near to f, with "quadratic sequence"

$$O_1 < O_2 < \cdots < O_r = O$$

then one finds, with

$$m_i = \text{maximal ideal of } O_i \,,$$
$$\text{ord}_O(m_i) = \min_{y \in m_i}(\text{ord}_O(y)) \,,$$
$$s_i = \text{virtual multiplicity of } H \text{ at } O_i$$

that

(**)
$$\text{ord}_O(\Theta_f(H)) = - \sum_{i=1}^{r} s_i \, \text{ord}_O(m_i) \,.$$

This formula determines $\Theta_f(H)$ since a divisor on Z is easily seen to be determined by its values at all the prime divisors on f (and $v(\Theta_f(H)) = 0$ for all v of first kind). We may remark here that in the text the integer $\text{ord}_O(m_i)$ is called "the multiplicity of O_i on a branch of lowest order passing through $O_1, O_2, \ldots, O_{r-1}, O$".

With the divisor D we can associate a divisor C on f, namely

$$C = C(D) = \sum_L \mathrm{ord}_L(D) \cdot L$$

where L runs through all integral curves on f and ord_L is the discrete valuation corresponding to L. The divisor

$$D_f = D - (\text{inverse image of } C \text{ on } Z)$$
$$= D - (\Theta_f(C))$$

may be called the "base part of D, with respect to f".

For any point O on a surface f', we define the *virtual multiplicity of D at O*, $s_O(D)$ in symbol, to be the integer $-\mathrm{ord}_O(D_{f'})$. (For a given point O, this integer does not depend on the choice of f'.) Then one shows that for almost all points O infinitely near to f, $s_O(D)$ vanishes, and, with

$$H = H(D) = \sum_O s_O(D) \cdot O \quad (O \text{ infinitely near to } f)$$

we have

$$\Theta_f(H) = D_f$$

i. e.

$$\Theta_f^{-1}(D) = C_H \qquad\qquad (C, H, \text{ as above}).$$

3. While we have dealt only with *divisors and base divisors*, the emphasis in the text is on *linear systems, with base conditions*. To relate the two, we first remark that any non-zero x in K defines a divisor (x) on Z whose corresponding object $\Theta_f^{-1}((x))$ on the surface f is just the usual divisor $\mathrm{div}_f(x)$ of x on f (with zero base divisor). It follows at once that two divisors D_1 and D_2 on Z are linearly equivalent (i. e. $D_1 - D_2 = (x)$ for some x) if and only if, for the corresponding $C_{i, H_i} = \Theta_f^{-1}(D_i)$ $(i = 1, 2)$ we have that C_1 is linearly equivalent to C_2 (in fact $C_1 - C_2 = \mathrm{div}_f(x)$) and $H_1 = H_2$.

We say naturally that $C_H \geqq 0$ if $\Theta_f(C_H) \geqq 0$, i. e. if

(a) $v(C) \geqq 0$ for all prime divisors v of first kind on f; i. e. C is a positive divisor on f, and

(b) $\mathrm{ord}_O(\Theta_f(C)) + \mathrm{ord}_O(\Theta_f(H)) \geqq 0$ for all O infinitely near to f.

In view of (**), (b) simply says that the divisor C "satisfies the base conditions imposed by H".

Thus, corresponding to $|D|$, the set of all positive divisors on Z linearly equivalent to D, there is (with $C_H = \Theta_f^{-1}(D)$) the set of all positive divisors on f which are linearly equivalent to C and satisfy the base conditions imposed by H; this is the set denoted by $|C|_H$ in the text. Our description of $|C|_H$, being in terms of divisors on Z, *is automatically invariant*. In other words, if $C'_{H'}$ is the transform of C_H on a birationally equivalent surface f', then $|C'|_{H'}$ consists precisely of all the transforms of members of $|C|_H$. So we have the notion of "transform of a linear system with base conditions".

4. Next we discuss "effective base divisors" and "proximity inequalities". For any C_H, the set of all x (including $x = 0$) such that $\operatorname{div}_f(x) + C_H \geqq 0$ forms a finite-dimensional vector space V over k. We assume that $V \neq (0)$, i. e. $|C|_H$ is not empty. Then the sheaf $V\mathcal{O}_Z$ is an invertible $\mathcal{O}_Z$-module, so that $V\mathcal{O}_Z = \mathcal{O}_Z(D)$ for some divisor D on Z. Let $C^*{}_{H^*} = \Theta_f^{-1}(D)$ be the corresponding object on f. The difference $C_H - C^*{}_{H^*}$ is then $\geqq 0$ and it depends only on $|C|_H$; it is called the "fixed part" (or "unassigned base") of $|C|_H$. $|C|_H$ is "reduced" if it has zero fixed part. For example, the linear system $|C^*|_{H^*}$, whose members are obtained from those of $|C|_H$ by subtracting off the fixed part, is a reduced linear system.

We say that H is an "effective" base divisor on f if there is a divisor C on f such that $|C|_H$ is a reduced linear system.

An effective base divisor is "simple" if it is not a sum of two other non-zero effective base divisors. It can be shown that there is a one-one correspondence $O \leftrightarrow H_O$ between points infinitely near to f and simple base divisors: if $O_1 < O_2 < \cdots < O_r = O$ is as usual, then the virtual multiplicity of H_O at O_i is $\operatorname{ord}_O(m_i)$ $(m_i = m(O_i),\ i = 1, 2, \ldots, r)$, and at all other infinitely near points is zero. Since H_O has virtual multiplicity one at O, the simple base divisors form a *free basis* for the group of all base divisors.

To gain more information about this situation we use the notion of proximity. For any two *distinct* points $O \subseteq P$ we say that P is "proximate" to O if the valuation ord_O is non-negative on P. If O is infinitely near to f, and H is a base divisor on f, we set

$$ e_O(H) = s_O(H) - \sum_P s_P(H) $$

where P runs through all points proximate to O, and $s_O(H)$, $s_P(H)$ are the virtual multiplicities of H at O, P respectively.

Theorem. *For any base divisor H and any point O infinitely near to f let H_O, $e_O(H)$ be as above. Then $e_O(H) = 0$ for almost all O, and*

$$ H = \sum_{\text{all } O} e_O(H) \cdot H_O . $$

Moreover, H is an effective base divisor if and only if $e_O(H) \geqq 0$ for all O (i. e. the virtual multiplicities of H satisfy the "proximity inequalities").

5. We mention, in closing, yet another approach, due to ZARISKI (2; also ZARISKI-SAMUEL, 1, Appendix 5), in terms of *complete ideals*. If $\mathscr{I}$ is any coherent sheaf of ideals on f, then $\mathscr{I}\mathcal{O}_Z$ is an invertible $\mathcal{O}_Z$-module i. e. $\mathscr{I}\mathcal{O}_Z = \mathcal{O}_Z(-D)$, where D is a divisor on Z which we denote $\operatorname{div}_Z(\mathscr{I})$. In this way we map the monoid of coherent ideals homomorphically into the group of divisors on Z. It is a fact that every divisor on Z can be represented in the form $\operatorname{div}_Z(\mathscr{I}) - \operatorname{div}_Z(\mathscr{J})$ for suitable $\mathscr{I}, \mathscr{J}$. The divisors on Z which correspond to *effective* base divisors on f are precisely those of the form $-\text{div}_Z(\mathscr{I})$ with $\mathscr{I}$ such that $\mathcal{O}_f : \mathscr{I} = \mathcal{O}_f$.

For given $\mathscr{I}$, there is a largest (in the sense of inclusion) coherent sheaf $\mathscr{I}'$ among those $\mathscr{J}$ such that $\text{div}_Z(\mathscr{I}) = \text{div}_Z(\mathscr{J})$. Such an $\mathscr{I}'$ is said to be "complete". The complete coherent ideals form a monoid with product $\mathscr{I}' * \mathscr{J}' = (\mathscr{I}'\mathscr{J}')'$. (Actually, f being non-singular, ZARISKI shows that $\mathscr{I}'\mathscr{J}' = (\mathscr{I}'\mathscr{J}')'$, i. e. *the product of complete ideals is complete*.) This monoid maps *injectively* into the group $\text{Div}(Z)$, and its image generates $\text{Div}(Z)$. Thus, *divisors on Z can be thought of as formal differences of complete ideals on f.*

The preceding theorem is a geometric counterpart of ZARISKI's theorem that *every complete ideal on f is in a* unique *way the product of simple complete ideals.*

Further remarks on § 1. ZARISKI and SCHILLING (1) prove by valuation-theoretic methods that on any surface F, an irrational pencil can have base points only at singular points of F, slightly generalizing the result stated in the text and extending it to char p. ZARISKI (5), (9), and (24) studied the 2 BERTINI theorems algebraically, and considered their extension to char p. It turns out that the 1st one is false in char p, except for instance for very simple linear systems like the system of hyperplane sections of a non-singular surface; but that the 2nd is true in the slightly weakened form — a reducible linear system is either the set of divisors $p^n D$, where D moves in an irreducible linear system, or else it is composed of the curves of a pencil.

Chapter III

Adjoint Systems and the Theory of Invariants

1. Complete linear systems of plane curves (CASTELNUOVO, 1). If $|C|_H$ is a complete linear system of plane curves of order n, defined by a set $H = \sum O_i^{s_i}$ of base points, the dimension r_n of $|C|_H$ is given, *for n sufficiently large*, say for $n \geq l$, by the *postulation formula* (CAYLEY, NOETHER) or by the *characteristic formula* (HILBERT):

$$r_n = \binom{n+2}{2} - k - 1,$$

where $k \leq \sum \frac{s_i(s_i+1)}{2}$ is an integer independent of n. If the proximity inequalities are satisfied, then $k = \sum \frac{s_i(s_i+1)}{2}$. For $n < l$ the *effective dimension* r'_n of $|C|_H$ is greater than its *virtual dimension* r_n evaluated according to the postulation formula. The difference $s = r'_n - r_n$ is called the *superabundance* of $|C|_H$. A system is *superabundant* if $s > 0$, *regular* if $s = 0$.

If $|C|_H$ is irreducible, of dimension r and of effective degree ν, and if H includes all the effective base points of the system taken with their effective multiplicities, then the following theorems hold (CASTELNUOVO, 1):

1. *The characteristic series g_ν^{r-1} of the system is complete.* Consequently $r = \nu - p + i + 1$, where i is the index of speciality of the series. 2. *The index of speciality i of the characteristic series is equal to the superabundance of the system.* As a corollary we have that *the characteristic series of an irreducible regular system is non-special.*

The following lemma is very useful in the applications: *If $|C^n|_H$ and $|C^{n+1}|_H$ are regular complete systems of curves of orders n and $n+1$ respectively, then the system of minimum dimension, which contains every reducible curve composed of a curve of $|C^{n+1}|_H$ and of a line, coincides with the complete system $|C^{n+2}|_H$* (CASTELNUOVO, 7, also ENRIQUES, 5, where this lemma is proved in the special case of the systems of adjoint curves of a given plane curve).

2. Complete linear systems of surfaces in S_3 (CASTELNUOVO, 7). Let Σ be a linear system of surfaces F_n of order n in S_3, without fixed components. We fix a pencil of planes $|\omega|$ such that no base curve of Σ lies in a plane ω of the pencil. The linear system of curves Γ_ω^n of order n, cut out by Σ on a *generic* plane ω of the pencil, defines completely its own set H_ω of (effective) base points and the corresponding complete system $|\Gamma_\omega^n| = |\Gamma_\omega^n|_{H_\omega}$. As ω varies in the pencil $|\omega|$, the set H_ω describes a set L of distinct or infinitely near algebraic curves, base curves of Σ. We say that a surface of order ν passes through L if it cuts out on a *generic* plane of the pencil $|\omega|$ a curve belonging to the system $|\Gamma_\omega^\nu|$. This incidence condition constitutes an indirect definition of the base curve L. It follows from this definition that if F_n and $\overline{F}_n$ are two surfaces of order n passing through L, then all the surfaces of the pencil determined by F_n and $\overline{F}_n$ pass through L. Hence the totality of surfaces F_n of a given order n passing through L constitutes a linear system. This system is denoted by $|F_n|_L$, or simply by $|F_n|$, and is called *the complete linear system relative to the base curve L*.

It can be proved that the surfaces of $|F_n|_L$ cut out also on a *generic plane $\overline{\omega}$ of the space* curves of the corresponding complete system $|\Gamma_{\overline{\omega}}^n|$. Hence the above definition of $|F_n|_L$ is independent of the particular pencil $|\omega|$ employed (CASTELNUOVO, 7).

The system $|F_n|$, although complete, does not necessarily cut out on ω the complete system $|\Gamma_\omega^n|$. The difference $\delta_n \geq 0$ between the dimension of $|\Gamma_\omega^n|$ and the dimension of the system Σ_ω^n cut out on ω by $|F_n|$ is called *the deficiency* of Σ_ω^n. Using the lemma of the preceding section it can be proved, however, that $\delta_n = 0$, *if n is sufficiently high*. We know that for sufficiently high values of n, say for $n \geq l$, the system $|\Gamma_\omega^n|$ is regular. We can therefore assign an integer $l_0 \geq l$ such that *for any $n \geq l_0$ the system $|F_n|$ cuts out on a generic plane ω a complete and regular system of curves*.

Let r_n' denote the effective dimension of $|F_n|$. Using the fact that $r_n' - r_{n-1}' - 1$ is equal to the dimension of the system Σ_ω^n, we find immediately the following relations: 1. $r_n' - r_{n-1}' - 1 = \varrho_n$, if $n \geq l_0$; 2. $r_n' - r_{n-1}' - 1 = \varrho_n - \delta_n$, if $l_0 > n \geq l$; 3. $r_n' - r_{n-1}' - 1 = \varrho_n - \delta_n + s_n$, if $l > n$, where s_n is the superabundance of $|\Gamma_\omega^n|$ and $\varrho_n = \binom{n+2}{2} - k - 1$ is its virtual dimension. From these recurrence relations we derive immediately the *postulation formula*, giving the *virtual dimension r_n* of $[F_n]$:

$$(1) \qquad r_n = \binom{n+3}{3} - kn + k' - 1,$$

where k is the postulation of the set H_ω and k' is another integral constant. As for the effective dimension r_n' we have:

$$(2) \qquad r_n' = r_n, \qquad n \geq l_0 - 1;$$

$$(2') \qquad r_n' = r_n + \sum_{i=n+1}^{l_0-1} \delta_i, \qquad l_0 - 1 > n \geq l - 1;$$

$$(2'') \qquad r_n' = r_n + \sum_{i=n+1}^{l_0-1} \delta_i - \sum_{j=n+1}^{l-1} s_j, \qquad l - 1 > n.$$

The system $|F_n|$ is said to be *regular* or *irregular* according as $n \geqq l_0 - 1$ or $n < l_0 - 1$. In other words, $|F_n|$ is regular if *all* the systems $|F_\nu|$, $\nu > n$, cut out on a generic plane ω complete and regular systems of curves; irregular, if the contrary is true. The difference $r'_n - r_n$ is called the *irregularity* of $|F_n|$. If $l_0 - 1 > n \geqq l - 1$, this difference is not negative and is equal to the sum of the deficiencies of the systems $\Sigma_\omega^{n+1}, \Sigma_\omega^{n+2}, \ldots$ For $n < l - 1$, there are two causes of irregularity, acting in opposite senses, and consequently the effective dimension is not necessarily greater than or equal to the virtual dimension.

The evaluation of the constant k' which occurs in (1) is a problem not solved in all generality. Its value is known, however, in several theoretically important cases. For instance, if L is a simple irreducible space curve of order m and genus p, possessing δ double points with distinct tangents and t triple points with non-coplanar principal tangents, then $k = m$ and $k' = 2t + p + \delta - 1$ (CASTELNUOVO, 2).

The preceding theory can be extended to complete systems of surfaces possessing *isolated base points*. In the most general case the behavior of the surfaces of the system at an isolated singular point O is fully described by assigning a set of points and of infinitesimal curves infinitely near O, through which the surfaces of the system are required to pass with assigned multiplicities. This manner of stating the conditions imposed by an isolated base singularity is in agreement with C. SEGRE's analysis of the composition of a singular point of an algebraic surface (I, 5). These conditions are analytically expressed by homogeneous relations between the coefficients of the equation of the surface (supposing that O is the origin of coördinates), and a particular solution can be obtained by putting equal to zero the coefficients of all the terms of degree lower than a certain constant. In other words, the conditions imposed by any isolated base singularity O are satisfied by cones of vertex O of a sufficiently high order. Using this fact, it is possible to prove also in the case of complete systems $|F_n|$ of surfaces with assigned base curves *and* isolated base singularities, that these systems, *for n sufficiently large*, cut out complete and regular systems of curves on a generic plane. The postulation formula (1) can then be readily extended to this more general type of complete systems of surfaces.

3. Subadjoint surfaces (ENRIQUES, 5). Let F be a surface of order n in S_3 with arbitrary singularities and let C denote a generic plane section of F. The surfaces φ_ν of a given order ν, which cut out on a *generic* plane ω adjoint curves Γ_ω^ν of order ν of the corresponding plane section C of F, are called *subadjoint surfaces* of F (of order ν). The polar surfaces of F are particular subadjoint surfaces φ_{n-1}; hence surfaces φ_ν certainly exist if $\nu \geqq n - 1$. Since the adjoint curves of a plane curve C are defined by base conditions at the multiple points of C, it follows that the surfaces φ_ν form a complete system $|\varphi_\nu|$ whose base curve L is made up of the multiple curves of F. It should be noted that $|\varphi_\nu|$ is free from assigned isolated base points; in particular, the isolated singularities of F (I, 6) do not impose any extra conditions on the subadjoint surfaces.

We shall denote by K_ν the curve of intersection, *outside* L, of a φ_ν with F. For particular surfaces φ_ν it may be necessary, however, to include in K some of the multiple curves of F to a certain multiplicity. Let L_1 be an irreducible s-fold curve of F, and let O be a *generic* point of L_1. The section C of F by a generic plane ω through O and a generic adjoint curve Γ^ν of C have at O an intersection multiplicity $\lambda = s(s-1)$ $+\sum s_i(s_i - 1) + \cdots$, where the summation is extended to all the infinitely near multiple points constituting the singularity of C at O. If Γ^ν is the section of ω with a given subadjoint surface φ_ν and if for this particular Γ^ν the intersection multiplicity with C at O is $\lambda' \geqq \lambda$, then L_1 must be counted $\lambda' - \lambda$ times as a component of K. This convention presupposes that the neighborhood of O on F consists of one irreducible branch. In the contrary case L_1 must be regarded, in the neighborhood of O, as a set of overlapping distinct curve elements, $L_{11}, L_{12}, \ldots$, one for each irreducible branch of F. To each curve element L_{1i} there will correspond one branch of C of origin O, and it is clear how the multiplicity σ_i of L_{1i} as a component of K should be defined. It is also clear that if F, although reducible in the neighborhood of a generic point of L_1, is irreducible along the whole curve L_1, then $\sigma_1 = \sigma_2 = \cdots$

From known properties of the adjoint curves of a plane curve C it follows that *any curve K_ν meets a generic plane section C of F in $2\pi - 2$ $+ rn$ points, where π is the genus of C and $r = \nu - (n-3)$, and that these points constitute a set of the linear series $g_{2\pi-2+rn}$ on C which is the sum of the canonical series $g_{2\pi-2}^{\pi-1}$ on C and of the r-tuple of the series cut out on C by the lines of its plane.* Enriques (5) has proved that this property of K_ν is characteristic for $r \geqq 0$: *if a curve K on F meets a generic C in a set of the above series $g_{2\pi-2+rn}, r \geqq 0$, then K is a K_ν, $\nu = n - 3 + r$, except when F is a ruled surface and $r < 2$.* This theorem is fundamental.

In stating the theorem of Enriques we have supposed that the above property of K holds with regard to the section of F by a generic plane of the space. Actually the proof of the theorem utilizes only the hypothesis that this property holds in regard to the section of F by the generic plane of a fixed pencil of planes, whose axis does not meet K. In the course of the proof, however, it is necessary to suppose that the section of F with *every* plane of the fixed pencil is irreducible. As a consequence the proof of the theorem as stated above fails when *every* pencil of plane sections of F contains reducible curves. By a known theorem of Kronecker-Castelnuovo the surface F is then either a ruled surface or a surface of Steiner.

If F is a ruled surface of genus π with ordinary singularities, there do not exist subadjoint surfaces of order $\nu = n - 3 \, (r = 0)$, and any $K_{n-2} \, (r = 1)$ is made up of $2\pi - 2 + n$ generators of F (ENRIQUES, 5; a, p. 122; PICARD and SIMART, a_2, p. 154). On the other hand any set of $2\pi - 2$ generators passing through the points of a canonical set of C gives a curve K satisfying the hypothesis of the theorem in the case $r = 0$. The above set of $2\pi - 2$ generators together with a plane section C gives a curve $\overline{K}$ of the theorem for the case $r = 1$, which is not a K_{n-2}. For $r \geqq 2$ the theorem holds also for ruled surfaces. That the theorem holds without exceptions also for STEINER surfaces follows immediately from known properties of these surfaces.

The curves $K_\nu, \nu = n - 3 + r$, form a linear system Σ_r. From the theorem of ENRIQUES it follows immediately that this system is complete. Moreover, *for r sufficiently high Σ_r is irreducible.* For the proof it is sufficient to observe that: 1) if there exists a K_ν, then $K_\nu + C$ is a $K_{\nu+1}$, and hence Σ_{r+1} is not composed of the curves of a pencil; 2) the surfaces φ_ν cut out on a generic plane the complete system of adjoint curves Γ_ν, if ν is sufficiently high; 3) the complete series $g_{2\pi-2+rn}, r \geqq 0$, is free from fixed points, and consequently Σ_r is free from fixed components.

4. Subadjoint systems of a given linear system (ENRIQUES, 5). We now proceed to give an invariantive formulation of the projective considerations of the preceding section. Let F be a surface free from singularities in an S_k or with ordinary singularities in S_3, and let $|C|_H = |C| \, (H = \sum O_j^{s_j})$ be a complete irreducible simple system on F, of degree n, of effective genus π, and of dimension $r \geqq 3$. We suppose that all the base points of $|C|$ are assigned with their effective multiplicities. We proceed to define the subadjoint systems of any system $|C|$ satisfying the above conditions.

By means of a generic web W contained in $|C|$ we transform F into a surface F^* of order n in S_3 whose plane sections C^* correspond to the curves of the web (II, 6). On F^* we may have arbitrary singularities, in particular isolated singularities corresponding to the fundamental curves of $|C|$. First let r be a sufficiently high integer, say $r \geqq r_0$, so that the system Σ_r^* cut out on F^* by the subadjoint surfaces φ_{n-3+r}^* enjoys the properties stated at the end of the preceding section. We call *subadjoint system of rank r* of $|C|$, denoting it by $|C_s^r|$, the *irreducible* system on F which corresponds to the system Σ_r^*. The set of base points of $|C_s^r|$ shall be denoted by H^r and shall consist of all the base points of $|C|$ which are effective base points of $|C_s^r|$, assigned with their effective multiplicities for $|C_s^r|$. In view of the results of the preceding section the system $|C_s^r|$ enjoys the following properties:

a. The curves C_s^r have $2\pi - 2 + rn$ *variable* intersections with the curves C of the web W and hence also with a generic curve of $|C|$. Consequently $(C_s^r \cdot C)_{H^r, H} = 2\pi - 2 + rn$.

b. The curves C_s^r cut out on a generic curve C of the web W *the complete series* $g_{2\pi-2+rn}$, the sum of the canonical series of C and the r-tuple of the characteristic series of $|C|$ on C. It follows that the curves C_s^r enjoy this property also in regard to the generic curve of any other generic web of $|C|$ having with W a generic (irreducible) pencil in common, and hence ultimately in regard to a *generic curve of* $|C|$.

c. The system $|C_s^r|$ with H^r as the assigned set of base points is *complete*.

d. Let us assign on a curve K a set G of base points consisting of base points of $|C|$ taken with arbitrary virtual multiplicities, provided, however, that K_G be an effective curve (II, 8, *Remark*). Let us suppose that the intersections of K and of a generic C, outside the assigned base sets[1] G and H respectively, constitute a set of the series $g_{2\pi-2+rn}$ mentioned in *b*. This implies in particular the numerical relation $(K \cdot C)_{G,H} = 2\pi - 2 + rn$. Under these conditions K *is either a* C_s^r *or may differ from a* C_s^r *at most by fundamental curves and by base points of* $|C|$.[2] In exact terms, there exist two sets Θ and Θ_1 of fundamental curves of $|C|$, without common components, such that Θ *is a component of* K and that the following relation holds:

$$(3) \qquad |K + \Theta_1| = |C_s^r + \Theta + \sum O_i^{t_i}|,$$

where $O_1, O_2 \ldots$ are base points of $|C|$. In this relation, as also through this and the following sections, any irreducible fundamental curve ϑ of a system $|C| = |C|_H$ shall carry with it an assigned set g of base points consisting of all the base points of $|C|$ which are on ϑ, taken with virtual multiplicities equal to their effective multiplicities for ϑ. If g is defined in this manner, then $(C \cdot \vartheta)_{H,g} = 0$.

The properties c. and d. follow from the fundamental theorem of the preceding section.

---

[1] If O is a base point of $|C|$, origin of a branch γ of a generic C, then the difference t between the effective and the virtual intersection multiplicities of K and γ at O is non-negative, since K_G is effective, and the point O, as origin of the branch γ, must be counted t times among the intersections of K and C.

[2] Let us apply to f a birational transformation, which does not possess fundamental curves on f and such that on the transformed surface $\bar{f}$ the transform $|\bar{C}|$ of $|C|$ is effectively free from base points (II, 7, *An application* etc.). In addition to the transforms of the fundamental curves of $|C|$, $|\bar{C}|$ may possess fundamental curves, which arise from base points of $|C|$. Let O_i be a base point of $|C|$ and let $L^{(i)}$ be the corresponding exceptional curve on $\bar{f}$. If all the points of a generic C, which follow *immediately* O_i, are also base points of $|C|$, then the irreducible component of $L^{(i)}$, which represents the immediate neighborhood of O_i, is a fundamental curve of $|\bar{C}|$. This curve may be expressed in the form $L^{(i)} + t_1 L^{(i+1)} + t_2 L^{(i+2)} + \cdots$, where $L^{(i+1)}$, $L^{(i+2)}$, $\ldots$ are exceptional curves corresponding to base points $O_{i+1}, O_{i+2}, \ldots$ following O_i and where $t_1, t_2, \ldots$ are convenient integers. A corresponding combination $O_i + O_{i+1}^{t_1} + O_{i+2}^{t_2} + \cdots$ may occur in (3) and can be considered as an *infinitesimal fundamental curve* of $|C|$.

Any curve such as K, with the assigned base set G on it, shall be called a *subadjoint curve of $|C|$ of rank r*.

Corollary. If Θ is any set of fundamental curves of $|C|$, then Θ is an accidental fixed component of the system $|C_s^r + \Theta|$.

e. The following relation holds:

$$(4) \qquad\qquad |C_s^{r+1}| = |C_s^r + C|, \qquad\qquad r \geqq r_0.$$

For the proof we observe that the generic curve of $|C_s^r + C|$ is irreducible and is a subadjoint curve of rank $r + 1$ of $|C|$. Hence, by (3), $|C_s^{r+1} + \Sigma O_i^{s_i}| = |C_s^r + C + \Theta_1|$, where Θ_1 is a set of fundamental curves of $|C|$. To prove (4) it is therefore sufficient to show that $|C_s^{r+1}|$ and $|C_s^r + C|$ have the same dimension and to observe that for each of these two systems the virtual multiplicities at the base points of $|C|$ coincide with the effective multiplicities. The series cut out on C by the curves of the system $|C_s^r + C|$ contains all sets made up of a set of the $g_{2\pi-2+rn}$ and of a characteristic set. Hence, by a lemma on linear series, due to CASTELNUOVO, and analogous to the lemma of section 1, this series is the complete series $g_{2\pi-2+(r+1)n}$. It follows that a generic C imposes on $|C_s^r + C|$ the same number of conditions as on $|C_s^{r+1}|$, namely $\pi - 1 + (r + 1)n$ conditions. By the preceding corollary the residual system $|C_s^r + \Theta_1|$ of $|C_s^{r+1}|$ with respect to $|C|$ possesses Θ_1 as a fixed component and is therefore of the same dimension as $|C_s^r|$. Consequently the two systems $|C_s^{r+1}|$ and $|C_s^r + C|$ have the same dimension, q. e. d.

It should be noted that (4) implies the following relation between the base sets: $H^{r+1} = H^r + H$.

We now use formally the relation (4) in order to define $|C_s^r|$ for $r_0 > r \geqq 0$: $|C_s^r| = |C_s^{r_0} + (r - r_0)C|$. For low values of r the system $|C_s^r|$ may be reducible. The properties (a.—e.) continue to hold for $r < r_0$, except that while the relation $(C_s^r \cdot C)_{H^r, H} = 2\pi - 2 + rn$ still holds (property a.), the $2\pi - 2 + rn$ intersections of C_s^r and C are not all necessarily variable intersections. In other words, $|C_s^r|_{H^r}$ may possess accidental base points or base points of accidental hypermultiplicity at some of the base points of $|C|$. Moreover, the series cut out by this system on a generic C may not be complete.

If $|C|$ possesses a pencil of unisecant curves (curves meeting the C's in *one* variable point), then F^* is a ruled surface, and therefore properties (a.—e.) do not hold for $r = 0$ or 1 (see preceding section). This exceptional case shall be excluded in the sequel.

The operation of *subadjunction* defined in this section is clearly invariant under birational transformations. If $|C|$ is any linear system on F satisfying the conditions stated at the beginning of this section and if F is birationally transformed into a surface $\overline{F}$, then the transformed system $|\overline{C}|$ will also satisfy the above conditions and *in general* $|\overline{C}_s^r|$ *will coincide with the transform of* $|C_s^r|$. An exception arises when $|C_s^r|$

possesses base points outside the base points of $|C|$ and fundamental for the transformation. According to our definition of $|C_s^r|$, such base points must be considered as accidental base points of $|C_s^r|$, and consequently the corresponding exceptional curves on $\overline{F}$ must be included in the transformed system as accidental fixed components counted to proper multiplicities (II, 7). This leads to the following *transformation law of subadjoint systems*: Let $\Theta_1, \Theta_2, \ldots$ *be the exceptional curves*[1] *of F which are transformed into points of $\overline{F}$ and which are fundamental for* $|C|$ *without being fundamental for* $|C_s^r|$. *Similarly, let* $\overline{\Theta}_1, \overline{\Theta}_2, \ldots$ *be the exceptional curves of $\overline{F}$ which correspond to points of F and which are fundamental for* $|\overline{C}|$ *without being fundamental for* $|\overline{C}_s^r|$. *Moreover, let* $(C_s^r \cdot \Theta_i) = \mu_i$ *and* $(\overline{C}_s^r \cdot \overline{\Theta}_i) = \overline{\mu}_i$. *Then the transform of the system* $|C_s^r + \sum \mu_i \Theta_i|$ *is the system* $|\overline{C}_s^r + \sum \overline{\mu}_i \overline{\Theta}_i|$.

5. The distributive property of subadjunction (ENRIQUES, 5). The operation of subadjunction enjoys a distributive property with respect to addition of linear systems which is fundamental for the theory.

Theorem I. *If $|C| = |A| + |B|$ and if all base points (resp. fundamental curves) of $|C|$ are also base points (resp. fundamental curves) of $|A|$, then the following relation holds:*

$$(5) \qquad |C_s^r + \Theta_1| = |A_s^r + (r+1)B + \Theta + \sum O_i^{s_i}|,$$

where Θ and Θ_1 are sets of fundamental curves of $|A|$, independent of r, and where $O_1, O_2, \ldots$ are base points of $|A|$.

Proof. Using the relation $|C_s^r| = |C_s^\varrho + (r - \varrho)C|$ and the hypotheses of the theorem, it is easily shown that the system $|C_s^r - (r+1)B| = |K^r|$ exists for sufficiently high values of r. If the surface F is transformed into a surface F^* of order n in S_3 by means of a generic web extracted from $|C|$ and containing a reducible curve $A + B$, it is seen that the subadjoint surfaces φ_{n-3+r}^* cut out in the plane of $A + B$ curves of order $n = \alpha + \beta$ ($n = (C^2)$, $\alpha = (C \cdot A)$, $\beta = (C \cdot B)$), which would be adjoint curves of $A + B$ except for the fact that they do not pass through the intersections of A and B. Consequently the curves C_s^r cut out on A sets of the series, the sum of the series cut out by $|B|$, the r-tuple of the series cut out by $|C|$ and the canonical series. This shows that *the curves K^r are subadjoint curves of rank r of $|A|$*, and from

---

[1] If the transformation possesses on $\overline{F}$ infinitely near fundamental points, it is necessary to introduce an exceptional curve for each of these points. For instance, if O and O_1 are fundamental points and if O_1 is infinitely near O, then to O there corresponds a reducible exceptional curve $\Theta = L + L_1$ and to O_1 there corresponds the exceptional curve $\Theta_1 = L_1$ (II, 6).

[2] We have defined the subadjoint systems only for such systems which satisfy certain conditions stated in the preceding section. It is therefore understood that in any statement involving the subadjoint systems of a system $|C|$, it is supposed that $|C|$ satisfies these conditions. In the present theorem this applies to the systems $|C|$ and $|A|$, while $|B|$ is arbitrary.

this (5) follows as a consequence of (3), section 4. Using (4) it is seen immediately that (5) holds also for low values of r, $r \geqq 0$, q. e. d.

Corollary 1. If $|B| = |kA|$, then $|C| = |(k+1)A|$ and (5) becomes $|C_s^r + \Theta_1| = |A_s^r + (r+1)kA + \Theta + \sum O_i^{s_i}|$. Since any fundamental curve of $|A|$ is also fundamental for any multiple of $|A|$, it follows that Θ_1 is a fixed component of $|C_s^r + \Theta_1|$ and therefore cannot occur, since for r sufficiently high $|A_s^r|$ is irreducible. We have then $|C_s^r| = |A_s^r + (r+1)kA + \Theta + \sum O_i^{s_i}| = |A_s^{(r+1)k+r} + \Theta + \sum O_i^{s_i}|$. The same reasoning shows that Θ cannot occur. Moreover, all the s_i's must vanish, since for each of the two systems $|C_s^r|$, $|A_s^{(r+1)k+r}|$ the virtual multiplicities of the points O_i coincide with the effective multiplicities. Hence we have the following relation:

$$(6) \qquad |(k+1)A|_s^r = |A_s^{(r+1)k+r}| = |A_s^r + (r+1)kA|.$$

Corollary 2. *With the additional hypothesis that the system* $|A|$ *has the same fundamental curves as* $|C|$ *and that* $|B|$ *does not posses fixed components which are fundamental for* $|A|$, *the formula* (5) *can be replaced by the following:*

$$(5') \qquad |C_s^r| = |A_s^r + (r+1)B + \Theta + \sum O_i^{s_i}|.$$

In fact, under the above conditions Θ_1 is a fixed component of $|C_s^r + \Theta_1|$ and therefore cannot occur, since $|A_s^r|$ is irreducible for r sufficiently high and since Θ and Θ_1 can be supposed to be without common components.

Corollary 3. *If in addition to the hypotheses of theorem 1 and of corollary 2 we suppose that the system* $|\varrho A - B|$ *does not possess, for* ϱ *sufficiently high, fixed components which are fundamental for* $|A|$, *then* (5') *can be further simplified:*

$$(5'') \qquad |C_s^r| = |A_s^r + (r+1)B + \sum O_i^{s_i}|.$$

In fact, the system $|C_s^r + \bar\varrho A - (r+1)B| = |A_s^r + \bar\varrho A + \Theta + \sum O_i^{s_i}| = |A_s^{r+\bar\varrho} + \Theta + \sum O_i^{s_i}|$, $\bar\varrho$ arbitrary, possesses Θ as fixed component and consequently Θ cannot occur, since for $\bar\varrho = (r+1)\varrho$ and for ϱ sufficiently high Θ cannot be a fixed component of $|\bar\varrho A - (r+1)B| = |(r+1)(\varrho A - B)|$.

We observe that if all the base points of $|B|$ are assigned with their effective multiplicities for $|B|$, and if $|A|$ does not possess infinitesimal fundamental curves (section 4, Footnote [2]), which are not fundamental for $|B|$ (and hence neither for $|C|$, so that this last condition is simply an extension of the first condition of corollary 2), then

$$(5''') \qquad |C_s^r| = |A_s^r + (r+1)B|.$$

Theorem 2. *Let* $|C| = |A| - \sum O_j^{r_j} - \sum P_i^{\mu_i}$, *where* $O_1, O_2, \ldots$ *are base points of* $|A|$ *and* $P_1, P_2, \ldots$ *are new base points; i. e. let* $|C|$ *be obtained from* $|A|$ *by raising the assigned multiplicities of some of the base points of* $|A|$ *and by imposing new base points* P_i (*see* II, 4). Then

$$(7) \qquad |C_s^r + \Theta| = |A_s^r - \sum O_j^{\nu_j(r+1)} - \sum P_i^{\mu_i(r+1)-1} + \Theta_1|,$$

where Θ *and* Θ_1 *are sets of fundamental curves of* $|C|$.[1] *If we suppose in addition that* $|C|$ *does not possess fundamental curves which are not fundamental for* $|A|$, *then*

$$(7') \qquad |C_s^r + \Theta| = |A_s^r - \sum O_j^{\nu_j(r+1)} - \sum P_i^{\mu_i(r+1)-1}|.$$

The proof of (7) follows from elementary projective considerations which show that the curves of $|A_s^r - \sum O_j^{\nu_j(r+1)} - \sum P_i^{\mu_i(r+1)-1}|$ are subadjoint curves of rank r of $|C|$. If the additional hypothesis of the theorem holds, then Θ_1 is a fixed component of $|A_s^r + \Theta_1|$ since it is made up of fundamental curves of $|A|$. Therefore Θ_1 is *a fortiriori* a fixed component of the system in the right-hand member of (7), and consequently cannot occur, since $|C_s^r|$ is irreducible for r sufficiently high.

Remark. It should be noted that Θ is a fixed component of $|C_s^r + \Theta|$ and hence does not occur in (7) or (7') if the system in the right-hand member of (7') is irreducible. This condition is satisfied if we replace the system $|A|$ by $|\varrho A|$, $\varrho > 1$, keeping fixed the integers ν_j and μ_i.

Theorem 3. *Let* $|C| = |A| + |B|$ *and let* $P_1, P_2, \ldots, P_k$ *be the base points of* $|C|$ *which are not base points of* $|A|$. *If for sufficiently high values of* ϱ *the system* $|\varrho A - \sum P_i|$ *does not possess fundamental curves which are not fundamental for* $|A|$ *or for* $|C|$, *and satisfies the conditions of section 4 (so that its subadjoint systems are defined), and if* $|B|$ *does not possess fixed components which are fundamental for* $|A|$, *then*

$$(8) \qquad |C_s^r| = |A_s^r + (r+1)B + \sum P_i + \Theta|,$$

where Θ *is a set of fundamental curves of* A, *including infinitesimal curves.*

Proof. We put $|\varrho A| = |\bar{A}|$, $|\varrho B| = |\bar{B}|$, $|\varrho C| = |\bar{C}|$, $|\bar{A} - \sum P_i| = |\bar{\bar{A}}|$, and $|\bar{C} - \sum P_i| = |\bar{\bar{C}}|$. We have then $|\bar{\bar{C}}| = |\bar{\bar{A}}| + |\bar{B}|$ and hence, by Corollary 2 of Theorem 1, $|\bar{\bar{C}}_s^r| = |\bar{\bar{A}}_s^r + (r+1)\bar{B} + \Theta|$, where Θ is a set of fundamental curves of $|A|$. By (7') and by the preceding remark we have $|\bar{\bar{A}}_s^r| = |\bar{A}_s^r|$, $|\bar{\bar{C}}_s^r| = |\bar{C}_s^r - \sum P_i|$. Consequently $|\bar{C}_s^r| = |\bar{A}_s^r + (r+1)\bar{B} + \sum P_i + \Theta|$, and from this (8) follows in view of (6).

6. Adjoint systems (Enriques, 2, 5; Noether, 2). The formulation of the distributive law of subadjunction is consistently complicated by the presence of fundamental curves of the linear systems which are being considered. This could have been expected, since in the definition of the subadjoint systems of a given system $|C|$ by means of the subadjoint surfaces of the transformed surface F^*, the isolated singularities of F^*, representing the fundamental curves of $|C|$, have not been taken into account. In order to eliminate the influence of the fundamental

---

[1] Θ may include infinitesimal fundamental curves of $|C|$.

curves it is, in general, necessary to consider a certain properly defined subsystem of $|C_s^r|$. For $r = 0$ we obtain the *adjoint system* of $|C|$, defined below, whose invariantive relationship with $|C|$ is particularly simple and is expressed in terms independent of the fundamental curves of $|C|$. For $r > 0$ nothing essentially new is obtained, in view of the relation (4).

First let $|C|$ be the system of hyperplane sections of F, free from singularities in S_r. We define the *adjoint system of* $|nC|$ (n—an arbitrary positive integer) as the subadjoint system of rank 0 of $|nC|$. We denote this system by $|(nC)'|$. We have $|(nC)'| = |(nC)_s^0| = |C' + (n-1)C|$ [formula (6)].

We shall call a system $|D|$ on F *non-singular* if it satisfies the conditions of section 4 and if in addition the surface $\overline{F}$ into which F is transformed by referring the curves D to the hyperplanes of an S_ϱ (ϱ is the dimension of $|D|$) *is free from singularities*. Let $H = \sum O_i^{s_i}$ be the set of base points of $|D|$. If n is sufficiently high, the system $|nC - \sum O_i^{s_i} - D|$ $= |D_1|$, virtually free from base points, exists, is irreducible, and is effectively free from base points (II, 5). We define *the adjoint system* $|D'|$ of $|D|$ as follows:

$$(9) \qquad |D'| = |(nC)' - D_1 - \sum O_i^{s_i-1}|.$$

This definition is obviously independent of n. The curves of D' are called *adjoint curves* of $|D|$. For the relationship between $|D'|$ and the subadjoint system $|D_s^0|$ of rank 0 of D, we have the following theorem: *If* $\Theta_1, \Theta_2, \ldots, \Theta_k$ *are the exceptional curves of* F *which correspond to* (*distinct or infinitely near*) *points* $P_1, P_2, \ldots, P_k$ *of* $\overline{F}$, *then*

$$(10) \qquad |D'| = |D_s^0 + \Theta_1 + \Theta_2 + \cdots + \Theta_k|,$$

i. e. the system $|D'|$ *possesses the exceptional curves* Θ_i *as fixed components*[1] *and the residual system coincides with the subadjoint system* $|D_s^0|$.

Proof. We consider the system $|C_1| = |nC - \sum O_i^{s_i}|$. If n is taken sufficiently high, the system $|C_1|$ does not possess fundamental curves and we will have, by (7'),

$$|C_{1s}^0| = |(nC)_s^0 - \sum O_i^{s_i-1}| = |(nC)' - \sum O_i^{s_i-1}|.$$

Let $|\overline{D}|$ denote the system of hyperplane sections of $\overline{F}$, and let $|\overline{C}_1|$ and $|\overline{D}_1|$ be the transforms of $|C_1|$ and $|D_1|$. These systems on $\overline{F}$ are irreducible and free from accidental base points (since the corresponding systems $|C_1|$ and $|D_1|$ on F are free from accidental base points and irreducible), and moreover $|\overline{C}_1|$ possesses at each point P_i a virtual *positive* multiplicity (equal to $(C_1 \cdot \Theta_i) > 0$). Using the relation (8) of the pre-

---

[1] See section 4, property d., corollary.

ceding section we find

$$|\bar{C}_{1_s}^r| = |\bar{D}_s^r + (r+1)\bar{D}_1 + \sum P_i|.$$

Now $|\bar{C}_{1_s}^r|$ and $|\bar{D}_s^r|$ are exactly the transforms of $|C_{1_s}^r|$ and $|D_s^r|$. This follows from the transformation law of subadjoint systems (section 4) and from the following remarks:

(a) If F is free from singularities, the generic hyperplane section of F on *any* point O of F has the same genus π as the generic hyperplane section of F. Consequently, the subadjoint system of a sufficiently high rank r of the system of hyperplane sections of F is *effectively free from base points*, since the complete series $g_{2\pi-2+rn}$ on a curve of genus π is free from fixed points.

(b) As a consequence of (a) it follows that if $|D|$ is a non-singular system on F, then the system $|D_s^r|$, for r sufficiently high, does not possess accidental base points (i. e. base points outside the base points of $|D|$), and that any total fundamental curve of $|D|$ (necessarily an exceptional curve of F) is also a total fundamental curve of $|D_s^r|$.

(c) The system $|C_1| = |nC - \sum O_i^{s_i}|$ is free from accidental base points or from base points of accidental hypermultiplicity, if n is sufficiently high.

We have consequently

$$|C_{1_s}^r| = |D_s^r + (r+1)D_1 + \sum \Theta_i|,$$

and in particular, for $r = 0$,

$$|D'| = |D_s^0 + \sum \Theta_i|, \quad \text{q. e. d.}$$

It should be noted that $|D'|$ possesses at each s-fold base point of $|D|$ the virtual multiplicity $s - 1$.

Let $|A_1|$ and $|A_2|$ be two non-singular systems and let $P_1^{(i)}, P_2^{(i)}, \ldots$ be the base points of $|A_i|$ ($i = 1, 2$) which are not base points of $|A_j|$ ($j \neq i$). From (9) we derive immediately the following *fundamental relation*:

(11) $$|(A_1 + A_2)'| = |A_1' + A_2 + \sum P_\alpha^{(2)}| = |A_2' + A_1 + \sum P_\beta^{(1)}|,$$

which expresses the *distributive property of adjunction*.

We are now in a position to define the adjoint system $|A'|$ of any system $|A|$ whatsoever, irreducible or not, of dimension ≥ 0. Let $|E|$ be a non-singular system with base points $P_1, P_2, \ldots$ We suppose for simplicity that none of the points P_i is an assigned base point of $|A|$. Replacing $|E|$, if necessary, by a multiple of $|E|$, we may assume that the system $|E - A - \sum O_j^{s_j}| = |D_1|$ exists, where $\sum O_j^{s_j}$ is the set of assigned base points of $|A|$. We define $|A'|$ as follows:

(12) $$|A'| = |E' - D_1 - \sum O_j^{s_j-1} - \sum P_\alpha|.$$

The set of assigned base points of $|A'|$ is clearly $\sum O_j^{s_j-1}$. This definition is independent of $|E|$. To see this it is sufficient, given any other non-singular system $|D|$, to consider the system $|E + D|$ and to apply (11). It is also seen immediately that *the fundamental relation* (11) *holds for any two systems* $|A_1|$ *and* $|A_2|$.

From (12) it follows that if $|A|$ possesses subadjoint systems, then the adjoint curves A' are subadjoint curves of $|A|$ of rank 0 *and hence cut out on a generic A canonical sets*. This implies the numerical relation

$$(13) \qquad (A \cdot A') = (A \cdot A')_{H, H'} = 2\pi - 2,$$

where π is the genus of $|A|$ and $H = \sum O_j^{y}$, $H' = O_j^{y-1}$.

If $|A|$ is an *arbitrary* system, of virtual genus π, the formula (13) still holds, as can be seen immediately by introducing an auxiliary non-singular system $|E|$ and by using the formulas (6) and (6') of II, 4. The function-theoretic complement of (13) to the effect that the curves A' cut out canonical sets on a generic A holds true for any irreducible system $|A|$ whose effective *multiple* base points are all assigned (i. e. of virtual genus = effective genus). To prove this it is sufficient to use the fundamental relation (11) and to observe that *if a variable irreducible curve A approaches a reducible curve* $A_1 + A_2$, *any canonical set of A approaches the sum of two sets* Γ_1 *and* Γ_2 *on* A_1 *and* A_2 *respectively, where* Γ_i *is a set equivalent to the sum of a canonical set on* A_i *and of the set cut out on* A_i *by* A_j $(j \neq i)$. This statement is nothing else than the function-theoretic complement of NOETHER's formula for the genus of a reducible curve (II, 4).

From the fact that the adjoint curves A' of a system $|A|$ cut out canonical sets on a generic A it follows that if $|A|$ possesses subadjoint systems $|A_s^r|$, then (section 4, property d)

$$(14) \qquad |A'| = |A_s^0 - \Theta| + \Theta_1,$$

where Θ and Θ_1 are sets of fundamental curves of $|A|$ and Θ_1 is a fixed component of $|A'|$. It is not difficult to see that A' *possesses as fixed components all the exceptional curves of F which are fundamental for* $|A|$ (this has been already proved for non-singular systems). In fact, if ϑ is such an exceptional curve, we consider a non-singular system $|E|$ for which ϑ is fundamental. Then ϑ is a fixed component of $|E'|$ and therefore also of $|A'| = |E' - D_1 - \sum O_j^{y-1} - \sum P_\alpha|$, where $|D_1| = |E - A - \sum O_j^{y}|$ [formula (12)], since in general (for instance, if we consider instead of $|E|$ a sufficiently high multiple of $|E|$) $|D_1|$ does not possess ϑ as a fixed component.

The relation (14) shows that on the surface $\overline{F}$ of order n in S_3, obtained by referring a web of $|A|$ to the system of plane sections, the adjoint system $|A'|$ is cut out by the subadjoint surfaces of order $n - 3$ satisfying certain additional base conditions at the isolated singularities of $\overline{F}$. The cumulative effect of these conditions is the subtraction of Θ

from $|A_s^0|$. These particular subadjoint surfaces of $\overline{F}$ are called *adjoint surfaces* (of order $n - 3$). If $|A|$ is a non-singular system, then $\overline{F}$ has only ordinary singularities and the adjoint surfaces coincide with the subadjoint surfaces. Concerning the conditions imposed by an isolated singularity on the adjoint surfaces, only partial results are available. The question has been treated by ENRIQUES in (5) and further elaborated in his treatise (ENRIQUES, a). We quote the following two theorems due to ENRIQUES:

If a fundamental curve $\overline{\Theta}$ occurs in Θ of (14), then $\overline{\Theta}$ is necessarily a proper fundamental curve, i. e. the effective genus of $|A_1| = |A - \overline{\Theta}|$ is less than the effective genus of $|A|$ (compare with the definition of a proper multiple point of a surface, I, 6).

If $|A_1|$ is irreducible, then the curves A' cut out on a generic A_1 sets of points equivalent to the sum of a canonical set on A_1 and of the set cut out on A_1 by $\overset{..}{\Theta}$.

The transformation law of adjoint systems: *If $|A_1|$, $|A_2|$ are corresponding systems on two surfaces F_1 and F_2, free from singularities, in birational correspondence, and if $\Theta_1^{(i)}, \Theta_2^{(i)}, \ldots$ $(i = 1, 2)$ are the exceptional curves of F_i which correspond to points of F_j $(j \neq i)$, not base points of $|A_j|$ (and which therefore are fundamental for $|A_i|$), then to the system $|A_1' - \sum \Theta_\alpha^{(1)}|$ corresponds the system $|A_2' - \sum \Theta_\beta^{(2)}|$.* For non-singular systems this theorem follows immediately from (10) and from the fact that the subadjoint systems of a non-singular system are absolutely invariant under birational transformations. Using (12) it is easy to prove the theorem for an arbitrary system $|A|$. The adjoint system $|A'|$ of a given system $|A|$ is consequently a *relative covariant* of $|A|$ (i. e. covariant to within exceptional curves).

If $|A|$ is an arbitrary linear system, of virtual characters n and π, and if n', π' are the virtual characters of its adjoint system $|A'|$, then

$$(15) \qquad\qquad n' = \pi + \pi' - 2.$$

For the proof it is sufficient to consider the system $|A + A'|$, of virtual genus $\pi + \pi' + (2\pi - 2) - 1 = 3\pi + \pi' - 3$ (13), whose adjoint system is $|2A'|$, and to calculate the intersection number of a $|A + A'|$ and a $|2A'|$. We obtain then $2(3\pi + \pi' - 3) - 2 = 2(2\pi - 2) + 2n'$, and from this follows (15).

We shall conclude this section by recalling several other definitions of adjoint systems. The adjoint surfaces $\varphi(x, y, z) = 0$ of a surface $f(x, y, z) = 0$ of order n, free from singularities at infinity, can be defined transcendentally by the condition that the double integral $\iint \frac{\varphi}{f_z'} \, dx\, dy$ (VII, 1) should be finite on the surface at every point at finite distance (CLEBSCH, NOETHER). If f has only ordinary singularities it can be verified

directly that $\varphi = 0$ must pass through the double curve and that consequently this definition is equivalent to the algebro-geometric definition
given in this section. The equivalence of the two definitions in the most
general case follows from the fact that *either definition implies the invariance of the adjoint surfaces φ_{n-4} of order $n - 4$ under birational transformations.* According to the transcendental definition this invariance follows
from the fact (CLEBSCH) that the integral $\iint \frac{\varphi_{n-4}}{f_z} \, dx \, dy$ is everywhere finite
on the surface f (including the points at infinity). In the algebro-geometric
definition the invariance of the surfaces φ_{n-4} follows directly from the
distributive property of adjunction and is stated explicitly in the next
section. The transcendental criterion provides in most cases the best if
not the only workable method for the determination of the behavior
of the adjoint surfaces at the isolated singularities. An interesting
result concerning singularities of a very general type has been obtained
by this method by HODGE (1).

Another definition of the adjoint system $|C'|$ of a system $|C|$ has
been given by ENRIQUES (9) and is based on the notion of the *Jacobian
system* $|C_j|$ of $|C|$. A Jacobian curve of a net is the curve of double
points of the curves of the net. It can be proved that the Jacobian
curves of the various nets extracted from $|C|$ are equivalent and therefore
are all contained in a complete linear system $|C_j|$. Each assigned s-fold
point of $|C|$ is considered as an assigned $(3s - 1)$-fold point of $|C_j|$.
It follows immediately from the definition that $|C_j|$ is a relative covariant
of $|C|$. The fundamental relation expressing the distributive property
of the Jacobian systems, and analogous to (11), is the following: $|(C+D)_j|$
$= |C_j + 3D + P_i|$, where $P_1, P_2, \ldots$ are the base points of $|D|$
which are not base points of $|C|$. The system $|C'|$ is defined as follows:
$|C'| = |C_j - 2C|$. From the above expression of $|(C + D)_j|$ we derive
immediately the fundamental relation (11) and also the equivalence
of the two definitions by observing that if $|C|$ is the non-singular system
containing totally the plane sections of a surface F with ordinary
singularities in S_3, then $|C_j|$ is cut out on F by adjoint surfaces of order
$n - 1$. This last statement follows from the fact that the Jacobian curve
of the net cut out on F by a bundle of planes of center O is cut out on F
by the polar surface of F with respect to O, and that the cuspidal points
of F, base points of the polar surfaces, must be regarded, as far as $|C_j|$
is concerned, as accidental base points.

SEVERI (10) defines $|C'|$ by means of a variable pencil $\{C\}$, extracted
from $|C|$, and a fixed arbitrary pencil $\{L\}$. If G is the curve of contacts
of a variable curve of $\{C\}$ with a variable curve of $\{L\}$, it can be shown
that as $\{C\}$ varies in $|C|$ the curve G varies in a linear system. The
adjoint system $|C'|$ is defined as the system $|G - C - 2L|$. The
equivalence of this definition with the preceding definition by means of

the Jacobian system follows from the relation $|(C+L)_j| = |G+C+L|$ (ENRIQUES, a).

7. The residue theorem in its projective form. The adjoint surfaces φ_ν of order $\nu = n - 3 + r$, $r \geqq 0$, of a surface F of order n in S_3 cut out on F the complete system $|C' + rC| = |[(r+1)C]'|$, where $|C|$ is the system of plane sections. It follows also that the surfaces φ_ν passing through a fixed curve D cut out on F outside D the complete system $|C' + rC - D|$. In particular, if $D = C$, we find that the adjoint surfaces of *any* order ν cut out complete systems. The residue theorem in its invariantive form (II, 4) can now be restated in its projective formulation: *The residual curves of a given D, with respect to the adjoint surfaces of a given order, are also residual curves of any other curve of $|D|$ with respect to the same adjoint surfaces.* This theorem is a generalization of the *Restsatz* of BRILL and NOETHER for algebraic plane curves.

It is clear that *any complete system on F can be cut out by adjoint surfaces of a sufficiently high order passing through a properly chosen curve D.*

We quote the following theorem, useful in applications: *If $|C|$ is the system of plane sections of F and if D is an arbitrary curve, then the system $|D + hC|$ cuts out on C a complete series, if h is sufficiently high* (CASTELNUOVO, 7). For the proof it is sufficient to observe that the systems $|D + C|, |D + 2C|, \ldots, |D + hC|, \ldots$ can be cut out on F by the adjoint surfaces of increasing orders passing through a fixed curve E independent of h, and to apply the results of section 2. It is well known that on a curve of genus π, complete series of a sufficiently high order $m (m \geqq 2\pi)$ are free from fixed points. This leads to the following corollary of the preceding theorem: *Let F be free from singularities in S_r, and let $|C|$ be a complete irreducible linear system on F, of dimension $r \geqq 3$ and simple. If D is an arbitrary curve on F and if h is sufficiently high, then the only possible fixed components of $|D + hC|$ are the fixed components of $|D|$ which are fundamental curves of $|C|$. Outside of these fixed components the system $|D + hC|$ is irreducible.*

8. The canonical system. From the fundamental relation (11) of section 6 it follows that given any system $|C| = |C|_H$, $H = \sum P_i^{a_i}$, the system

$$(16) \qquad |K| = |C' - C - \sum P_i|,$$

virtually free from base points, is independent of $|C|$ and is therefore uniquely determined on F. The system $|K|$, if it effectively exists, is called the *canonical system*. In the contrary case $|K|$ is defined by (16) as a virtual system (II, 8) and could be referred to as the *virtual canonical system*.

The transformation law of adjoint systems (section 6) leads immediately to the following *transformation law of the canonical system*:

If F_1 and F_2 are two surfaces in birational correspondence, whose canonical systems are $|K_1|$ and $|K_2|$ respectively, and if $\Theta_1^{(i)}$, $\Theta_2^{(i)}$, ... are the exceptional curves of F_i $(i = 1, 2)$ which correspond to points of F_j $(j \neq i)$, then the transform of $|K_1 - \sum \Theta_\alpha^{(1)}|$ is $|K_2 - \sum \Theta_\beta^{(2)}|$. Consequently, the canonical system of a surface F is a *relative functional invariant of F*.

Since the exceptional curves of F which are fundamental for $|C|$ are fixed components of $|C'|$, and since any exceptional curve of F is fundamental for some system $|C|$, it follows that *all the exceptional curves of F are fixed components of the canonical system*, if this system effectively exists. The system obtained from $|K|$ by neglecting these fixed components is called the *pure canonical system*. We shall denote it by $|K^*|$. *This system is absolutely invariant under birational transformations.* Its dimension, virtual genus, and virtual degree are denoted by $p_g - 1$, $p^{(1)}$, and $p^{(2)}$ respectively and are naturally *absolute numerical invariants* of F. The invariants p_g and $p^{(1)}$ are called respectively the *geometric genus* (Flächengeschlecht) and the *linear genus* (Kurvengeschlecht) of F (NOETHER, 2; ENRIQUES, 2, 5, 9).

On a surface F of order n in S_3 the canonical system $|K^*|$ is cut out by the adjoint surfaces of order $n - 4$. The invariance of these surfaces under birational transformations was first stated by CLEBSCH and proved algebraically by NOETHER (2).

If $|K|$ is a virtual system, then $p_g = 0$, while $|K^*|$ and $p^{(1)}$, $p^{(2)}$ are not defined. However, irrespective of the effective existence of $|K|$, it is possible to consider its virtual characters: its virtual genus $\bar{p}^{(1)}$ (denoted by ω in CASTELNUOVO-ENRIQUES, 2) and its virtual degree $\bar{p}^{(2)}$. The characters $\bar{p}^{(1)}$ and $\bar{p}^{(2)}$ are *relative invariants* of F. They are, however, not independent. In fact, if $|C|$ is a system virtually free from base points, of characters n and π, and if n' and π' are the characters of $|C'|$, then from $|C'| = |C + K|$ it follows that

$$(17) \qquad \bar{p}^{(1)} = \pi' - 3\pi + n + 3,$$
$$(17') \qquad \bar{p}^{(2)} = n' + n - 4\pi + 4,$$

and consequently, in view of (15),

$$(18) \qquad \bar{p}^{(2)} = \bar{p}^{(1)} - 1.$$

Solving for n' and π' and using (18) we find the formulas:

$$(19) \qquad n' = 4(\pi - 1) - n - 1 + \bar{p}^{(1)},$$
$$(19') \qquad \pi' = 3(\pi - 1) - n + \bar{p}^{(1)},$$

which express the virtual characters of $|C'|$ in terms of the virtual characters of $|C|$ and $\bar{p}^{(1)}$.

The relative invariance of $\bar{p}^{(1)}$ can be formulated exactly as follows: *Let F_1 and F_2 be two surfaces in birational correspondence and let $\bar{p}_1^{(1)}$ and $\bar{p}_2^{(1)}$ be their corresponding invariants. If there are e_i exceptional curves* on F_i $(i = 1, 2)$ *which correspond to points of* F_j $(j \neq i)$, *then*

(20) $$\bar{p}_1^{(1)} + e_1 = \bar{p}_2^{(1)} + e_2.$$

Proof: Let $\Theta_1^{(1)}, \Theta_2^{(1)}, \ldots, \Theta_{e_1}^{(1)}$ be the e_1 exceptional curves of F_1. We have $[\Theta_i^{(1)}] = 0$, $(\Theta_i^{(1)2}) = -1$ (II, 6, 7). Moreover, it is not difficult to see that $(\Theta_i^{(1)} \cdot \Theta_j^{(1)}) = 0$, if $i \neq j$. This is obvious whenever $\Theta_i^{(1)}$ and $\Theta_j^{(1)}$ correspond to two points of F_2 at a finite distance from each other. If the two points are infinitely near this is still true and follows from the considerations developed in II, 6. Denoting then by ϱ_1 the virtual genus of the system $|\bar{K}_1| = \left| K_1 - \sum\limits_{i=1}^{e_1} \Theta_i^{(1)} \right|$, we find $\bar{p}_1^{(1)} = \varrho_1 + \sum\limits_{i=1}^{e_1} (\bar{K}_1 \cdot \Theta_i^{(1)}) - e_1$ $= \varrho_1 + \sum\limits_{i=1}^{e_1} (K_1 \cdot \Theta_i^{(1)})$. Let $|C_1|$ be the (non-singular) system on F_1 which corresponds to the hyperplane sections of F_2, and let $P_1^{(1)}, P_2^{(1)}, \ldots, P_{e_2}^{(1)}$ be its base points, corresponding to the exceptional curves $\Theta_1^{(2)}, \Theta_2^{(2)}, \ldots, \Theta_{e_2}^{(2)}$ of F_2. Since $|C_1'| = \left| C_{1s}^0 + \sum\limits_{i=1}^{e_1} \Theta_i^{(1)} \right|$ [formula (10), section 6], and since each $\Theta_i^{(1)}$ is fundamental for $|C_{1s}^0|$ (section 6, p. 62), we deduce that $(C_1' \cdot \Theta_i^{(1)}) = -1$. From the definition of K_1 (16) it follows that

(21) $$\bar{p}_1^{(1)} = \varrho_1 - e_1 - \sum\limits_{i=1}^{e_1} \sum\limits_{j=1}^{e_1} s_{ij}^{(1)},$$

where $s_{ij}^{(1)}$ is the assigned multiplicity of $\Theta_i^{(1)}$ at $P_j^{(1)}$. In a similar manner we find that

(21') $$\bar{p}_2^{(1)} = \varrho_2 - e_2 - \sum\limits_{i=1}^{e_1} \sum\limits_{j=1}^{e_1} s_{ij}^{(2)},$$

where $\varrho_2 = \left[K_2 - \sum\limits_{i=1}^{e_2} \Theta_i^{(2)} \right]$ and $s_{ij}^{(2)}$ is the assigned multiplicity[1] of $\Theta_i^{(2)}$ at the point $P_j^{(2)}$ which corresponds to $\Theta_j^{(1)}$. It is obvious that $s_{ij}^{(1)} = s_{ji}^{(2)}$ and also that, in view of the transformation law of the canonical system stated above, $\varrho_1 = \varrho_2$. Hence from (21) and (21') it follows that $\bar{p}_1^{(1)} + e_1 = \bar{p}_2^{(1)} + e_2$, q. e. d.

It follows from (20) that *if a surface* F *possess a finite number* e *of exceptional curves, then* $\bar{p}^{(1)} + e$ *is an absolute invariant of* F. The number of exceptional curves of F is certainly finite if $p_g > 0$, because these curves are then fixed components of the effectively existing canonical system $|K|$. *In this case the absolute invariant* $\bar{p}^{(1)} + e$ *is nothing else than the linear genus of* F:

(22) $$p^{(1)} = \bar{p}^{(1)} + e.$$

This statement follows from (21) in conjunction with a theorem of CASTELNUOVO-ENRIQUES (2) to the effect that if the number of exceptional curves of a surface is finite, then they are all necessarily of the

---

[1] We recall that each exceptional curve Θ of a surface F carries with it a set of assigned base points such that with respect to this set $(\Theta^2) = -1$ and $[\Theta] = 0$. If Θ is of the first kind, then Θ is virtually free from base points (II, 6).

1st kind (IV, 4). If we then apply (21) to the system $|K^*| = \left| K - \sum_{i=1}^{e} \Theta_i \right|$, we find $\bar{p}^{(1)} = p^{(1)} - e$, since all the assigned multiplicities s_{ij} are zero (see the preceding footnote).

From the proof of (20) it follows incidentally that if Θ is an exceptional curve of the 1st kind, then

(23) $$(K \cdot \Theta) = -1,$$

and if $|K|$ effectively exists, then

(23') $$(K^* \cdot \Theta) = 0.$$

The relation (22) can now be assumed as *the definition of the linear genus* $p^{(1)}$ for surfaces whose geometric genus p_g is zero and which possess a finite number of exceptional curves.

For a surface F with infinitely many exceptional curves CASTELNUOVO (7) proposes to define $p^{(1)}$ as *the maximum value of* $\bar{p}^{(1)}$ *on surfaces birationally equivalent to* F. However, by a fundamental result of CASTELNUOVO-ENRIQUES (2), F belongs then to the class of ruled surfaces (IV, 4) and so this exceptional case is relatively unimportant.

Another relative invariant of F, whose behavior under birational transformations is exactly the opposite of that of $\bar{p}^{(1)}$, is the ZEUTHEN-SEGRE *invariant I*. Let $\{C\}$ be an irreducible linear pencil of curves, of genus π and degree n, and let δ be the number of curves in $\{C\}$ which are of genus $\pi - 1$. Then $I = \delta - n - 4\pi$. It can be shown that this expression is independent of the pencil $\{C\}$ (ZEUTHEN, 1, C. SEGRE, 2). A topological proof will be given in VI, 10. If F_1 is birationally transformed into a surface F_2 and if we suppose, for simplicity, that the transformation introduces on F_2 only one new exceptional curve $\Theta^{(2)}$, corresponding to a point P_1 or F_1, but that no exceptional curve of F_1 is transformed into a point of F_2, we see immediately that for the transformed pencil $\{C_2\}$ on F_2 we have $\delta_2 = \delta_1 + 1$. In fact, to the curve C_1 of $\{C_1\}$ passing through P_1 there corresponds in $\{C_2\}$ a curve $C_2 = C_2^* + \Theta^{(2)}$, having a double point at the intersection of the two components. Consequently $I_2 = I_1 + 1$ (whereas $\bar{p}_2^{(1)} = \bar{p}_1^{(1)} - 1$). More generally, if e_1 exceptional curves of F_1 are transformed into points of F_2 and e_2 exceptional curves of F_2 correspond to points of F_1, then $I_1 - e_1 = I_2 - e_2$. It follows that the *expression* $I + \bar{p}^{(1)}$ *is an absolute invariant of* F. NOETHER (2) has shown that this invariant is related to the arithmetic genus p_a of F (IV, 1) by the formula $I + \bar{p}^{(1)} = 12 p_a + 9$.

The invariant I can be calculated also in terms of the characters of any irrational pencil $\{C\}$. We recall that such a pencil is necessarily of degree 0 (II, 1). Let π be the genus of a generic C and let ϱ be the genus of the pencil. If δ denotes as before the number of curves in $\{C\}$ which are of genus $\pi - 1$, then $I = \delta + 4(\pi - 1)(\varrho - 1) - 4$ (CASTELNUOVO-ENRIQUES, 2).

9. The pluricanonical systems (ENRIQUES, 5). The multiples $|K_i|$ $= |iK|$ of the canonical system $|K|$ are called *pluricanonical systems*. The dimension of $|K_i|$, increased by 1, is called the *i-genus* of F and is denoted by P_i. In particular, $P_1 = p_g$, P_2 is the *bigenus*, etc. The plurigenera $P_2, P_3, \ldots$ are absolute invariants of F. The virtual genus Π_i and the virtual degree N_i of $|K_i|$ can be expressed in terms of $\bar{p}^{(1)}$ as follows (CASTELNUOVO, 7):

$$(24) \qquad \Pi_i = \frac{i(i+1)}{2}(\bar{p}^{(1)} - 1) + 1, \qquad N_i = i^2(\bar{p}^{(1)} - 1).$$

In the same paper (also CASTELNUOVO-ENRIQUES, 2) CASTELNUOVO gave a lower limit for P_i by using the theorem of RIEMANN-ROCH (IV, 2). He finds that

$$(24') \qquad P_i \geqq p_a + \frac{i(i-1)}{2}(p^{(1)} - 1) + 1,$$

provided $p^{(1)} > 1$ (see IV, 4, p. 86). This inequality gives a sufficient condition that the *i*-canonical system $|K_i|$, a priori virtual, should effectively exist.

The plurigenera are of special importance in the theory of surfaces for which $p_g = 0$, because then the canonical system $|K|$ does not exist, but there may exist the system $|iK|$ (and hence $P_i > 0$) for sufficiently high values of i. A very simple example is furnished by a surface of 6th order in S_3 passing doubly through the edges of a tetrahedron (ENRIQUES, 5). Quadrics passing through the 6 edges do not exist, hence $p_g = 0$. There exists however one (and only one) surface of order 4 passing doubly through the edges of the tetrahedron, namely the tetrahedron itself. Hence, $P_2 = 1$. Other examples of surfaces for which $p_g = 0$, $P_2 > 0$ and whose linear genus $p^{(1)}$ is *positive* have been given by CASTELNUOVO (5), GODEAUX (4), CAMPEDELLI (1). The plurigenera play an essential role in the classification of surfaces according to the particular values of their invariants, dealt with in several important papers by CASTELNUOVO and ENRIQUES [see the report of H. GEPPERT (a)].

Appendix to Chapter III

By David Mumford

Re: § 3. If F is a surface in $\mathbf{P}^3$, and $\pi : F_1 \to F$ is its normalization, let $\mathfrak{C} \subset \mathcal{O}_F$ be the sheaf of conductors of the affine rings of F_1 over the affine rings of F, and let $\mathfrak{C}_0$ denote the inverse image of $\mathfrak{C}$ in $\mathcal{O}_{\mathbf{P}^3}$. Since the local rings of F are Cohen-MacCauley rings, the associated points of $\mathfrak{C}$ (the points corresponding to the associated prime ideals of $\Gamma(U, \mathfrak{C})$, for various affines $U \subset F$) are all of codimension 1. Hence in view of the arithmetic characterization of the adjoints of plane curves (Gorenstein, 1), it follows that the *subadjoint surfaces* for F of degree n are the surfaces $\varphi = 0$, where $\varphi \in \Gamma(\mathbf{P}^3, \mathfrak{C}_0 \cdot \mathcal{O}_{\mathbf{P}^3}(n))$.

Since the conductor of $\Gamma(\pi^{-1}(U), \mathcal{O}_{F_1})$ over $\Gamma(U, \mathcal{O}_F)$ is an ideal in $\Gamma(\pi^{-1}(U), \mathcal{O}_F)$ also, we see that we get a sheaf of conductors $\mathfrak{C}_1 \subset \mathcal{O}_{F_1}$ too and

$$\pi_* \, \mathfrak{C}_1 = \mathfrak{C},$$
$$\pi^* \, \mathfrak{C} = \mathfrak{C}_1.$$

Similarly all the associated points of $\mathfrak{C}_1$ have codimension 1 so $\mathfrak{C}_1 = \mathcal{O}_{F_1}(-\Delta)$ for some positive divisor Δ. Now for any normal surface G, we will denote by ω_G be the sheaf of rational 2-forms on G, regular in codimension 1, hence regular at all non-singular points of G. Equivalently, ω_G can also be characterized as the double dual of Ω_G^2, the Kähler 2-forms on G. Then the fundamental isomorphism, on which the theory of adjoint systems rests, is:

$$\omega_{F_1} \cong \mathcal{O}_{F_1}((n-4)\, H - \Delta)$$

where $n = $ degree of F and H stands for a plane section of F or its inverse image on F_1. This isomorphism is defined by means of the residue:

$$\Omega_{\mathbf{P}^3}(F) \xrightarrow{\ \text{Res.}\ } \Omega_{k\,(F)/k}^2 = \Omega_{k\,(F_1)/k}^2,$$

$$a\,\frac{dx \wedge dy \wedge dz}{f} \mapsto a\,\frac{dx \wedge dy}{\partial f/\partial z} \bmod (f, df).$$

The essential point is that $a\,\dfrac{dx \wedge dy}{\partial f/\partial z}$ is regular on F_1 in codimension 1 if and only if $a \in \mathfrak{C}_0$. Once this is proven, since $\Omega_{\mathbf{P}^3}(F) \cong \mathcal{O}_{\mathbf{P}^3}(n-4)$ it follows that residue defines a map from $\mathcal{O}_{F_1}((n-4)\,H - \Delta)$ to ω_{F_1} which is an isomorphism in codimension 1; since both sheaves are isomorphic to their double duals, it is an isomorphism everywhere. For a proof of the essential point and a discussion of this isomorphism, see ZARISKI (32), § 15. In the language of divisors, if we write $\omega_{F_1} = \mathcal{O}_{F_1}(K_{F_1})$ where K_{F_1} is the *canonical divisor class*, the conclusion can be expressed:

$$K_{F_1} \equiv (n - 4) H - \Delta .$$

The great importance of subadjoint surfaces is due to the fact that they are the easiest way to construct *complete* linear systems on F_1. In fact, using the exact sequences:

$$0 \to \mathcal{O}_{\mathbf{P}^3}(m - 4) \xrightarrow[\text{eq. of } F]{\text{mult. by}} \mathfrak{C}_0 \cdot \mathcal{O}_{\mathbf{P}^3}(n + m - 4) \to \mathfrak{C} \cdot \mathcal{O}_F(n + m - 4) \to 0$$

and

$$\mathfrak{C} \cdot \mathcal{O}_F(n + m - 4) \cong \pi_* \, \mathcal{O}_{F_1}(K_{F_1} + m H)$$

it follows that

$$H^0(\mathfrak{C}_0 \cdot \mathcal{O}_{\mathbf{P}^3}(n + m - 4)) \to H^0(\mathcal{O}_{F_1}(K_{F_1} + m H))$$

is surjective, i. e., the subadjoint surfaces of order $n + m - 4$ cut out on F_1 the complete linear system $|K_{F_1} + m H|$.

Re: §§ 4, 5, 6. We may express ENRIQUES' definition of subadjoint systems of a complete irreducible simple linear system $|C|$ on a non-singular surface F, as follows:

i) let F_1 be the non-singular surface obtained from F by blowing up the base points of $|C|$,

ii) let F_2 be the normal surface obtained by blowing down on F_1 the exceptional curves of $|C|$,

iii) let $F^* \subset \mathbf{P}^3$ be a generic projection of F_2, i. e., the image of F by a generic web in $|C|$.

We get the diagram:

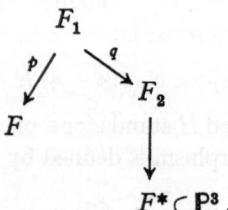

Since the base points of $|C|$ are assigned with their effective multiplicities, $|C|$ becomes a linear system on F_1 free of both assigned and effective base points. "Lift" the sheaf ω_{F_2} and the divisor K_{F_2} to F_1 as follows:

let $\tilde{\omega}_{F_2}$ = the double dual of $\omega_{F_2} \otimes_{\mathcal{O}_{F_2}} \mathcal{O}_{F_1}$

= the double dual of $\Omega_{F_2}^2 \otimes_{\mathcal{O}_{F_2}} \mathcal{O}_{F_1}$

and let

$$\tilde{\omega}_{F_2} = \mathcal{O}_{F_1}(\tilde{K}_{F_2}) .$$

Then by definition $|C_s^r|$ is the complete linear system on F_1:

$$|\tilde{K}_{F_s} + (r+1)\,C|$$

without *assigned* base points as a linear system on F_1.

(Of course, it can be transformed by the rules in Chapter II into a complete linear system with assigned base points on F.)

On the other hand, the adjoint system of $|C|$ is just the complete linear system on F_1:

$$|C'| = |K_{F_1} + C|$$

without *assigned* base points as a linear system on F_1. The definition given in § 6 is the same as this one by virtue of the *transformation law* for K_F under quadratic transformations:

If F is a non-singular surface and $\pi: F_1 \to F$ is the quadratic transformation with center $x \in F$, then

$$K_{F_1} = \pi^{-1}(K_F) + E$$

where $E = \pi^{-1}(x)$.

Note that modulo details about base points and exeptional curves, these definitions of subadjoint and adjoint linear systems make their "distributive" property obvious. The Italians took this apparently roundabout course of defining adjoint systems and proving distributivity before defining the canonical divisor class because they preferred linear systems of effective divisors and regarded the canonical divisor class as rather "unreal" because it involved "virtual divisors".

The fact that $|C'|$ is related to $|C_s^0|$ by the formula

(*) $\qquad\qquad\qquad |C'| = |C_s^0 - \Theta| + \Theta_1$

where Θ, Θ_1 are exceptional curves of q and Θ_1 is a fixed component of $|C'|$ [formula (14), p. 63], follows from the observation that

$$q_* \, \omega_{F_1} \subset \omega_{F_s}$$

hence

$$q^* (q_* \, \omega_{F_1}) \subset \tilde{\omega}_{F_s}.$$

If the double dual of $q^*(q_* \, \omega_{F_1})$ equals $\omega_{F_1}(-\Theta_1)$, then Θ_1 is a sum of exceptional curves for q which must be fixed components of every linear system $|K_{F_1} + q^{-1} D|$, for every divisor D on F_2. For instance Θ_1 is a fixed component of $|C'|$.

Now $\omega_{F_1}(-\Theta_1) \subset \tilde{\omega}_{F_s}$. Define the positive divisor Θ by:

$$\omega_{F_1}(-\Theta_1) = \tilde{\omega}_{F_s}(-\Theta).$$

Then since the linear system $|C'| - \Theta_1$ is defined by the sections of $\omega_{F_1}(C - \Theta_1)$, and the linear system $|C_s^0 - \Theta|$ is defined by the sections of $\tilde{\omega}_{F_s}(C - \Theta)$, (*) follows.

The essentially birational-invariant nature of the adjoint system $|C'|$ can be illustrated by rephrasing the definition yet again, as follows:

(I) Let v be any prime divisor of the function field K/k of transcendence degree 2. Let F be any non-singular model of K in which the center C of v is 1-dimensional and non-singular. Let

$\Omega_k^2(v)$ = vector space of differentials $\omega \in \Omega_{K/k}^2$ such that ω is regular
    on F except for a simple pole on C.

Then $\Omega_k^2(v)$ is *independent of the model F*.

(II) If $|C|$ is any irreducible linear system on a non-singular surface F, with base points assigned with their effective multiplicity, and if $C \in |C|$ is sufficiently generic so that C is singular only at the base points of $|C|$ and has an r-fold point at each r-fold base point, then

$|C'|$ = linear system of divisors of zero of the differentials $\omega \in \Omega_k^2(v)$
    where v = prime divisor centered at C.

(III) Notice too that there is an invariant *residue map:*

$$\text{Res}: \Omega_k^2(v) \to \Omega_v^1$$

where Ω_v^1 is the space of everywhere regular differentials on the curve C which is the center of v.

Re: § 8. The invariants $\bar{p}^{(1)}$ and I introduced here are essentially the Chern numbers of the tangent sheaf T_F (or of the cotangent sheaf Ω_F^1). The Chern classes and the Chern numbers have an exceedingly large literature devoted to them. I am not competent to give a history of their development but very roughly, after the publication of the first edition of this book, the following took place: on the one hand, the Italian school, and notably Severi, Todd, Eger, and B. Segre (13), developed a general theory of Chern classes in the algebraic case; simultaneously, Chern and may other topologists and differential geometers developed the concept of Chern classes in the topological case. Finally, Grothendieck (3), unaware that the Italian school had developed a general algebraic theory, showed how Chern classes could be defined by a simple and general procedure applicable in the algebraic and topological cases. In any case,

$$\bar{p}^{(1)} = \deg(c_1^2) + 1 \ ,$$

$$I = \deg c_2 - 4$$

where $c_i = i$th Chern class of the tangent sheaf T_F. And Noether's famous formula $\bar{p}^{(1)} + I = 12 p_a + 9$ is, of course, the special case of the Riemann-Roch theorem:

$$(c_1^2) + (c_2) = 12(\chi(\mathcal{O}_F)) \ .$$

For a recent analysis of Todd's original definition, see Porteous (1).

Chapter IV

The Arithmetic Genus and the Generalized Theorem of Riemann-Roch

1. The arithmetic genus p_a. Let F be a surface of order n with ordinary singularities in S_3. The effective dimension of the complete system $|\varphi_{n-4}|$ of adjoint surfaces of F of order $n-4$ is $p_g - 1$. Cayley (1) and Noether (2) have first shown by examples that the *virtual dimension* of $|\varphi_{n-4}|$ evaluated according to the postulation formula (III, 2) may be less than $p_g - 1$. Thus, for instance, if F is a ruled surface of genus p, it is found (Cayley, 1) that $p_a = -p$, while $p_g = 0$, because in this case $|\varphi_{n-3}|$ and hence a fortiriori $|\varphi_{n-4}|$ does not exist (III, 3). The virtual dimension of $|\varphi_{n-4}|$ increased by 1 is called the *arithmetic genus* of F and is denoted by p_a (Noether, 2). From the well known fact that the adjoint curves of order $\geqq n-3$ of a plane irreducible curve of order n form a regular system, it follows in view of III, 2, formula (2'), that $p_g \geqq p_a$ and that

$$(1) \qquad\qquad p_g - p_a = \omega_{n-3} + \omega_{n-2} + \cdots,$$

where ω_{n-3+i} is the deficiency of the system of curves cut out by $|\varphi_{n-3+i}|$ on a generic plane. The difference $q = p_g - p_a$ is called *the irregularity* of F and F is said to be a *regular* or an *irregular* surface according as $q = 0$ or $q > 0$.

If the double curve of F is of order m, genus p, and possesses t triple points, then $p_a = \binom{n-1}{3} - m(n-4) + 2t + p - 1$ (III, 2). If the double curve is reducible, then p should be evaluated by means of Noether's formula for the genus of a reducible curve. The definition of p_a can be extended to surfaces with arbitrary singularities. Its evaluation requires in every case the knowledge of the postulation formula for the system of adjoint surfaces of a given order.

Zeuthen (1) was the first to prove, under certain restrictions concerning the singularities of F, that p_a is an absolute invariant under birational transformations. A complete proof was given by Enriques (5) by means of considerations which threw a new light upon the invariantive character of the irregularity q and hence also of p_a. We proceed to sketch Enriques' proof.

a. Let $|C|$ be a complete irreducible system of effective genus π. The series cut out on a generic C by its adjoint system $|C'|$ is totally contained in the canonical series $g_{2\pi-2}^{\pi-1}$, and if it is not complete it will have a certain deficiency, which we shall denote by $\delta(C)$. A generic C imposes then $\pi - \delta(C)$ conditions on the curves C' constrained to contain C, and therefore the residual system $|C' - C|$ has the dimension $r' - \pi + \delta(C)$, where r' is the dimension of $|C'|$. But this residual system coincides, to within the virtual multiplicities of the base points, with the canonical system $|K|$. Consequently,

(2) $r' = $ dimension of $|C'| = p_g - 1 + \pi - \delta(C)$.

It is seen immediately that (2) holds also if the series cut out by $|C'|$ on C is not defined, i. e. if $|C'|$ does not exist (and hence $p_g = 0$) or if $|C|$ is of dimension 0 and C is a fixed component of $|C'|$, provided that we consider then the above series as a series of dimension -1 and consequently put $\delta(C) = \pi$.

b. Let $|C| = |C_1| + |C_2|$, where each system is supposed to be irreducible, and let π, π_1, π_2 be their effective genera. The system $|C'|$ cuts out on a generic C_2 a series of order $2\pi_2 - 2 + i$, where $i = (C_1 \cdot C_2)$ [see formula (11), III, 6], which is therefore non-special. It follows that C_2 imposes at most $\pi_2 - 1 + i$ conditions on the curves C' constrained to contain it, say $\pi_2 - 1 + i - \varepsilon$ conditions, $\varepsilon \geqq 0$. Since the residual system $|C' - C_2|$ coincides, to within the virtual multiplicities of the base points, with $|C_1'|$, it follows that the dimension of $|C_1'|$ equals the dimension of $|C'|$ diminished by $(\pi_2 - 1 + i - \varepsilon)$. Hence, by a., $\pi_1 - \delta(C_1)$ $= \pi - \delta(C) - (\pi_2 - 1 + i - \varepsilon)$, or finally, since $\pi = \pi_1 + \pi_2 + i - 1$,

(3) $\delta(C) = \delta(C_1) + \varepsilon,$ $\varepsilon \geqq 0.$

The above proof can be extended very easily to the case in which $|C_2|$ is reducible and to the case in which $|C_1|$ is obtained from $|C|$ by increasing the assigned multiplicities of some of the base points of $|C|$, or by imposing new base points. We have therefore the following theorem:

If an irreducible system $|C_1|$ is partially contained in an irreducible system $|C|$, then $\delta(C_1) \leqq \delta(C)$.

Corollary. If $|C|$ is irreducible, then

$$\delta(C) \leqq \delta(2C) \leqq \delta(3C) \leqq \cdots.$$

c. We now suppose that F is in S_3, is of order n and has ordinary singularities. Let C be a generic plane section of F and let $|C|$ be the corresponding complete system. The system $|(rC)'|, r \geqq 1$, is cut out by the adjoint surfaces φ_{n-4+r}. If we denote by ϱ_{n-4+r} the dimension of $|\varphi_{n-4+r}|$, we find by (2)

$$\varrho_{n-4+r} - (r-1)(r-2)(r-3)/6 = p_g - 1 + \pi_r - \delta(rC),$$

where π_r is the effective genus of $|rC|$, and consequently

$$\varrho_{n-3+r} - \varrho_{n-4+r} = \pi + rn - 1 + \frac{(r-1)(r-2)}{2} - \delta((r+1)C) + \delta(rC),$$

since $\pi_{r+1} = \pi_r + \pi + rn - 1$. On the other hand $\varrho_{n-3+r} - \varrho_{n-4+r}$ $= \bar{\varrho}_{n-3+r} + 1 - \omega_{n-3+r}$ (III, 2), where $\bar{\varrho}_{n-3+r}$ is the dimension of the complete system of the adjoint curves of order $n - 3 + r$ of the plane section C and where ω_{n-3+r} denotes, as usual, the deficiency of the system cut out by $|\varphi_{n-3+r}|$ on the plane of C. From the well known expression of $\bar{\varrho}_{n-3+r} \left(= \pi + rn - 2 + \dfrac{(r-1)(r-2)}{2}, \, r \geqq 0 \right)$ we derive that

(4) $$\delta((r+1)C) - \delta(rC) = \omega_{n-3+r}.$$

If r is sufficiently high, then $\omega_{n-3+r} = 0$. On the other hand we have obviously $\delta(C) = \omega_{n-3}$. Using (1) we find that if r is sufficiently high, then

(5) $$\delta(rC) = \omega_{n-3} + \omega_{n-2} + \cdots = p_g - p_a = q,$$

i. e. *the non-decreasing deficiencies $\delta(C), \delta(2C), \ldots, \delta(rC), \ldots$ reach a maximum, and this maximum is equal to the irregularity of F.*

If now $|D| = |D|_H$ $(H = \sum O_i^{s_i})$ is an arbitrary irreducible system, the system $|rC - D - \sum O_i^{s_i}|$ is irreducible for r sufficiently high. From b. we deduce that $\delta(D) \leqq \delta(rC)$. Hence, *the deficiency $\delta(D)$ of the series cut out on a generic curve D of an irreducible system $|D|$ by its adjoint system $|D'|$ is never greater than the irregularity q of F, and there exist systems $|D|$ for which $\delta(D) = q$.* This result proves that q and hence also p_a are absolute invariants of F, q. e. d.

We observe that $\delta(D) = q$, whenever $|D|$ is a sufficiently high multiple of an irreducible system $|E|$ of dimension $\geqq 2$. In fact, if $|C|$ is any system, it is possible to find h sufficiently high so that $|hE|$ should partially contain $|C|$ and that consequently $\delta(hE) \geqq \delta(C)$. In particular, if we suppose that $|E|$ is the system of plane sections of a surface F^* of order n in S_3, birationally equivalent to F and possessing *arbitrary* singularities, we deduce that the arithmetic genus p_a of F can be evaluated directly on F^*, as the virtual dimension of the system of its adjoint surfaces of order $n - 4$.

2. The theorem of RIEMANN-ROCH on algebraic surfaces. We recall that by the classical theorem of RIEMANN-ROCH in the theory of algebraic curves the dimension r of a complete linear series g_n^r on a curve of genus p can be expressed in terms of the characters of the series and of p. If i is the *index of speciality* of the g_n^r, i. e. if $i - 1$ is the dimension of the series $g_{2\pi-2-n}^{i-1}$, the difference between the canonical series $g_{2\pi-2}^{\pi-1}$ and the g_n^r, then the theorem of RIEMANN-ROCH says that $r = n - p + i$.

The analogous problem in the theory of surfaces consists in expressing, if possible, the dimension of a complete linear system $|D|$ on a surface F in terms of the characters of $|D|$—its virtual degree, virtual genus and another character analogous to the index of speciality of a linear series. Partial results have been stated by NOETHER (5) and obtained (especially for regular surfaces) by ENRIQUES (2). The complete solution is due to CASTELNUOVO (6, 7). The proof has been considerably simplified by SEVERI (1), who also extended the final result to reducible (SEVERI, 5) and virtual (SEVERI, 6) systems.

The *index of speciality* i of a linear system $|D|$ is defined as the dimension, increased by 1, of the system $|K - D|$, where $|K|$ is the canonical system of F. In performing the operation $K - D$ we must regard $|D|$ as virtually free from base points. If $|K - D|$ does not exist, we put $i = 0$. The system $|D|$ is *special* or *non-special* according as $i > 0$ or $i = 0$. If $p_g = 0$, every system on F is non-special.

The theorem of RIEMANN-ROCH on an algebraic surface:

If $|D|$ is a complete linear system on F, of virtual genus π, of virtual degree n and of index of speciality i, then

(6) $$r \geqq n - \pi + p_a + 1 - i.$$

We outline below the simplified proof given by SEVERI (1).

a. We first observe that the results of the preceding section already prove (6) for a very large class of non-special linear systems. In fact, if $|D|$ is the adjoint system of an irreducible system $|C|$, then by (2) and by the inequality of ENRIQUES, $\delta(C) \leqq q$, it follows that $r \geqq p_a + \bar{\pi} - 1$, where $\bar{\pi} = [C]$, and consequently [III, 6, formula (15)], $r \geqq n - \pi + p_a + 1$. Moreover, $i = 0$ in this case, because $|D| = |C'|$ partially contains the canonical system $|K| = |C' - C - \sum P|$ and therefore is not partially contained in $|K|$. Hence, the theorem of RIE-MANN-ROCH is thus proved for any *sufficiently ample* system $|D| = |D|_H (H = \sum P_i)$, i. e. *such that* $|C| = |D - K - \sum P_i|$ *exists and is irreducible* (ENRIQUES, 2 and a). We point out that if $\delta(C) = q$, then it follows from the proof that $r = n - \pi + p_a + 1$.

We shall call the system $|D|$ *regular*, if it is non-special and if $r = n - \pi + p_a + 1$. *The adjoint system $|C'|$ of an irreducible system $|C|$ is regular if and only if $|C'|$ cuts out on a generic C a series of maximum deficiency q.* In view of the results of the preceding section it follows that *if $|C|$ is a sufficiently high multiple of an irreducible system of dimension $\geqq 2$, then $|C'|$ is regular* (CASTELNUOVO, 7). A stronger theorem due to PICARD (13) and SEVERI (15) will be found in VII, 7.

b. Let F be in S_3, of order n and with ordinary singularities, and let $|C|$ be the complete system of its plane sections. Let r_h, n_h, π_h be the characters of $|C_h| = |hC|$. If h is sufficiently high, the system $|A| = |C_h - K|$, where $|K|$ is the canonical system, exists and is irreducible. Hence, by a., $r_h \geqq n_h - \pi_h + p_a + 1$. Let s be an integer high enough that $\delta(C_s) = q$ [see section 1, formula (5)]. We can take h so high that $|A|$ shall partially contain $|C_s|$. We will have then $\delta(A) \geqq \delta(C_s)$, hence $\delta(A) = q$, and since $|C_h| = |A'|$ it follows, by a., *that $|C_h|$ is a regular system and that* $r_h = n_h - \pi_h + p_a + 1$.

c. The central point of SEVERI's proof is the following

Lemma. *The surfaces F_m of any order m which pass through the double points of the curve C_h, the intersection of F with a generic surface F'_h of order h, cut out on C_h a complete series* (SEVERI, 1).

A well known theorem of CASTELNUOVO (2), valid for any irreducible space curve $L(= C_h)$, states that the series mentioned in the above lemma is complete *if m is sufficiently high*. In our case C_h is however a special curve, in the sense that it is the complete intersection of two surfaces, F and F'_h. The lemma of SEVERI adds to CASTELNUOVO's theorem the very necessary information that in this case the theorem holds also if m is arbitrary. The above lemma leads immediately to important consequences. The double points of C_h are at the intersection of F'_h with the double curve of F. Hence, if $h > n - 4$ the surfaces F_{n-4} ($m = n - 4$) of the lemma, if they exist, necessarily contain the double curve of F, i. e. *are adjoint surfaces* φ_{n-4} *of F.* The series cut out on C_h by these surfaces $F_{n-4}(= \varphi_{n-4})$ is the difference of the series cut out on C_h by the adjoint surfaces φ_{n-4+h} and of the series cut out by the surfaces of order h, i. e. *the difference of the canonical series on C_h and of the characteristic series of $|C_h|$ on C_h.* Conversely, if this series difference exists, i. e. if *the characteristic series of $|C_h|$ is special,* the surfaces F_{n-4} must exist, since they must cut out on C_h that series difference of dimension $\geqq 0$. Hence we have the following result:

If $h > n - 4$, the adjoint surfaces φ_{n-4}, if they exist (i. e. if $p_g > 0$), cut out on a generic C_h a complete series. If $p_g = 0$, the characteristic series of $|C_h|$ is non-special.

d. We now are in position to prove the theorem of RIEMANN-ROCH, as stated above, for any system $|D|$, irreducible or not, of dimension $\geqq 0$. We know from III, 7 that the system $|D + C_h| = |D + hC|$ is irreducible if h is sufficiently high (here $|C|$ is the system of plane section of F, or more generally any non-singular system on F). Moreover, if h is sufficiently high, this system is sufficiently ample in the sense stated in a. Hence, if R, N, Π are the characters of $|D + C_h|$, we may suppose that $R \geqq N - \Pi + p_a + 1$. Denoting by ϱ the dimension of the series cut out by $|D + C_h|$ on a generic C_h, we have $r = R - \varrho - 1$. This series is a g_{n_h+s}, where $s = (D \cdot C_h)$. If j denotes its index of speciality, then $\varrho \leqq n_h + s - \pi_h + j$. From this inequality and from the relation $r = R - \varrho - 1$, we find, using the formulas of NOETHER for the genus and for the degree of a reducible curve, that

$$r \geqq n - \pi + p_a + 1 - j.$$

Now it follows directly from the final result stated in c., that j, the index of speciality of the series g_{n_h+s}, is also equal to the number of independent canonical curves K which pass through the set of s intersections of a D with a C_h. If h is taken sufficiently high, s will be so large that all these canonical curves will have to contain D. Hence $j = i$, and $r \geqq n - \pi + p_a + 1 - i$, q. e. d.

The expression $n - \pi + p_a + 1 - i$ is referred to as the *virtual dimension* of $|D|$, and the difference $r - (n - \pi + p_a + 1 - i) = s$ is called the *superabundance* of $|D|$. We have $r = n - \pi + p_a + 1 - i + s$, and the system is regular if and only if $i = s = 0$. The difference $s - i$ is called the *irregularity* of $|D|$.

If $|D|$ is the canonical system $|K|$, then $r = p_g - 1$, $i = 1$, $n = \bar{p}^{(2)}$, $\pi = \bar{p}^{(1)}$, and $n - \pi = -1$ [III, 8, formula (18)]. Hence $s = q$: *the superabundance of the canonical system equals the irregularity of the surface.*

If $|D| = |A - B|$ is a *virtual* system, its index of speciality i is defined as the dimension of the system $|K - D| = |K + B - A|$, increased by 1. In the case of virtual systems i may be greater than p_g. It is readily seen that the RIEMANN-ROCH theorem holds also for virtual systems, by repeating practically *verbatim* the reasoning in d. The inequality $r \geqq n - \pi + p_a + 1 - i$ is to be interpreted in this case as *an arithmetic criterion for the effective existence of an a priori virtual system*:

If D is a curve, a priori virtual, of characters n, π, i, a sufficient condition that the system $|D|$ be effective (i. e. that it contain effective curves) is that $n - \pi + p_a + 1 - i$ be non-negative (SEVERI, 6).

A curve D whose characters satisfy the inequality $n - \pi + p_a + 1 - i \geqq 0$ is said to be *arithmetically effective*.

3. The deficiency of the characteristic series of a complete linear system. It is well known that in the plane (and hence also on any rational surface) the characteristic series of any complete irreducible system of curves is complete (III, 1). A proof given by ENRIQUES (2), although incomplete, constituted nevertheless a strong indication that this theorem is true on any regular surface. In projective language, the theorem says that the hyperplane sections of a regular *normal surface* are *normal curves* (p. 37). On the other hand, examples of irregular surfaces for which this theorem is not true have been known since the early stages of the theory. For instance, C. SEGRE has shown that for any value of $p \geqq 0$ there exist ruled surfaces in S_3, of order n and genus p (and hence of irregularity p), which are obtained by projection from a normal surface of the same order in a space of dimension $n - 2p + 1$, whereas the generic plane section of such a surface, being of genus p. is the projection of a normal curve in a space of $n - p$ (or more) dimensions. Hence the deficiency of the characteristic series of the complete system, totally containing the system of the plane sections of such a ruled surface, is at least p.

The question was thoroughly investigated and completely solved by CASTELNUOVO (7), who arrived at the following fundamental result:

Theorem of Castelnuovo. *The deficiency of the characteristic series of any complete irreducible linear system on an algebraic surface is not greater than the irregularity of the surface, and there exist systems for which this deficiency equals the irregularity.*

The original proof of Castelnuovo is somewhat complicated. The first step of the following proof, in the course of which other noteworthy results are arrived at incidentally, constitutes the simplification contributed by Severi (1). The rest of the proof is as in Castelnuovo.

a. *If $|C_h|$ is a sufficiently high multiple of the system of plane sections of F (or of any non-singular system on F), the deficiency of the characteristic series of $|C_h|$ equals the irregularity of F.* This statement follows from Severi's lemma quoted in the preceding section. In fact, let δ_h be the deficiency of the characteristic series $g_{n_h}^{r_h-1}$ of $|C_h|$. Then the complete series to which this series belongs is of dimension $r_h + \delta_h - 1$. It follows by the ordinary Riemann-Roch theorem that $\pi_h - n_h + r_h + \delta_h - 2$ is the dimension of its residual series with respect to the canonical series on C_h, i. e. of the difference between the series cut out on C_h by the adjoint surfaces φ_{n-4+h} and by the surfaces of order h. By section 2, c., this complete residual series is cut out on C_h by the ∞^{p_g-1} canonical curves. Hence $\pi_h - n_h + r_h + \delta_h - 2 = p_g - 1$. If $p_g = 0$ this relation still holds, because then, again by section 2, c., the characteristic series of $|C_h|$ is non-special and hence $r_h + \delta_h - 1 = n_h - \pi_h$. Since $r_h = n_h - \pi_h + p_a + 1$ (section 2, b.), we find $\delta_h = p_g - p_a = q$, q. e. d.

b. Let $|A_1|$ and $|A_2|$ be two complete irreducible systems, of dimension r_1, r_2 and of degree n_1, n_2 respectively. The complete system $|A| = |A_1 + A_2|$ cuts out on a generic A_1 a series $g_{n_1+i}^{r-r_2-1}$, where $i = (A_1 \cdot A_2)$ and where r is the dimension of $|A|$. The series cut out on A_1 by the curves A passing through the i intersections of A_1 with a fixed A_2 is a $g_{n_1}^{r-r_2-\varepsilon-1}$, where $0 \leqq \varepsilon \leqq i$. This series totally contains the characteristic series $g_{n_1}^{r_1-1}$ of $|A_1|$. Consequently, if δ_1 denotes the deficiency of the characteristic series of $|A_1|$, we have $\delta_1 \geqq r - r_1 - r_2 - \varepsilon$, *where the equality holds, if $|A|$ cuts out on A_1 a complete series.* Similarly, if δ_2 is the deficiency of the characteristic series of $|A_2|$, then $\delta_2 \geqq r - r_1 - r_2 - \varepsilon$. Hence, *if $|A|$ cuts out on A_1 a complete series, then $\delta_1 \leqq \delta_2$.*

c. The theorem on the irregularity of the residual system. *Let $|E|$ and $|A|$ be two complete irreducible systems, such that $|E_1| = |E - A|$ exists and is irreducible. If j and δ denote the index of speciality and the deficiency of the series cut out by $|E|$ on a generic A, then*

(7)   *irregularity of $|E_1|$ = irregularity of $|E| + (\delta - j)$.*

Since δ and j are given, we know the number of conditions which a generic A imposes on the curves E, and hence also the dimension of $|E_1|$ in terms of δ, j and of the dimension of $|E|$. The theorem follows then immediately from the definition of the irregularity of a system and from simple calculations.

d. Let $|A|$ be an arbitrary complete irreducible system and let $|C_h|$ be the system considered in a. If h is sufficiently high, the system $|E| = |A + C_h|$ can be regarded as the adjoint system of the effectively existing and irreducible system $|A + C_h - K - \sum O_i|$, where $O_1, O_2, \ldots$ are the base points of $|A|$. This last system, if h is sufficiently high, contains any assigned multiple $|C_s|$ of $|C|$. If we take s so high that the adjoint system of $|C_s|$ is regular (section 2, a.), it will follow a fortiriori (section 1) that $|E|$ is regular. We now apply (7) by putting $|E_1| = |C_h|$. Since both $|E|$ and $|E_1|$ are in this case regular systems, it follows that $\delta = j$. *If h is sufficiently high*, the series cut out by $|E|$ on A is non-special, and therefore *the system $|A + C_h|$ cuts out on A a complete series*. In view of b., the deficiency of the characteristic series of $|A|$ is therefore not greater than the deficiency of the characteristic series of $|C_h|$, i. e., by a., not greater than the irregularity q of F, q. e. d.

It is not difficult to see that from the theorem of CASTELNUOVO follows readily the theorem of RIEMANN-ROCH (CASTELNUOVO, 7). It is sufficient to replace in a. the system $|C_h|$ by an arbitrary irreducible system $|D|$ and to repeat the same reasoning. If δ is the deficiency of the characteristic series of $|D|$, we find, as in a., that the dimension of its residual series with respect to the canonical series is $\pi - n + r + \delta - 2$, where π, n, r are the characters of $|D|$. However we cannot affirm now, as we did in the case of $|C_h|$, that this dimension equals $p_g - 1$, because of the following two circumstances: (1) $|D|$ may be special; if i is its index of speciality, then the canonical curves cut out on a generic D a series, not of dimension $p_g - 1$, but of dimension $p_g - i - 1$; (2) this series may not be complete. We may write then the following inequality: $\pi - n + r + \delta - 2 \geqq p_g - i - 1$, or $r \geqq n - \pi + p_g + 1 - \delta - i$, *where the equality holds if and only if the canonical system cuts out on D a complete series.* Since by the theorem of CASTELNUOVO $\delta \leqq p_g - p_a$, we deduce that $r \geqq n - \pi + p_a + 1 - i$, q. e. d.

From this mode of deriving the RIEMANN-ROCH theorem it becomes apparent incidentally *under what conditions the superabundance of $|D|$ is zero.* In order that the equality $r = n - \pi + p_a + 1 - i$ should hold, it is necessary and sufficient that the two inequalities used in the course of the proof should become equalities, i. e. that we should have $r = n - \pi + p_g + 1 - \delta - i$, and $\delta = p_g - p_a$. Hence, *the superabundance of $|D|$ is zero if and only if*: (1) *the canonical system cuts out on a generic D a complete series*; (2) *the characteristic series of $|D|$ has the maximum deficiency q.* If $p_g = 0$, the condition (1) must be replaced by the following: (1') *the characteristic series of $|D|$ is non-special* (CASTELNUOVO, 7).

The theorem on the irregularity of the residual system proved in c. has been applied by CASTELNUOVO (7) toward the determination of the superabundance of sufficiently high multiples of a given irreducible system $|C|$, ∞^2 at least. We quote the following theorems:

If $|D|$ is an arbitrary linear system, then the superabundance of $|D + hC|$ remains constant if h is sufficiently high.

If i_h and δ_h are respectively the index of speciality and the deficiency of the series cut out by $|hC|$ on a generic C, and if s is the constant super-abundance of $|hC|$ when h is high, then $\sum_{h=1}^{\infty}(i_h - \delta_h) = s + p_a$ (it is clear that for sufficiently high values of h, $i_h = 0$, and by [III, 7] also $\delta_h = 0$).

If $|C|$ is free from proper fundamental curves, then $s = 0$, i. e. any sufficiently high multiple of $|C|$ is a regular system.

4. The elimination of exceptional curves and the characterization of ruled surfaces (CASTELNUOVO-ENRIQUES, 2). If L is an exceptional curve on a surface F free from singularities, then $(L^2) = -1$, $[L] = 0$. If in addition L is free from assigned base points, then it is of the 1st kind (II, 6). Using RIEMANN-ROCH's theorem it is possible to show that conversely *if $(L^2) = -1$ and $[L] = 0$ and if L is irreducible, then it is an exceptional curve*[1]. For the proof we use an auxiliary regular non-singular system $|E|$ of characters n, π, r, and we consider the system $|E_i| = |E + iL|$, $0 \leq i \leq m$, where $m = (E \cdot L)$. If r_i denotes the virtual dimension (section 2) of the *non-special* system $|E_i|$, it is easily found that $r_i = r_{i-1} + m - i + 1$. Since $(E_i \cdot L) = m - i$ and since therefore the curve L imposes at most $m - i + 1$ conditions on the curves E_i, it follows that r_i coincides with the effective dimension of $|E_i|$, provided the same is true of $|E_{i-1}|$ and r_{i-1}. Since this is true if $i = 0$ ($|E_0| = |E|$), it follows that $|E_i|$ is a regular system of effective dimension $r_i = r_{i-1} + m - i + 1$ and therefore irreducible. In particular, the system $|E_m|$ possesses L as a total fundamental curve and the residual system $|E_m - L| = |E_{m-1}|$ is immediately seen to be of effective degree one less than the effective degree of $|E_m|$, q. e. d. (see II, 6).

If L is virtually free from base points and if we take for $|E|$ a sufficiently high multiple of the system of hyperplane sections of F, then $|E_m|$ will be effectively free from base points, and therefore L is an exceptional curve of the 1st kind.

The system $|E_m|$ is non-singular and its characters n_1, π_1 are connected to the characters n, π of $|E|$ by the relation $n_1 - (2\pi_1 - 2) = n - (2\pi - 2) + m > n - (2\pi - 2)$. If we transform F into a surface F_1 of order n_1 by means of the system $|E_m|$, and if the transformed surface F_1 (on which to L there corresponds a point) still possesses exceptional curves, it will be possible to construct a linear system on F_1

---

[1] A characterization of reducible exceptional curves is given in the joint paper by Dr. BARBER and myself, quoted in III, 6.

whose characters n_2, π_2 satisfy the inequality $n_2 - (2\pi_2 - 2) > n_1 - (2\pi_1 - 2)$, etc. We have then the following theorem:

If it is not possible to transform F into a surface free from exceptional curves, there exist on F non-singular regular systems, of characters n, π, for which the difference n $-$ (2π $-$ 2) is arbitrarily high, in particular > 0.

As an immediate consequence we have the following theorem: *If not all the plurigenera of F are zero, then it is possible to transform birationally F into a surface free from exceptional curves.*

In fact, if $P_i \neq 0$, the system $|i(C' - C)| = |iK|$ effectively exists $(|C|$ arbitrary) or consists of a curve of order zero $(|iC'| = |iC|)$. It follows that if $|C|$ is irreducible, of dimension > 0, and of characters n, π, the intersection number $(iK \cdot C) = i(2\pi - 2 - n)$ is never negative. The characters n, π of any such system $|C|$ satisfy therefore the inequality $n \leqq 2\pi - 2$, q. e. d.

We observe that for the existence on F of linear systems for which the difference $n - (2\pi - 2)$ is arbitrarily high it is sufficient that there should exist on F a linear system for which this difference is > 0, *because $(C^2) - 2[C] + 2$ is an additive numerical function of curves* (II, 5). We are thus confronted with the following question: *What can be said of the class of surfaces F which possess linear systems, sufficiently general (for instance, irreducible and of dimension > 0), whose characters n, π satisfy the inequality n > 2π $- 2$?* This question has been investigated by CASTELNUOVO and ENRIQUES in (2) with the view of classifying the surfaces on which *the process of successive adjunctions applied to a sufficiently general system $|C|$ terminates after a finite number of steps*, i. e. such that in the sequence $|C|, |C'|, |C''| = $ adjoint system of $|C'|$, $|C'''|, \ldots$ there exists a term $|C^i|$ which is not an effective system. These authors have shown the complete identity of these two classes of surfaces. The result is as follows:

If F possesses a linear system $|D|$ whose characters n, π satisfy the inequality n > 2π $- 2$ and if no exceptional curve of F is a fixed component of $|D|$, then the process of successive adjunctions applied to any system $|C|$ on F terminates after a finite number of steps. Conversely, if the process of successive adjunctions applied to a given system $|C|$ of dimension $\geqq 1 - p_a$ terminates after a finite number of steps, then there exists a system $|D|$ satisfying the above conditions.

Proof. For the proof of the first part of the theorem we first observe that from the existence of *all* the successive adjoint systems of a given system $|C|$ follows the existence of all the successive adjoint systems of *any* system partially containing $|C|$. Hence it is sufficient to prove the theorem for irreducible systems $|C|$ of dimension $\geqq 1$. We may also suppose that $|D|$ is virtually free from base points without destroying the inequality $n > 2\pi - 2$. Let N and Π be the characters of $|C|$, and let i be an integer high enough that the characters N_i and Π_i

of the system $|C_i| = |C + iD|$ satisfy the inequality $N_i - 2\Pi_i$ $+ 2 (= N - 2\Pi + 2 + i(n - 2\pi + 2)) > 0$. If $|C_i^j|$ denotes the jth adjoint system of $|C_i|$, then $|C_i^j - C_i + \Sigma O_\alpha| = |j(C_i' - C_i)|$, where the summation is extended to all the assigned base points of $|C_i|$ of virtual multiplicities $\leqq j - 1$, and hence $(C_i^j \cdot C_i) \leqq 2\Pi_i$ $- 2 - (j - 1)(N_i - 2\Pi_i + 2)$. We choose j high enough that $(C_i^j \cdot C_i) < 0$. If $|C_i^j|$ exists, it must then have in common with $|C_i|$, and hence also with $|D|$, common irreducible fixed components of negative virtual degree. Let L be such a fixed component, of characters ν, ϱ, and let n_1, π_1 be the characters of $|D_1| = |D - L|$. Since L is by hypothesis not an exceptional curve of the first kind, the case $\nu = -1, \varrho = 0$ is excluded, and consequently $\nu - 2\varrho + 2 \leqq 0$. From $n - 2\pi + 2 = (n_1 - 2\pi_1 + 2) + (\nu - 2\varrho + 2)$, it follows that $n_1 - 2\pi_1$ $+ 2 \geqq n - 2\pi + 2 > 0$. If $|D_1|$ still possesses fixed components of negative virtual degree, we proceed as before. Hence, we may suppose that $|D|$ is free from fixed components of negative virtual degree, and we may then conclude that $|C_i^j|$, and a fortiriori $|C^j|$, does not exist if j is sufficiently high.

For the proof of the second part of the theorem, we observe that, as a consequence of the hypothesis, p_g must vanish; hence $p_a \leqq 0$. Let $p_a = -p$, when $p \geqq 0$, and let $|C|$ be the given system of dimension $\geqq 1 + p$. We may suppose that F can be birationally transformed into a surface free from exceptional curves, because in the contrary case the existence of a system such as $|D|$ follows from the preceding theorem. We shall therefore assume that F is free from exceptional curves.

Let $|C^i|$ be the last existing adjoint system of $|C|$. If π_i is its genus, the virtual dimension of $|C^{i+1}|$ equals $\pi_i + p_a - 1 = \pi_i - p - 1$ (section 2, a.), and consequently we have $\pi_i < p + 1$. Since $\pi \geqq p + 1$, we must have for some s $(0 < s \leqq i)$, $\pi_{s-1} > \pi_s$. If n_s denotes the virtual degree of $|C^s|$, we have $n_s = \pi_s + \pi_{s-1} - 2$ and hence $n_s - 2\pi_s$ $+ 2 > 0$, q. e. d.

Using the above theorem CASTELNUOVO and ENRIQUES arrive at the fundamental result of the quoted paper (2), to the effect that *if on a surface F the process of successive adjunction terminates after a finite number of steps, the surface is birationally equivalent to a ruled surface.* In the proof of this theorem different cases have to be considered, especially the two cases $\bar{p}^{(1)} \leqq 1$, $\bar{p}^{(1)} > 1$, and in each case it is shown that F must possess a linear system of a type which, by other well known theorems, is characteristic for ruled surfaces, i. e. (a) a pencil of rational curves (NOETHER, 3; ENRIQUES, 6); or (b) a linear pencil of elliptic curves with at least one simple or double base point (CASTELNUOVO-ENRIQUES, 3); or (c) more generally, an irreducible system of genus $p > 2$ and of dimension $r \geqq 3p - 5$.

As a consequence of the above results, it follows that *the only surfaces, which cannot be birationally transformed into surfaces free from exceptional curves, are those which are birationally equivalent to ruled surfaces.* A ruled surface possesses infinitely many exceptional curves of the second kind, its generators (of virtual degree 0). *Any surface, which does not belong to the class of ruled surfaces, possesses at most a finite number of exceptional curves, all of the first kind, and such that no two of them have a point in common*[1].

In view of this result and from III, 8, it follows that the linear genus $p^{(1)}$ is defined for any surface F which does not belong to the class of ruled surfaces. It is not difficult to see that *always* $p^{(1)} \geq 1$. To prove this we observe that if $|C|$ is any system on F, of genus π, and if π' and π'' are the genera of $|C'|$ and $|C''|$, then the following relation holds: $\pi - \pi' = \pi' - \pi'' + \bar{p}^{(1)} - 1$ [formulas (15), (19) and (19') of the preceding chapter]. If F is free from exceptional curves, then $\bar{p}^{(1)} = p^{(1)}$. If $p^{(1)} < 1$ and if we take for $|C|$ the system of hyperplane sections of F, then the genera π', π'', ..., $\pi^{(i)}$, ... of the successive adjoint systems of $|C|$, beginning with a proper $\pi^{(i)}$, would form a decreasing succession and hence, by the preceding proof, F would be birationally equivalent to a ruled surface.

The consideration of the two possible cases, $p^{(1)} > 1$, $p^{(1)} = 1$, leads to interesting conclusions (CASTELNUOVO-ENRIQUES, 2). If $p^{(1)} > 1$, then from the formula (24') of III, 9 it follows that for sufficiently high values of i the plurigenera P_i are positive and assume arbitrarily high values. If $p^{(1)} = 1$ the quoted formula may not hold. There exist surfaces for which $p^{(1)} = 1, p_a = P_i = 1$ (for instance the general surface of order 4) or $p^{(1)} = 1, p_a = 0, P_i = 0, 1$, according as i is odd or even (for instance the surface of order 6 passing doubly through the edges of a tetrahedron).

We have quoted in III, 8 CASTELNUOVO's definition of $p^{(1)}$ for ruled surfaces, as the maximum of $\bar{p}^{(1)}$. It is proved that this maximum exists (CASTELNUOVO, 7, CASTELNUOVO-ENRIQUES, 2). In the second of the papers just quoted the authors show that *if F is a ruled surface of genus $p \geq 1$ then $p^{(1)} = -8(p-1) + 1$ and hence $p^{(1)} < 0$, except when $p = 1$.* If F is rational, $p^{(1)} = 10$. The problem of characterizing the ruled surfaces by a finite number of vanishing plurigenera has been solved by ENRIQUES (11): *a surface F is birationally equivalent to a ruled surface if $P_4 = P_6 = 0$.*

In CASTELNUOVO-ENRIQUES (2) *a necessary and sufficient condition that a surface F be rational is found to be $p_a = P_2 = 0$.*

Another criterion for ruled surfaces, based on the consideration of the arithmetic genus, has been obtained by CASTELNUOVO (11). The

---

[1] See p. 40.

point of departure is a theorem of de FRANCHIS (3) to the effect that *a necessary and sufficient condition that a surface possess an irrational pencil of curves is that it possess two independent* PICARD *integrals of the first kind* (VII, 1) *which are functionally dependent.* CASTELNUOVO shows that *this condition is satisfied for all surfaces whose genera* p_g, p_a *are such that* $p_g \geq 2(p_a + 2)$. An immediate consequence of this result is the following theorem, proved by DE FRANCHIS (3) in a different manner: *if* $p_a < -1$, *then F possesses an irrational pencil of curves.* Let ϱ be the genus of the pencil and let π be the genus of a generic curve of the pencil. *Let us suppose that F is not birationally equivalent to a ruled surface.* Then we have necessarily $\pi \geq 1$. If we eliminate I between NOETHER's relation $I + \bar{p}^{(1)} = 12 p_a + 9$ and the relation $I = \delta + 4(\pi - 1)(\varrho - 1)$ (III, 8) and if we observe that in view of our hypothesis it is permissible to suppose that F is free from exceptional curves and that consequently $\bar{p}^{(1)} = p^{(1)}$, we obtain the relation

$$\delta + 4(\pi - 1)(\varrho - 1) + p^{(1)} = 12 p_a + 13.$$

If $p_a < -1$, the right-hand member is negative, while the left-hand member is positive, since $\delta \geq 0$, $\pi \geq 1$, $\varrho \geq 1$ and since by a previous result stated above, $p^{(1)} \geq 1$. From this contradiction it follows that the quoted theorem of DE FRANCHIS can be replaced by the following more precise statement: *Any surface of arithmetic genus* $p_a < -1$ *is birationally equivalent to a ruled surface* (CASTELNUOVO, 11).

Appendix to Chapter IV

By DAVID MUMFORD

The RIEMANN-ROCH theorem has a vast literature devoted to it by now, and in this appendix I only want to point out a few of these developments. An extensive report on many of its ramifications can be found in HIRZEBRUCH (2), 3rd edition, in the appendix of SCHWARZENBERGER. To begin with, it should be pointed out that the RIEMANN-ROCH theorem, as treated in the book, is now treated as 3 separate theorems: let X be a non-singular projective variety, and let D be a divisor on X:

Part I: A formula for $\chi(\mathscr{O}_X(D))$ in terms of the intersection numbers of D and the CHERN classes of T_X. This is a special case of the result which called the RIEMANN-ROCH theorem now.

Part II: Construction of a non-degenerate pairing between $H^i(\mathscr{O}_X(D))$ and $H^{n-i}(\Omega_X^n(-D))$. This is known as SERRE's duality theorem. When $n = 1$, (resp. $n = 2$) it shows

$$\chi(\mathscr{O}_X(D)) = \dim H^0(\mathscr{O}_X(D)) - \dim H^0(\mathscr{O}_X(K_X - D)),$$

$$(\text{resp. } \chi(\mathscr{O}_X(D)) \leqq \dim H^0(\mathscr{O}_X(D)) + \dim H^0(\mathscr{O}_X(K_X - D))).$$

Part III: The vanishing theorem that if H is a hyperplane section of X and $n \gg 0$,

$$H^1(\mathscr{O}_X(D + n H)) = (0), \quad i > 0$$

hence $\chi(\mathscr{O}_X(D + n H)) = \dim H^0(\mathscr{O}_X(D + n H))$. This is needed to check that $\chi(\mathscr{O}_X(D))$ is the virtual dimension of $|D|$, plus one.

The formula of Part I was proven first by HIRZEBRUCH (2), by topological methods; then by GROTHENDIECK [his proof was published by BOREL and SERRE (1)] and by WASHNITZER (2) by algebraic methods. See also GROTHENDIECK et al. (6), 1966—67. The duality theorem in Part II was proven by SERRE (1) analytically, and by SERRE (2) algebraically. This 2nd paper, known as FAC, is a watershed in the development of algebraic geometry; in it, he introduced the cohomology of sheaves into algebraic geometry for the first time and proved also the vanishing theorem. See also the report of ZARISKI (22). The duality theorem was then greatly generalized by GROTHENDIECK (4) [most of GROTHENDIECK's work was published by HARTSHORNE (1)]. There are still problems, especially in finding local proofs of the local assertions about residues, but TATE (4) seems to point the way to a more complete understanding of this theory. To complete this list of references, ZARISKI himself in (18) and (32) has given 2 proofs of the RIEMANN-ROCH theorem for surfaces which are more classical in their approach. Actually the classical methods of analyzing linear systems have always been essentially homological in nature and the difference between the classical and modern proofs is often little more than a difference in notation.

Re: § 2. The RIEMANN-ROCH formula (6) given on p. 78, comes out from cohomology groups as follows:

Let F be a non-singular projective surface and D a divisor on F.

i) the modern formula states:

$$\chi(\mathcal{O}_F(D)) = \frac{(D \cdot D + c_1)}{2} + \frac{1}{12}((c_1^2) + (c_2))$$

where $c_i = i$th CHERN class of T_F.

ii) Using the fact that $c_1 \equiv -K_F$ and subtracting the same formula with $D = 0$, we get:

$$\chi(\mathcal{O}_F(D)) = \frac{(D \cdot D - K_F)}{2} + \chi(\mathcal{O}_F) .$$

iii) By definition

$$p_a = \chi(\mathcal{O}_F) - 1 = \text{artihmetic genus of } F ,$$
$$n = (D^2) = \text{virtual degree of } |D| ,$$
$$\pi = \frac{(D \cdot D + K_F)}{2} + 1 = \text{virtual genus of } |D|$$

so $\chi(\mathcal{O}_F(D)) - 1 = n - \pi + p_a + 1 = $ right hand side of (6).

iv) Finally, by SERRE duality:

$$\chi(\mathcal{O}_F(D)) - 1$$
$$= \underbrace{[\dim H^0(\mathcal{O}_F(D) - 1]}_{\dim |D| \text{ or } r} - \underbrace{[\dim H^1(\mathcal{O}_F(D))]}_{\text{superabundance}} + \underbrace{[\dim H^0(\mathcal{O}_F(K_F - D))]}_{\text{index of speciality } i}$$

hence $r + i \geq n - \pi + p_a + 1$.

The lemma of SEVERI (c., p. 79) is really the focal point of the proof. ZARISKI analyzed it very carefully in (18), generalizing it to arbitrary normal projective varieties X in the form: if D is any divisor on X, then the trace of the linear system $|D|$ on a generic hypersurface section H_n of degree n is complete if $n \gg 0$. Equivalently, this means that:

$$H^0(\mathcal{O}_X(D)) \to H^0(\mathcal{O}_{H_n}(D \cdot H_n))$$

is surjective. This is now called the lemma of ENRIQUES-SEVERI-ZARISKI. SERRE (2, p. 270) proved this by showing that in fact:

$$H^1(\mathcal{O}_X(D - n H)) = (0) , \quad n \gg 0 ,$$

$H = $ hyperplane section, D a CARTIER divisor.

Re: §§ 1, 3. Enriques' theorem that the irregularity of F is the maximum deficiency of the trace of $|C'|$ on C, and Castelnuovo's theorem that the irregularity is also the maximum deficiency of the trace of $|C|$ on C illustrate how basically homological the classical work on linear systems was. Recall the concept of the *satellite* of a functor: If $F : \mathfrak{C} \to$ (Ab) is a covariant additive functor from an abelian category to the category of abelian groups, its satellite $F^* : \mathfrak{C} \to$ (Ab) is defined as follows: For all $A \in \mathrm{Ob}(\mathfrak{C})$, consider the category R_A of all short exact sequences

$$0 \to A \to B \to C \to 0 \text{ in } \mathfrak{C}$$

where morphisms are diagrams:

$$
\begin{array}{ccccccccc}
0 & \to & A & \to & B_1 & \to & C_1 & \to & 0 \\
  &   & \| &   & \downarrow & & \downarrow & & \\
0 & \to & A & \to & B_2 & \to & C_2 & \to & 0 .
\end{array}
$$

It is possible to form a well-defined limit:

$$F^*(A) = \varinjlim_{R_A} F(C)/\mathrm{Im} F(B) .$$

Apply this with $\mathfrak{C} =$ category of coherent sheaves of $\mathcal{O}_F$-modules, and $F = \Gamma$, the global section functor. For all divisors D, $\mathcal{O}_F(D) \in \mathrm{Ob}(\mathfrak{C})$ and it is easy to see that the exact sequences:

$$0 \to \mathcal{O}_F(D) \to \mathcal{O}_F(D + C) \to \mathcal{O}_C(D \cdot C + C^2) \to 0$$

where C is a positive divisor, are cofinal in the category $R_{\mathcal{O}_F(D)}$. Therefore

$$\Gamma^*(\mathcal{O}_F(D)) = \varinjlim_{\text{positive } C} \Gamma(\mathcal{O}_C(D \cdot C + C^2))/\mathrm{Im}\,\Gamma(\mathcal{O}_F(D + C)) .$$

On the other hand, Enriques and Castelnuovo consider the dimension of $\Gamma(\mathcal{O}_C(D \cdot C + C^2))/\mathrm{Im}\,\Gamma(\mathcal{O}_F(D + C))$ where $D = K_F$ in Enriques' case, and $D = 0$ in Castelnuovo's case. In both cases, they find that for all sufficiently ample C, this "deficiency" is constant and equal to $p_g - p_a$. Cohomologically, the point is that if C is ample enough

$$\Gamma(\mathcal{O}_C(D \cdot C + C^2))/\mathrm{Im}\,\Gamma(\mathcal{O}_F(D + C)) \cong \Gamma^*(\mathcal{O}_F(D)) \cong H^1(\mathcal{O}_F(D)) .$$

The fact that $H^1(\mathcal{O}_F(K_F))$ and $H^1(\mathcal{O}_F)$ have the *same* dimension is due to Serre duality.

Enriques' characterization of the irregularity as the maximum deficiency of adjoint systems has the advantage that, when applied to linear systems with possible base points, it implies immediately that the irregularity is a *birational* invariant. In fact, in the notation introduced in the appendix to Chapter 3, Enriques shows:

$$\text{irreg of } F = \max_{\substack{\text{prime divisors} \\ v \text{ of } k(F)}} \dim [\Gamma(\Omega_C^1)/\mathrm{res}\,\Omega_{k(F)}^2(v)]$$

where C is the non-singular curve which is the center of v on all sufficiently blown up models of $k(F)$.

The birational invariance of the arithmetic genus p_a was also proven in all characteristics, for surfaces, by ZARISKI and MUHLY (1), as a consequence of (i) the factorization of birationals maps between surfaces into quadratic transformations and their inverses, and (ii) a direct computation that, in a quadratic transformation, the arithmetic genus does not change[1].

Re: § 4. Almost all the material in this section has been reworked and extended to char p in recent years. ENRIQUES (11) published a book summarizing his results on classification of surfaces. ZARISKI (24), (25) showed that all non-ruled surfaces have "minimal non-singular models", i. e., models which are dominated by every other non-singular model. He also proved in (26) CASTELNUOVO's theorem in all characteristic characterizing rational surfaces by $p_a = P_2 = 0$. KODAIRA (8) and ŠAFAREVITCH (2) worked out in all rigor ENRIQUES' classification of surfaces into

  a) ruled surfaces,
  b) 2-dimensional abelian varieties,
  c) KUMMER surfaces, i. e., $K_F \equiv 0$ and $p_g = p_a = 1$,
  d) surfaces with pencils of elliptic curves,
  e) surfaces with $|n K_F|$ irreducible and simple for $n \gg 0$.

In particular, they show that $P_4 = P_6 = 0$ or "adjunction terminating" characterize ruled surfaces. MUMFORD (6) has extended these results to char p. CASTELNUOVO's result that $p_a < -1$ implies F ruled does not seem to have been analyzed in char p, however. It would be interesting to know, in char 0 as well as char p, exactly what values the 2 CHERN numbers (c_1^2), (c_2) can take in class (e) in ENRIQUES' classification.

---

[1] The proof in this paper in the 3-dimensional case is incomplete because no factorization theorem is known. MATSUMURA (1) and HIRONAKA (4) showed, however, by spectral sequences, that in all dimensions and characteristics, a strong resolution theorem implies that $\dim H^i(\mathcal{O}_x)$ is a birational invariant.

Chapter V

Continuous Non-linear Systems

1. Definitions and general properties. In II, 1 we have accepted for temporary purposes a definition of an algebraic system Σ of curves C on a surface F, as a system cut out on F by an algebraic system Σ' of hypersurfaces. It will serve as a preliminary clarifying remark if we point out immediately why that definition is not sufficiently general. The base loci of the given system Σ' of hypersurfaces determine a complete linear system of hypersurfaces of the same order as the hypersurfaces of Σ'. This linear system cuts out on F a linear system of curves of the same order as the curves C and containing Σ. Hence Σ *is totally contained in a linear system of curves*. However—and this is a fundamental point of the theory, which will be discussed in section 3 of this chapter—there exist surfaces (notably, irregular surfaces) which carry algebraic systems of curves not contained in linear systems. The simplest example is given by surfaces carrying an *irrational pencil* Σ of curves. Obviously, such a pencil (supposing for simplicity that the curves of the pencil are irreducible) cannot be contained in a linear system of dimension r, where r is necessarily $\geqq 2$, because the curves of the pencil are of virtual degree zero (II, 1). If the pencil is reducible, it is seen immediately that its curves are composed of the curves of another irrational pencil, and the statement that Σ is not contained in a linear system is essentially equivalent to the statement that an irrational involution of sets of points on an algebraic curve cannot be contained in a linear series. If the curves of Σ are irreducible, then a stronger statement holds to the effect that *any curve C_1 of Σ is isolated with respect to linear equivalence*. In fact, each curve of Σ is a fundamental curve of $|C_1|$ (since $(C \cdot C_1) = 0$), and consequently $|C_1|$ cannot be of dimension $\geqq 1$.

In order to set up, in the most general case, the equations of an algebraic system Σ, ∞^r, of curves C on a surface F in S_ϱ $(x_0, x_1, \ldots, x_\varrho)$, we project the curves of Σ onto a plane $S_2(y_0, y_1, y_2)$ from a generic $S_{\varrho-3}$. We obtain then in S_2 an algebraic system $\overline{\Sigma}$, ∞^r, of curves $\overline{C}$, such that the generic curve $\overline{C}$ of $\overline{\Sigma}$ is the *simple* projection of the corresponding curve C of Σ. Now in S_2 the system $\overline{\Sigma}$ can be given by two algebraic equations:

(1) $$f(y_0, y_1, y_2; t_0, t_1, \ldots, t_{r+1}) = 0,$$

(1') $$\psi(t_0, t_1, \ldots, t_{r+1}) = 0,$$

where the t's are homogeneous parameters. We can then write the parametric equations of a generic curve C of Σ in the following form:

(2) $$\sigma x_i = \varphi_i(y_0, y_1, y_2; t_0, t_1, \ldots, t_{r+1}), \qquad i = 0, 1, \ldots, \varrho,$$

where the φ's are homogeneous polynomials in the two sets of variables y and t. The equations (1), (1') and (2) constitute together the complete algebraic representation of Σ.

The curves of Σ are in $(1 - 1)$ algebraic correspondence with the points of the r-dimensional variety (1'). If this variety is irreducible, then also the system Σ is said to be *irreducible*. The system Σ is *rational*, if the variety (1') is rational, i. e. if it can be put into birational correspondence with a linear space S_r.

We may also define irreducible algebraic systems on F by considering an algebraic irreducible correspondence between F and an irreducible algebraic variety ψ (I, 1). If in this correspondence to a generic point P of ψ there corresponds a curve C on F, then, as P varies on ψ, C varies in an algebraic system Σ. For special points P of ψ, fundamental for the correspondence, the curve C may become indeterminate, while special curves C may acquire multiple components, of which it is necessary to determine the multiplicity. On this subject see the papers of van der Waerden (1, 2, 5) and Severi (26), dealing with the foundations of algebraic geometry.

Instead of the term "algebraic system" we shall often use the more expressive term "continuous system". The totality of curves C of a given order on F distribute themselves into a finite number of irreducible algebraic systems, and any continuous family of curves of the given order belongs to one of these algebraic systems.

A rational system Σ is always totally contained in a linear system (Enriques, 4). For the proof we follow a reasoning employed by Humbert (2) in a similar connection. It is sufficient to consider the case $r = 1$, because any two curves C of a rational system Σ, ∞^r, belong to a rational sub-system, ∞^1, of Σ — the image of a straight line in S_r. The variety (1') is in this case a rational curve Γ, $\psi(t_0, t_1, t_2) = 0$. Let ν—*the index of* Σ—denote the number of curves C which pass through a generic point of F ($\nu = 1$, if Σ is a pencil). Let us suppose for simplicity that F is in $S_3(x_0, x_1, x_2, x_3)$. A generic point $P \equiv (x)$ of F determines a set of ν curves C in Σ and hence a set of ν points on Γ. As P varies on F, this set varies in a series, ∞^2, of sets of points on Γ. Since on a rational curve all sets of ν points are equivalent, it follows that this series is cut out on Γ, outside of fixed points, by an algebraic system of curves $\lambda_0 \varphi_0(t_0, t_1, t_2) + \cdots + \lambda_k \varphi_k(t_0, t_1, t_2)$, where $\sigma \lambda_i = f_i(x_0, x_1, x_2, x_3)$ is a rational integral homogeneous function of the coordinates x_j of the point P on F. It is then immediately seen that Σ is cut out on F, outside of fixed loci, by the algebraic system of surfaces

(3) $\varphi_0(t_0, t_1, t_2) f_0(x_0, x_1, x_2, x_3) + \cdots + \varphi_k(t_0, t_1, t_2) f_k(x_0, x_1, x_2, x_3) = 0$,

and is consequently totally contained in a linear system.

Remark. A delicate point of the proof consists in showing that along any k-fold component of a generic curve C of Σ the corresponding surface (3) has exactly a k-fold contact with F.

If Σ is irreducible and if E is a variable curve of another irreducible continuous system, the intersection number $(C \cdot E)$ remains constant as C and E vary in their respective systems[1]. In particular, the *virtual degree* $n = (C^2)$ of C remains constant as C varies in Σ. If $r \geqq 1$, then n is also the intersection number of two variable curves of Σ.

We now suppose that the generic C is irreducible and that Σ is irreducible, of dimension $r \geqq 1$. Let (t^0) be a generic *simple* point of the variety $(1')$ and let C_0 be the corresponding curve of Σ. To any analytic (in particular algebraic) arc L on the variety $(1')$, issued from the point (t^0), there corresponds in Σ an analytic (in particular algebraic) subsystem Σ_1, ∞^1, containing C_0. As C varies in Σ_1 and approaches C_0, its intersections with C_0 approach a definite set G of n points on C_0. A set like G is called *a characteristic set on C_0*. Elementary considerations of differential character show that the set G depends only on the tangent to the arc L at the point (t^0). It follows that the characteristic sets G on C_0 form a rational variety, in $(1 - 1)$ correspondence with the tangential directions of the variety $(1')$ about (t^0). Moreover, it is easily seen that this series is involutorial, i. e. that $r - 1$ generic points of C_0 determine the characteristic set to which they belong. Consequently, we are dealing with a *linear series* g_n^{r-1}. This series is called *the characteristic series of Σ on C_0* (SEVERI, 4).

In this manner we are able to attach to every continuous non-linear irreducible system of irreducible curves, of dimension $\geqq 1$, a *linear* series defined on a generic curve of the system. This is a point of fundamental importance for the theory.

We use the plane representation of Σ, given by the equations (1), $(1')$, (2), and we suppose for simplicity that F is in S_3. Let $u_1, u_2, \ldots, u_r$ be local analytic parameters on the variety $(1')$ in the neighborhood of and vanishing at (t^0). Substituting into (1), we find the equation

$$\bar{f}(y_0, y_1, y_2; u_1, u_2, \ldots, u_r) = 0,$$

---

[1] This is a consequence of the *principle of the conservation of the number*. For a precise formulation and proof of this principle see SEVERI (26) and VAN DER WAERDEN (5). It also follows from the fact that the number $(C \cdot E)$ is the KRONECKER index of the cycles C, E and consequently remains invariant when the cycles are deformed (LEFSCHETZ, e).

which is analytic in the parameters u_i and which represents the curves C of Σ in the neighborhood of $\overline{C}_0$. The curves $\overline{C}$ possess a certain number δ of variable double points, corresponding to the bisecants of C through the center of projection O, and moreover have a certain number s of variable contacts with the branch curve $\overline{D}$ of the surface F, represented upon the multiple plane (y) (apparent contour of F viewed from O). *The characteristic series of Σ on C_0 corresponds, by projection, to the series cut out on C_0 by the linear system* $\sum \lambda_i \left(\dfrac{\partial \overline{f}}{\partial u_i} \right)_{u_i=0}$, *outside the δ double points of C_0 and its s contacts with $\overline{D}$* (ENRIQUES, 10). These last $\delta + s$ points are seen to be base points of the above mentioned linear system.

The characteristic series of Σ on a generic C, completed if necessary, coincides with the virtual characteristic series on C, defined in II, 8 (SEVERI, 4). This theorem is obvious if C belongs to a linear system of dimension ≥ 1 contained in Σ. To prove the theorem in general, we may suppose that Σ is ∞^1 and that the variety $(1')$ is therefore a curve Γ. We take on Γ a g_ν^2 free from fixed points. The ∞^2 reducible curves $E = C_1 + C_2 + \cdots + C_\nu$ which correspond to the sets of this series form a rational variety and, by the preceding theorem of ENRIQUES, are therefore contained in a linear system $|E|$. As one of the variable components of E, say C_1, approaches the fixed curve C, the curve $C_2 + \cdots + C_\nu$ approaches a curve of the system $|E - C|$. It follows that the characteristic set of Σ on C is a set of the series, the difference of the two series cut out on C by $|E|$ and by $|E - C|$ respectively, q. e. d.

2. Complete continuous systems and algebraic equivalence. The extension of the relation of equivalence of curves (from linear equivalence to algebraic equivalence) is due to SEVERI (4, 5, 10, 13, 16, 17). The definitions of algebraic equivalence vary according to the manner of defining complete continuous systems (ALBANESE, 1, 4; LEFSCHETZ, a). The interrelations between the various types of algebraic equivalence have been investigated by ALBANESE (1). In the following exposition the types of algebraic equivalence are considered in the order of increasing generality.

An irreducible continuous system Σ of curves C is said to be *complete in the restricted sense*, if it is not contained in an *irreducible* system of curves of the same order as C. Two curves C_1, C_2 belonging to an irreducible continuous system are *algebraically equivalent in the restricted sense*, in symbols $C_1 \parallel\!\!\!| C_2$ (ALBANESE, 1). As in the case of linear systems, we may also here impose a set H of base points and consider complete continuous systems with respect to H. However, since such base points may be replaced by exceptional curves, we shall only consider continuous systems virtually free from base points.

A given curve C does not always determine uniquely the complete continuous system Σ (in the restricted sense) to which it belongs. A very simple example was given by Albanese (1). The system $|A|$, ∞^3, of the plane sections of a cubic elliptic cone F, is a linear system whose characteristic series is complete, and hence $|A|$ is complete not only as a linear system but also as a continuous system (in the restricted sense). It can also be shown that the system, ∞^3, of reducible curves $D + R_1 + R_2 + R_3$, where D is the infinitesimal curve represented by the vertex of F and where R_1, R_2, R_3 are any three generators of F, is complete in the restricted sense. These two systems have in common the curves $D + R_1 + R_2 + R_3$ made up of D and of 3 *coplanar* generators.

As a consequence, the transitive property does not hold for the present type of algebraic equivalence, i. e. from $A \parallel\!\parallel B$ and $B \parallel\!\parallel C$ does not follow necessarily $A \parallel\!\parallel C$.

It is fairly obvious, however, that the uniqueness of the complete continuous system can be asserted for the *generic* curve C of a given complete continuous system Σ (i. e. for all curves of Σ, not belonging to one or more subsystems of Σ of lower dimension than Σ). In fact, if a generic curve C of Σ belonged to an irreducible algebraic system Σ_1 not contained in Σ, then as C varied in Σ the system Σ_1 would describe a continuous irreducible system containing Σ, and this is impossible.

In particular, Σ must contain the complete linear system $|C|$ determined by its *generic* curve C. The dimension of $|C|$, as long as C is generic, has a certain constant value r. For particular curves C of Σ the dimension may be greater than r, and then $|C|$ may or may not be contained in Σ. The second possibility is illustrated by the preceding example. The first possibility can be illustrated by the following example. Let F possess an irrational pencil of genus p. The system Σ, ∞^p, of all sets of p curves of the pencil is complete in the restricted sense. For a generic curve C of Σ the dimension of $|C|$ is zero. However, Σ also contains linear systems of dimension >0, corresponding to those sets of p curves which determine in the pencil a g_p^r, $r \geqq 1$. Other, less trivial, examples can be found in Albanese (1).

The preceding considerations lead to a significant extension of the notion of a complete continuous system (Severi, 4, 5, 10). Given a continuous system Σ, complete in the restricted sense, we consider the system Σ' generated by the *complete* linear systems $|C|$ as C varies in Σ. *The system Σ' may not be irreducible as a totality of curves, but is irreducible as a totality of linear systems.* It may, however, belong to a more inclusive system irreducible as a totality of linear systems (see the first of the two examples quoted in this section). *A continuous system of curves C, irreducible as a totality of complete linear systems, is called complete* (without further specifications) *if it is not contained in another continuous system of the same nature.* Such a system will be denoted by $\{C\}$. Any two curves of $\{C\}$ are said to be *algebraically equivalent in the effective sense*, briefly *effectively equivalent*, in symbols, $C_1 \equiv C_2$

(SEVERI). It is clear that if $C_1 \parallel C_2$, then also $C_1 \equiv C_2$, but not conversely.

We know from III, 7 and IV, 3, d., that if C is a given curve of $\{C\}$ and if $|E|$ is the system of hyperplane sections of F, then there exists an integer λ_0, such that for any $\lambda \geq \lambda_0$ the system $|C + \lambda E|$ is irreducible and regular. It can be easily verified, by making the proof of this statement, that $\lambda_0 = \lambda_0' + \lambda_0''$, where λ_0' and λ_0'' are defined as follows: for any $\lambda \geq \lambda_0'$ the system $|C + \lambda_0' E|$ should be irreducible (and hence λ_0' depends only on the order of C); the system $|\lambda_0'' E|$ should partially contain the canonical system. It follows that as C varies in $\{C\}$, it is possible to assign a fixed value of λ_0 in such a manner, that the system $|C + \lambda_0 E|$ remains regular for an *arbitrary* (and not only for the generic) curve C of $\{C\}$. The continuous system generated by this variable linear system is constituted of linear systems, all of the same dimension, and its therefore irreducible not only as a totality of linear systems but also as a totality of curves. We have therefore the following result (ALBANESE, 1):

If $C_1 \equiv C_2$ then $C_1 + \mu E \parallel C_2 + \mu E$, where $|E|$ is the system of hyperplane sections of F and μ is a sufficiently high integer. ALBANESE has also shown that this theorem remains valid when $|E|$ is *any irreducible system of virtual positive degree and free from fundamental curves.*

As in the case of continuous systems complete in the restricted sense, so also here *it is not possible to assert the uniqueness of the complete continuous system $\{C\}$ containing a given curve C*. Examples of distinct systems $\{C\}$ having curves in common have been given by ALBANESE (1). It follows that from $A \equiv B$ and $B \equiv C$ does not necessarily follow $A \equiv C$. However, the uniqueness of $\{C\}$ can be asserted for a very general class of curves C (see section 3).

It is clear that if $A \equiv B$, then $A + C \equiv B + C$, C arbitrary. The converse is not always true, as can be shown by examples (ALBANESE, 1). Hence, *with regard to effective equivalence subtraction is not always a single-valued operation*. This suggests a further extension of the notion of algebraic equivalence. *Two curves A and B shall be said to be algebraically equivalent in the virtual sense, or simply algebraically equivalent, in symbols $A = B$* (LEFSCHETZ, ALBANESE), *whenever there exists a curve C such that $A + C \equiv B + C$.* It follows from the definition that: (1) if $A = B$ and $B = C$, then $A = C$; (2) if $A = B$, then $A - C = B - C$. The relation $A = B$ has a meaning also if A, or B, or both A and B are virtual curves. The totality of all curves, effective or virtual, which are algebraically equivalent to a given curve A is called by ALBANESE (1) a *complete virtual continuous system* and denoted by V_A. It is clear that V_A is uniquely determined by A, and that if B is any curve of V_A, then V_A coincides with V_B. The virtual system V_A may contain a finite number of complete continuous systems (irreducible as totalities of linear systems) of effective curves.

If $A = B$, the virtual curve $\Gamma = A - B$ is said to be *equivalent to zero*, $\Gamma = 0$. All curves equivalent to zero are algebraically equivalent. We have $(\Gamma^2) = 0, [\Gamma] = 1$, and $(C \cdot \Gamma) = 0$, C arbitrary.

It follows from the definition that if $A = B$, then it is either possible to deform A into B in the domain of effective algebraic curves, or this can be done only after adding to both A and B a convenient curve C. In both cases the curves A and B represent two homologous 2-cycles on the surface ($A \backsim B$, see VI, 12). The more restrictive condition that A and B be deformable into each other in the domain of effective algebraic curves signifies that there exists a finite sequence of effective curves $C_1, C_2, \ldots, C_k$, such that $A \parallel\mid C_1 \parallel\mid C_2 \parallel\mid \ldots \parallel\mid C_k \parallel\mid B$. If this is possible, the curves A and B are said by ALBANESE (1) to be *chainwise equivalent*. He gave examples of chainwise equivalent curves A, B which yet are not effectively equivalent $(A \not\equiv B)$.

3. The completeness of the characteristic series of a complete continuous system. Let Σ be a complete continuous system of curves in the restricted sense, on a surface F, and let R be the dimension of Σ. If n is the virtual degree of a generic curve C of Σ and if r is the dimension of the complete linear system $|C|$, then the characteristic series g_n^{r-1} of $|C|$ on C is totally contained in the characteristic series g_n^{R-1} of Σ on C (since $|C|$ belongs to Σ). In view of CASTELNUOVO's theorem on the deficiency of the characteristic series of a complete linear system (IV, 3), it follows that $R \leqq r + q$, where q is the irregularity of F. Hence, *on a surface of irregularity q every complete continuous system consists of at most ∞^q distinct complete linear systems.* This proof (SEVERI, 4) leads immediately to the following theorem of ENRIQUES (7), who has derived it in a different manner: *on a regular surface every continuous system is totally contained in a linear system*; in other words, every complete continuous system is a linear system.

As a corollary we have (see section 1) that *if a surface possesses an irrational pencil of curves, it is irregular*—a remark due to CASTELNUOVO (5).

Much more difficult is the proof of the converse of the theorem of ENRIQUES, i. e. *that every irregular surface possesses continuous systems not contained in linear systems*. This qualitative characterization of irregular surfaces, completed by the quantitative specification to the effect that *an algebraic surface of irregularity q possesses complete continuous systems consisting of ∞^q distinct linear systems*, is a fundamental result of the theory of surfaces, due to the combined efforts of several geometers (HUMBERT, CASTELNUOVO, ENRIQUES, SEVERI, PICARD, POINCARÉ). It is interesting to note the analogy with a corresponding theorem of the theory of algebraic curves: on a curve of genus p the sets of n points $(n \geqq p)$ form a continuous system, ∞^n, consisting of ∞^p linear series g_n^{n-p}.

A proof of the above fundamental result has been first proposed by ENRIQUES (10). Immediately after, another proof was proposed by SEVERI (7). SEVERI himself has later pointed out that neither proof is entirely rigorous (SEVERI, 20). Nevertheless, in view of the intrinsic interest of the considerations involved, it is worthwhile to review these proofs in order to analyse the assumption on which they are based and for which as yet an algebro-geometric proof is not available.

We first take up ENRIQUES' proof. Let F be a surface in S_3, of genera p_g, p_a, and of order ν, and let D denote a plane section of F. We consider on F a regular linear system $|C|$ of characters n, π, r, virtually free from base points. Then $r = n - \pi + p_a + 1$ and the characteristic series of $|C|$ has the deficiency $q = p_g - p_a$. Let O be a generic point of the space, K the contact curve of F with the tangent cone to F drawn from O, $\overline{K}$ the section of this cone by an arbitrary plane π. If we project F on to the plane π from O, F is represented on the ν-fold covered plane π with $\overline{K}$ as the branch curve, and the system $|C|$ is projected into a continuous system $\overline{\Sigma}_1$, ∞^r, of plane algebraic curves $\overline{C}$, of order $m = (C \cdot D)$, possessing a certain number d of variable double points and $s = (C \cdot K)$ variable contacts with $\overline{K}$. The characteristic series g_n^{r-1} on a given C is projected into the characteristic series g_n^{r-1} of $\overline{\Sigma}_1$ on the corresponding curve $\overline{C}$. This last series is cut out on $\overline{C}$ by a linear system ∞^{r-1} of curves of order m, passing through the d double points of $\overline{C}$ and its s contacts with $\overline{K}$, and is contained in the complete series g_n^{r-1+q} cut out on $\overline{C}$ by the totality of (adjoint) curves of order m defined by the above base conditions.

In the $(\nu, 1)$ correspondence between π and F the ν-fold covered curve $\overline{C}$ is reducible, consisting of C and of another, generally irreducible, component L, in $(1, \nu - 1)$ correspondence with $\overline{C}$. Let us suppose that $\overline{\Sigma}_1$ is contained in a more inclusive irreducible complete system $\overline{\Sigma}$ of curves of order m, possessing d variable double points and s variable contacts with $\overline{K}$. *The ν-fold covered curve $\overline{C}$ of $\overline{\Sigma}$ degenerates in the same manner as the ν-fold covered curve of $\overline{\Sigma}_1$.* This can be seen: (1) either by observing that the ν-fold covered curve $\overline{C}$ of $\overline{\Sigma}$ possesses on F the same number $\nu d + s$ of double points as the ν-fold covered curve of $\overline{\Sigma}_1$ and by applying ENRIQUES' principle of degeneration (II, 5); or (2) from the fact that as a curve $\overline{C}$ of $\overline{\Sigma}$ approaches (without acquiring new singularities) a curve $\overline{C}_1$ of $\overline{\Sigma}_1$, the branch points of the ν-fold covered curve $\overline{C}$ (its simple intersections with $\overline{K}$) approach *without coincidences* the branch points of the ν-fold covered curve $\overline{C}_1$ and that consequently the monodromy groups of the two ν-fold covered curves are isomorphic. *The curves C on F in $(1 - 1)$ correspondence with the curves $\overline{C}$ of $\overline{\Sigma}$ form an irreducible continuous system Σ containing the linear system $|C|$.*

It remains, however, to justify our assumption that $\overline{\Sigma}_1$ itself is not complete, i. e. does not coincide with $\overline{\Sigma}$. To this effect ENRIQUES makes use of a reasoning tending to prove that *the characteristic series of the complete system $\overline{\Sigma}$ is the complete series g_n^{r-1+q}*. From this it would follow not only that $|C|$ is contained in a complete continuous system Σ of higher dimension than $|C|$, but also that Σ is of dimension $r + q$ and therefore consists of ∞^q linear systems. It is this part of ENRIQUES' proof that has been recognized later as insufficient.

A similar situation arises in the proof given by SEVERI, in which no use is made of the multiple plane. Let Σ be a complete (in the restricted sense) continuous system of curves C, free from singularities, and let $|E|$ be a sufficiently general linear system, such that the series cut out by $|E|$ on a generic C is complete (for instance, a sufficiently high multiple of the system of hyperplane sections of F). As C varies in Σ, the linear system $|D| = |E - C|$ varies or remains fixed, as matters may be, and generates a continuous system Σ_1, irreducible as a totality of linear systems. Let $\delta = (D \cdot C)$ and let V be the continuous system of all reducible curves $D + C$, where C is any curve in Σ and D is any curve of $|E - C|$. From ENRIQUES' principle of degeneration it follows immediately that V, obviously irreducible as a totality of curves, coincides with the algebraic system of the curves E possessing δ double points, or with one of its irreducible components if this system is reducible. The curves of Σ_1 infinitely near to a generic curve $C + D$ of Σ_1 are curves E passing through the δ intersections of C with D, and it is clear that these curves, forming a linear system, cut out on C characteristic sets of Σ on C. *If we admit that this linear system coincides with the complete system of the curves E passing through the δ intersections of C with D,* it will follow, since $|E|$ cuts out on C a complete series, that *the characteristic series of Σ on C is complete.*

In both proofs we have, at the start, a linear system of curves (in ENRIQUES' proof the system of all plane curves of order m, in SEVERI's proof the system $|E|$ on F), and then on the curves of this system a certain number of algebraic non-linear conditions is imposed (d variable double points and s variable contacts with $\overline{K}$ in the first case, δ variable double points in the second case). Both proofs make use of the assumption that in the algebraic system of curves thus obtained (or, if this system is reducible, in one of its irreducible components) the linear system of curves infinitely near to a generic curve of the system is *complete*. SEVERI's criticism (20) is to the effect that the available algebro-geometric proof of this assumption fails *if the characteristic series of the considered continuous system Σ is special.*

Let us examine, for the sake of concreteness, the second of the proofs given above. If we refer the curves of $|E|$ to the points of a linear space S_r (r is the dimension of $|E|$), the ∞^{r-1} curves of $|E|$ possessing a double point (in non-assigned position) will be represented in S_r by an irreducible algebraic hypersurface M. The tangent hyperplane at a generic point of M is the image of the linear subsystem of $|E|$ whose curves pass simply through the double point of the corresponding curve E. The system V of the curves $C + D$ is represented by a multiple subvariety V of M. Let $C_0 + D_0$ be a generic but fixed curve of V, E_0 its image point on M, and $P_1^0, \ldots, P_\delta^0$ the double points of $C_0 + D_0$. The totality of all curves E possessing δ double points and very near $C_0 + D_0$ is represented in S_r by the complete intersection of δ analytic branches M_i of M, of origin E_0, where M_i $(i = 1, 2, \ldots, \delta)$ represents the curves E very near E_0 and possessing one double point near P_i^0. It is not difficult to show that each M_i is a branch *of first order* (a *linear branch*). From the theory of systems of simultaneous analytic equations it is well known that their complete intersection is composed of one or more irreducible analytical branches, each of dimension $\geqq r - \delta$. Since V is complete, one of these branches is on V and represents the complete neighborhood of E_0 in V.

What is the tangent space S_ϱ (ϱ is the dimension of V) of V at E_0? It is well known that if the tangent hyperplanes of the branches M_i at E_0 are independent (i. e. if the Jacobian of the corresponding δ analytical equations in r variables is of rank δ at E_0), then the complete intersection of the branches M_i is irreducible, linear, of dimension $r - \delta$ and possesses as a tangent space the intersection $S_{r-\delta}$ of the δ tangent hyperplanes just mentioned. Therefore in this case $\varrho = r - \delta$ and S_ϱ coincides with $S_{r-\delta}$. If, however, the above δ hyperplanes are dependent, they intersect in an $S_{r-\sigma}$, $\sigma \leqq \delta$, and S_ϱ either coincides with or is a subspace of $S_{r-\sigma}$. Hence $r - \delta \leqq \varrho \leqq r - \sigma$.

If we recall that the tangent hyperplane of M_i at E_0 is the image of the linear system, ∞^{r-1}, of the curves E passing through the double point P_i^0 of $C_0 + D_0$ and that therefore the intersection of these hyperplanes is the image of the complete system of the curves E passing through the double points of $C_0 + D_0$, we conclude that *the completeness of the linear system of the curves $C + D$ infinitely near $C_0 + D_0$ can be asserted only if the δ double points of $C_0 + D_0$ impose independent conditions on the curves E.* If we suppose, without loss of generality, that the series cut out by $|E|$ on C_0 is non-special, and if we recall that the series cut out on C_0 by the curves E passing through the double points of $C_0 + D_0$ is the characteristic series of Σ on C_0, we see that the above condition is equivalent to the condition that the *characteristic series of Σ on C_0 be non-special.* On the contrary, if this series is special, then the points P_i^0 certainly do not impose independent conditions on the curves E and *a priori* S_ϱ may not coincide with $S_{r-\sigma}$.

We point out that on surfaces for which $p_g > 0$ it is this second alternative which always takes place. If, however, $p_g = 0$, then the characteristic series of Σ is non-special, provided Σ is sufficiently general, for instance such that the system $|C|$ determined by a generic curve of Σ is regular (IV, 3).

The above method of analytical branches ("metodo di falde analitiche") has been devised by SEVERI and applied by him to the study of continuous systems of plane curves (see VIII, 4).

A rigorous proof of the existence on a surface of irregularity q of continuous systems consisting of ∞^q linear systems was given for the first time by POINCARÉ (8, 9) (see VII, 5). POINCARÉ's proof is analytic and has been subsequently simplified by SEVERI (20) and by LEFSCHETZ (a). The system obtained by POINCARÉ as an immediate result of his method is an irreducible system, ∞^q, whose generic curve is isolated in the sense of linear equivalence. The value of the construction of such a system is greater than that of mere example; indeed it is an essential step in the theory of continuous systems. We proceed to show that from the existence of *one* continuous system made up of ∞^q linear systems, the existence of infinitely many analogous systems and the completeness of the characteristic series of sufficiently general complete continuous systems follow with relative simplicity by means of the RIEMANN-ROCH theorem.

Let $\{A\}$ be the continuous system of POINCARÉ and let $|C|$ be any complete linear system of arithmetically effective curves, i. e. of characters n, π, i such that $n - \pi + p_a - i + 1 \geqq 0$ (IV, 2). If A and A_1 are any two curves of $\{A\}$, then $A \equiv A_1$, i. e. $A_1 - A$ is algebraically equivalent to zero. From this it follows (section 2) that the a priori virtual curve $C + A_1 - A$ has the same virtual characters n, π as C. As for its index of speciality, it obviously cannot be greater than the index of speciality of the particular curve $C + A_1 - A_1 = C$ obtained by letting A coincide with A_1. Consequently $C + A_1 - A$ is arithmetically effective and therefore, by SEVERI's criterion (IV, 2), the linear system $|C + A_1 - A|$ is of dimension $\geqq 0$. We fix A_1 and let A vary in $\{A\}$. The linear system $|C + A_1 - A|$ generates then a continuous system $\{C\}$ (irreducible as a totality of linear systems), consisting of ∞^q distinct linear systems (and therefore complete). We thus have the following theorem (SEVERI, 5): *On a surface of irregularity q every complete linear system $|C|$ of arithmetically effective curves belongs to a complete continuous system $\{C\}$ consisting of ∞^q distinct linear systems.*

We observe that if the generic C is irreducible, then the characteristic series of the *generic* system $|C|$ contained in $\{C\}$ has necessarily the maximum deficiency q and that *the characteristic series of $\{C\}$ on a generic C is complete.*

It is not difficult now to show that *the system $|C|$ of the preceding theorem cannot belong to two distinct complete continuous systems* (irreducible as totalities of curves). For this it is sufficient to prove that any system Σ, irreducible as a totality of curves and containing a curve C_0 of $|C|$, belongs to the system $\{C\}$ of the preceding theorem. Let C_1 be a generic but fixed curve of $\{C\}$ and let B denote a variable curve of Σ. The system $|\bar{C}| = |C_1 + C_0 - B|$ is of dimension ≥ 0 (since $C_1 + C_0 - B$ is arithmetically effective) and, as B varies in Σ, generates a continuous system Σ', irreducible as a totality of linear systems. The system Σ' contains $|C_1|$ and consequently (section 2) is contained in $\{C\}$, since C_1 is a generic curve of $\{C\}$. It follows that for *any* curve B of Σ there exists a curve $\bar{C}$ in $\{C\}$ such that $|B| = |C_1 + C_0 - \bar{C}|$. As $\bar{C}$ varies in $\{C\}$, the system $|C_1 + C_0 - \bar{C}|$, of dimension ≥ 0, generates a continuous system Σ'', irreducible as a totality of linear systems and containing $|C_1|$. Therefore Σ'', and hence also $|B|$, where B is *any* curve of Σ, belong to $\{C\}$, q. e. d.

Corollary. *Any complete continuous system* $\{C\}$, *whose generic curve is arithmetically effective, consists of* ∞^q *distinct linear systems. If the generic C is irreducible, then the characteristic series of* $\{C\}$ *is complete* (SEVERI).

Remark. If $\{C\}$ is reducible as a totality of linear systems, it consists of systems $\Sigma_1, \Sigma_2, \ldots$, irreducible as totalities of curves. It should then be understood that by the characteristic series of $\{C\}$ on a particular C, belonging to a system Σ_i, is meant the characteristic series of Σ_i on C. Among the systems Σ_i there is one and only one, say Σ_1, which is composed of ∞^q linear systems. *The preceding corollary asserts only the completeness of the characteristic series of* Σ_1 *on its generic curve.*

From the preceding theorems concerning arithmetically effective curves it follows that *for these curves effective equivalence coincides with algebraic equivalence* (ALBANESE, 1). In fact, let A and B be arithmetically effective curves, and let $A \equiv B$. Then there exists a curve D such that $A + D \equiv B + D$. We may suppose that both D and $A + D$ are arithmetically effective. The systems $\{A\}, \{D\}, \{A + D\}$ are then uniquely determined and consist each of ∞^q linear systems. As D varies in $\{D\}$ the system $|A + D|$ generates the system $\{A + D\}$. Hence, for a given D, there exists a curve $\bar{D}_0$ in $\{D\}$ such that $|A + D| = |B + \bar{D}_0|$. But $\{A\}$ is obviously generated by the system $|A + D - \bar{D}|$ as $\bar{D}$ varies in $\{D\}$. Consequently $|B|$ is in $\{A\}$ and $A \equiv B$, q. e. d.

For the proof of the completeness of the characteristic series in all its generality two steps in different directions are necessary: (1) to extend the theorem to curves which are not arithmetically effective; (2) to prove it for continuous systems, complete in the restricted sense. These extensions are due to B. SEGRE (5). He has proved that *the characteristic series of any continuous system, complete in the restricted sense, on a generic curve of the system, is complete.* In the same paper SEGRE also proves the following theorem, which is in part stronger than the preceding one:

If C is an irreducible curve whose multiple points are all assigned and if the virtual characteristic series of C effectively exists, then C belongs to at least one continuous system, complete in the restricted sense, whose characteristic series on C is complete. A weaker but more expressive form of this theorem is the following: *The effective existence of the virtual characteristic series of an irreducible curve C is a sufficient (and obviously necessary) condition for the existence of an infinite irreducible continuous system containing C.*

4. The variety of Picard. It is well known that on a curve of genus p the ∞^p linear series g_n^{n-p} ($n \geqq p$ and fixed) form an algebraic p-dimensional variety independent of n and hence birationally determined. It is the Jacobi *variety J* of the curve. The fundamental result of the preceding section suggests the following generalization to surfaces. If a complete continuous system $\{C\}$ on F is sufficiently general, for instance if the generic C of the system is arithmetically effective, it consists of ∞^q distinct linear systems. The q-dimensional variety of these linear systems is algebraic and irreducible. We shall denote it by V_q. Given two systems $\{C\}$ and $\{D\}$ of arithmetically effective curves, it is possible to set up a birational correspondence, depending on q arbitrary parameters, between the linear systems of $\{C\}$ and $\{D\}$, as follows: We fix arbitrarily a system $|\bar{C}|$ in $\{C\}$ and a system $|\bar{D}|$ in $\{D\}$ and we consider as the homologue of a system $|C|$ the system $|D| = |\bar{D} + C - \bar{C}|$. It follows that *the variety V_q is birationally determined, being independent of the particular system $\{C\}$, and that it admits a transitive continuous abelian group of ∞^q birational transformations into itself* (Castelnuovo, 9). V_q is called the *variety of* Picard attached to the surface F. Picard (4, 5) and Painlevé (1) have shown that any algebraic variety V_q satisfying the above conditions is *abelian*, i. e. that the coördinates of a variable point of V_q can be expressed as $2q$-ply periodic functions of q parameters. Using this property of V_q Castelnuovo (9) has been able to show that the surface F possesses q distinct simple integrals of total differentials of the first kind (see VII, 3, p. 163).

5. Equivalence criteria. These criteria enable us to recognize when two given curves A and B are equivalent, either linearly ($A \equiv B$) or algebraically ($A \equiv B$ or $A = B$), and have been established by Severi in several papers (10, 11, 12, 13). They are in part transcendental (and for linear equivalence generalize to algebraic surfaces the ordinary theorem of Abel) and in part algebro-geometric. We shall deal in this section with the algebro-geometric criteria, referring for the transcendental criteria to VII, 3, p. 164, VII, 4. p. 168 and VII, 7, p. 180.

Criterion 1. *Let Σ be an irreducible (rational or irrational) pencil of curves C, free from reducible curves. If two curves A_1 and A_2, virtually free from base points, cut out equivalent sets of points on every curve of Σ, then* *they are linearly equivalent to within curves of* Σ, *i. e.* $A_1 + \sum\limits_{i=1}^{\alpha'} C_i' \equiv A_2$ $+ \sum\limits_{j=1}^{\alpha''} C_j''$. *In particular, if* Σ *is rational (i. e. linear), then* $A_1 \equiv A_2 + \lambda C$, *where* λ *is an integer. If in addition the linear pencil* Σ, *virtually free from base points, is of positive degree (i. e. has at least one base point), then* $A_1 \equiv A_2$ (SEVERI, 11).

Proof (ENRIQUES, a, p. 171). Let x be a variable point on a generic C and let $y(x)$ be the rational function on C whose zeros and poles are the intersections of C with A_1 and A_2 respectively. The function $y(x)$ is determined to within a constant factor. Let L be an arbitrary fixed curve on F, such that $(L \cdot C) = h > 0$, and let $x_1, x_2, \ldots, x_h$ be the intersections of L and C. The function $y_1(x) = y(x)/(y(x_1) + \cdots + y(x_h))$ *is rationally determined on every* C and therefore defines *a rational function* Y *on* F (since Σ is a pencil). The curve of zeros of Y consists of the curve A_1 and of the curves C_i' $(i = 1, 2, \ldots, \alpha')$ of Σ which pass through the intersections of L with A_2. Similarly, the polar curve of Y consists of A_2 and of those curves C_j'' $(j = 1, 2, \ldots, \alpha'')$ of Σ which meet L in a set of points for which $\sum\limits_{i=1}^{h} y(x_i) = 0$. Hence $A_1 + \sum\limits_{i=1}^{\alpha'} C_i' \equiv A_2 + \sum\limits_{j=1}^{\alpha''} C_j''$. If Σ is linear, then any two curves C are equivalent and hence $A_1 \equiv A_2 + \lambda C$. We have $(A_1 \cdot C) = (A_2 \cdot C) + \lambda(C^2)$ and also, by hypothesis, $(A_1 \cdot C) = (A_2 \cdot C)$. Consequently, if $(C^2) > 0$, then $\lambda = 0$, q. e. d.

If Σ contains reducible curves, then A_1 and A_2 are equivalent to within curves of Σ *and* components of the reducible curves of Σ.

If we replace the pencil Σ by an algebraic system, ∞^1, we have the following more general

Criterion 2. *If* A_1 *and* A_2 *cut out equivalent sets on every curve* C *of an irreducible algebraic system* Σ, ∞^1, *free from reducible curves, and of degree* >0, *then* $A_1 \equiv A_2$.

The proof is the same as that of the preceding criterion, except that now, if Σ is of index ν (see section 1), $y_1(x)$ defines a ν-valued function on F and it is therefore necessary to introduce the function $Y_1(x) = y_1^{(1)}(x) + y_1^{(2)}(x) + \cdots + y_1^{(\nu)}(x)$.

If instead of having two curves A_1 and A_2 we have an irreducible algebraic system S of curves A (so that we deal *a priori* with curves which are already algebraically equivalent in the restricted sense), the following stronger criterion holds:

Criterion 3. *If the curves of* S *cut out equivalent sets on one irreducible curve of a linear pencil of degree* >0, *then* S *is contained in a linear system* (SEVERI, 10). A transcendental proof will be given in VII, 3, p. 165.

We now pass to criteria of algebraic equivalence.

Criterion 4. *If A_1 and A_2 are two curves of the same order and if $(A_1^2) = (A_1 \cdot A_2) = (A_2^2) = n > 0$, then there exists a positive integer λ, such that $\lambda A_1 \equiv \lambda A_2$* (Severi, 13).

Proof. Let $[A_i] = \alpha_i$ $(i = 1, 2)$, $\alpha_2 \geqq \alpha_1$, and let $|E|$ be the system of hyperplane sections of F. By hypothesis, $(E \cdot A_1) = (E \cdot A_2)$, and this relation continues to hold if we replace $|E|$ by any multiple of $|E|$. Hence we may suppose that $|E|$ is non-special and of virtual dimension $r > 0$, and that also the system $|E_t| = |E + tA_1 - tA_2|$ is non-special[1], for any value of the positive integer t. The curves of the systems $|E|, |E_1|, |E_2|, \ldots$ are all of the same order, and the virtual dimension of $|E_t|$ is $r + (\alpha_2 - \alpha_1)t \geqq r$. It follows in the first place that $|E_t|$ is effective, and moreover that $\alpha_2 = \alpha_1$, since the dimension of $|E_t|$ cannot increase indefinitely as t increases. If the number of distinct systems $|E_t|$ is finite, then there exists an integer λ such that $\lambda A_1 \equiv \lambda A_2$. If there are infinitely many distinct systems $|E_t|$, we observe that the curves E_t, being all of the same order, belong to a finite number of complete continuous systems in the restricted sense. Let Σ be one of these systems containing curves of infinitely many systems $|E_t|$. Let $|E_{i_1}|, |E_{i_2}|, (i_2 > i_1)$ be two systems $|E_t|$ having curves in Σ, and let i_2 be taken large enough that $|(i_2 - i_1)A_1| = |\lambda A_1|$ is non-special. Using the hypothesis $(A_1^2) > 0$ it is then immediately seen that if D is any curve of Σ, then the system $|C| = |E_{i_1} + \lambda A_1 - D|$ is effective. As D varies in Σ the system $|C|$ varies in a complete continuous system $\{C\}$ (irreducible as a totality of linear systems) containing $|\lambda A_1|$ $(D \to E_{i_1})$ and $|\lambda A_2|$ $(D \to E_{i_2})$, q. e. d.

Remark 1. If the condition $n > 0$ is not fulfilled, then we consider the curves $\overline{A}_1 = A_1 + tE$, $A_2 = \overline{A}_2 + tE$. If t is sufficiently high, then $(\overline{A}_1^2) = (\overline{A}_1 \cdot \overline{A}_2) = (\overline{A}_2^2) > 0$, and consequently $\lambda A_1 + \lambda t E \equiv \lambda A_2 + \lambda t E$, i. e. *if the condition $n > 0$ is not fulfilled, we can only assert that $\lambda A_1 = \lambda A_2$* (section 2). In this last form our equivalence criterion can be immediately extended also to virtual curves A_1, A_2.

Remark 2. If $n > 0$ and if one of the curves A_1, A_2 is free from components of virtual negative degree (in particular, if one of these curves is irreducible), then it is not necessary to use the hypothesis that A_1 and A_2 are of the same order (this hypothesis follows then *a posteriori* from the other hypotheses of the theorem). As system $|E|$ we may take in this case the system $|E| = |k(A_1 + A_2)|$, where k is taken high enough to satisfy the condition $(E \cdot A_1) [= (E \cdot A_2) = 2kn] > (K \cdot A_1)$, $(K \cdot A_2)$, where $|K|$ is the canonical system. Then it is immediately

---

[1] That E_t is non-special can be seen as follows: If $|K|$ is the canonical system, we may suppose that $(E \cdot K) < (E^2)$. If $|E_t|$ is special, then $|E + tA_1|$ is partially contained in $|K + tA_2|$, i. e. $|E + tA_1 + L| = |K + tA_2|$, where $|L|$ is an effective system. Intersecting with E we find $(E^2) + (L \cdot E) = (K \cdot E)$, hence $(L \cdot E) < 0$, and this is impossible, since $|E|$ is irreducible and of dimension > 0.

seen that all the systems $|E_t| = |E + t(A_1 - A_2)|$ are non-special. In fact, if $|E_t|$ is special, then $|E + tA_1 + L| = |K + tA_2|$, where L is an *effective* curve. Intersecting with A_1 and A_2 we find $(L \cdot A_1) < 0$ and $(L \cdot A_2) < 0$, and this is impossible since one of the curves A_1, A_2 is free from components of virtual negative degree. It remains to observe that if k is sufficiently high, then the virtual dimension of $|E|$ is positive. The rest of the proof is unaltered.

6. The theory of the base and the number ϱ of PICARD (SEVERI, 13, 16, 17). *Definition.* Given a set of (effective or virtual) curves $C_1, C_2, \ldots,$ C_l on F, they are said to be *algebraically dependent* if there exist integers $\lambda_1, \lambda_2, \ldots, \lambda_l$, not all zero, such that $\lambda_1 C_1 + \lambda_2 C_2 + \cdots + \lambda_l C_l = 0$.

Let $(C_i \cdot C_j) = n_{ij}$ and let m_i be the order of C_i.

Theorem 1. *A necessary and sufficient condition that the curves C_i $(i = 1, 2, \ldots, l)$ be algebraically dependent is that the matrix $\left\| \begin{matrix} n_{ij} \\ m_j \end{matrix} \right\|$ be of rank $< l$* (SEVERI, 13).

Proof. If $\lambda_1 C_1 + \cdots + \lambda_l C_l = 0$, then intersecting with C_i and with a hyperplane section E, we find

$$(4) \qquad \begin{cases} \displaystyle\sum_{j=1}^{l} \lambda_j n_{ij} = 0, \\[2mm] \displaystyle\sum_{j=1}^{l} \lambda_j m_j = 0, \end{cases} \qquad (i = 1, 2, \ldots, l)$$

and hence the condition of the theorem is necessary. Conversely, let the matrix $\left\| \begin{matrix} n_{ij} \\ m_j \end{matrix} \right\|$ be of rank $< l$, and let $\lambda_1, \lambda_2, \ldots, \lambda_l$ be a set of integers, not all zero, satisfying (4). The virtual curve $C = \lambda_1 C_1 + \lambda_2 C_2 + \cdots$ $+ \lambda_l C_l$ satisfies then the following conditions: (1) $(C^2) = 0$, (2) $(C \cdot E)$ $= 0$. The virtual curve Γ which is algebraically equivalent to zero also satisfies these conditions, and moreover $(C \cdot \Gamma) = 0$. Hence, by the criterion 4. of the preceding section, Remark 1, $\lambda C = \lambda \Gamma$, i. e. $\lambda C = 0$, q. e. d.

While necessary references to PICARD integrals will be found in Chapter VII, we give below the theorem which establishes the connection between SEVERI's theory of algebraic dependence and the transcendental theory of PICARD and from which the existence of a base for algebraic curves on F follows as a corollary.

Theorem 2. *A necessary and sufficient condition that given effective curves $C_1, C_2, \ldots, C_l$ be algebraically dependent is that there exist on F a PICARD integral of the 3rd kind having the curves C_i as the only logarithmic curves* (SEVERI, 13).

Proof. We first prove that the condition is necessary. Let $\lambda_1 C_1 + \lambda_2 C_2$ $+ \cdots + \lambda_l C_l = 0$. The coefficients λ_i cannot all have the same sign, since the curves C_i are effective. Let us suppose that $\lambda_i \geqq 0$, when $i = 1, 2, \ldots, t$,

$\lambda_i < 0$, when $i = t+1, \ldots, l$. We put $A_1 = \sum\limits_{i=1}^{t} \lambda_i C_i$, $A_2 = -\sum\limits_{i=t+1}^{l} \lambda_i C_i$, where A_1 and A_2 are therefore effective curves. From $A_1 = A_2$ we deduce (see section 2, p. 97), that there exists an effective curve D such that $A_1 + D \parallel A_2 + D$, i. e. that there exists an irreducible algebraic system Σ, ∞^1, containing the curves $\overline{A}_1 = A_1 + D$ and $\overline{A}_2 = A_2 + D$. The proof now follows along the lines of a very fruitful reasoning employed by HUMBERT (2). Let $\varphi(\xi, \eta) = 0$ be the non-homogeneous equation of a curve whose points are in $(1-1)$ correspondence with the curves of Σ (see section 1), and let $a_1 = (\xi_1^0, \eta_1^0)$, $a_2 = (\xi_2^0, \eta_2^0)$ be the points of φ which correspond to the curves $\overline{A}_1$ and $\overline{A}_2$. We consider a normal abelian integral $\tilde{\omega}(\xi, \eta)$ of the 3rd kind attached to φ, possessing at a_1 and a_2 logarithmic singularities with polar periods $+1$ and -1 respectively. If (ξ_i, η_i) $(i = 1, 2, \ldots, \nu$, where ν is the index of Σ) are the ν points of φ which correspond to the ν curves of Σ passing through a point (x) of F, then it is immediately seen that the function $J(x) = \sum\limits_{i=1}^{\nu} \tilde{\omega}(\xi_i, \eta_i)$ is a PICARD integral of the 3rd kind attached to F, possessing C_i as a logarithmic curve with polar period λ_i. As for the auxiliary curve D, it is seen that its polar period is zero, and hence it is not a logarithmic curve of $J(x)$.

We now prove that the condition is sufficient. Let $J(x)$ be a PICARD integral of the 3rd kind attached to F, possessing $C_1, C_2, \ldots, C_l$ as the only logarithmic curves with polar periods $\lambda_1, \lambda_2, \ldots, \lambda_l$ respectively. If D is an irreducible curve on F meeting each C_i in a finite number of points, forming a set Γ_i, then $J(x)$ defines on D an abelian integral of the 3rd kind possessing at each point of the set Γ_i a logarithmic singularity with polar period λ_i. Since the sum of the polar periods must be equal to 0, we have

$$\lambda_1 (C_1 \cdot D) + \cdots + \lambda_l (C_l \cdot D) = 0.$$

If we take for D a hyperplane section E of F, we find

$$\lambda_1 m_1 + \cdots + \lambda_l m_l = 0,$$

where m_i is the order of C_i. We now put $|D| = |C_j + \mu E|$ and we observe that for μ sufficiently high $|C_j + \mu E|$ is irreducible. We then find

$$\sum_{i=1}^{l} \lambda_i n_{ij} = 0, \qquad\qquad j = 1, 2, \ldots, l,$$

where $n_{ij} = (C_i \cdot C_j)$. It follows that the matrix $\left\| \begin{matrix} n_{ij} \\ m_j \end{matrix} \right\|$ is of rank $< l$ and that therefore (theorem 1) the curves C_i are algebraically dependent, q. e. d.

In view of theorem 2, the existence of a base for algebraic curves on F now follows as a corollary from the following

3. **Theorem of PICARD.** *There exists an integer $\varrho \geqq 0$, such that given $\varrho + 1$ arbitrary curves on F there always exists a PICARD integral of the 3rd kind of which they are the only logarithmic curve. On the contrary, if the number of the given curves is $\leqq \varrho$, such an integral in general (i. e. when the given curves are arbitrary) does not exist* (PICARD and SIMART, a_2, Chap. 9). As a corollary we have the following

4. **Fundamental theorem of the base.** *There exist on F ϱ algebraically independent curves $C_1, C_2, \ldots, C_\varrho$, and every other curve C on F is algebraically dependent on these ϱ curves, i. e. is connected to these curves by a relation of the type*

$$\lambda C + \lambda_1 C_1 + \cdots + \lambda_\varrho C_\varrho = 0,$$

where λ and the λ_i's are integers and where $\lambda \neq 0$ (SEVERI, 13).

It is clear that, conversely, from the fundamental theorem of the base follows the theorem of PICARD, in view of theorem 2. A direct topological proof of the theorem of the base will be given in VI, 12.

The ϱ curves $C_1, C_2, \ldots, C_\varrho$ are said to form a *base*. If $\overline{C}_1, \overline{C}_2, \ldots, \overline{C}_\varrho$ form another base, then $\lambda_i \overline{C}_i = \sum_{j=1}^{\varrho} \lambda_{ij} C_j$, where $\lambda_i \neq 0$ and $|\lambda_{ij}| \neq 0$ $(i, j = 1, 2, \ldots, \varrho)$. Conversely, if these inequalities hold, then $\overline{C}_1, \overline{C}_2, \ldots, \overline{C}_\varrho$ form a base. Let $(C_i \cdot C_j) = n_{ij}$, $(\overline{C}_i \cdot \overline{C}_j) = \overline{n}_{ij}$. Then $(\lambda_1 \lambda_2 \ldots \lambda_\varrho)^2 \, |\overline{n}_{ij}| = |\lambda_{ij}|^2 \, |n_{ij}|$ and consequently, if the determinant $|n_{ij}|$ of one base is $\neq 0$, then the same is true for any other base. It is always possible to find a base $C_1, C_2, \ldots, C_\varrho$ such that one of the curves C_i, say C_1, is a hyperplane section. Then $m_i = n_{1i}$, where m_i is the order of C_i, and since the matrix $\left\| \begin{matrix} n_{ij} \\ m_j \end{matrix} \right\|$ is of rank ϱ, it follows that $|n_{ij}| \neq 0$. On the other hand, if ϱ curves C_i are such that the determinant $|n_{ij}|$ of their intersection numbers is different from zero, they are independent and therefore form a base. We have then the following

Theorem 5. *A necessary and sufficient condition that ϱ curves $C_1, C_2, \ldots, C_\varrho$ form a base is that the determinant $|n_{ij}| = |C_i \cdot C_j|$ be different from zero* (SEVERI, 13).

Let F be birationally transformed into a surface F' and let us suppose that the transformation creates e' new exceptional curves on F', while no exceptional curve of F is transformed into a point of F'. If $C_1, C_2, \ldots, C_\varrho$ form a base on F, then it is seen that their transforms C_i' and the above e' exceptional curves form a base on F'. Hence, if ϱ' is the number of PICARD relative to F', then $\varrho' = \varrho + e'$. This shows that ϱ *is a relative invariant of F.* If F and F' possess a finite number of exceptional curves of the 1st kind, say e and e' respectively, then $\varrho + e' = \varrho' + e$, and hence $\varrho - e$ *is an absolute invariant of F* (SEVERI, 13).

If F belongs to the class of ruled surfaces, it possesses a linear system $|E|$ of dimension $\geqq 3$ and a pencil Σ of curves C which are unisecants of $|E|$. A base on F is given by a curve E, a curve C and by the exceptional curves of the 1st kind which are components of curves C of Σ (SEVERI, 2). In particular, if the curves C are straight lines, then $\varrho = 2$.

The evaluation of ϱ for a given surface presents in general grave difficulties. For hyperelliptic surfaces, see BAGNERA-DE FRANCHIS (1).

LEFSCHETZ (5, p. 359) has determined the value of ϱ for surfaces variable in certain linear systems $|E|$ in S_3. He finds that if $|E|$ is simple, irreducible, and ∞^3 at least, and if the base locus of the system consists of k curves which are independent on a generic E and do not contain any multiple base point of the system, then the generic E contains only curves which are cut out by surfaces passing through the base curves, and for it $\varrho = k + 1$. A closely related result is the first complete proof (LEFSCHETZ, a, p. 108) of NOETHER's surmise that the general surface of order 4 in S_3 contains only curves which are complete intersections with other surfaces of the space. This is proved by LEFSCHETZ for surfaces of any order $m > 3$.

We point out explicitly that in SEVERI's definition of ϱ the surface F is supposed to be free from singularities. If F possesses isolated singularities, the invariantive definition of ϱ requires that they be treated as infinitesimal curves. For instance, on a KUMMER surface a base is given by one of its 16 conics and by the 16 double points, each considered as a curve of virtual degree -2; hence $\varrho = 17$ (SEVERI, 16).

If $C_1, C_2, \ldots, C_\varrho$ form a base on F and if D and E are any two algebraic curves on F, then $\lambda D = \sum \lambda_i C_i$, $\mu E = \sum \mu_i C_i$ and $\lambda \mu (C \cdot D) = \sum n_{ij} \lambda_i \mu_j$, where $n_{ij} = (C_i \cdot C_j)$. This important relation (SEVERI) is a generalization of the formula of BEZOUT. For every base there is an associated bilinear form $\sum n_{ij} x_i y_j$ which permits us to compute the number of intersections of any two curves on F. The result is complete especially in the case of ruled surfaces (SEVERI, 2) since then the base is known.

ALBANESE (1) has shown that it is possible to choose a base $C_1, C_2, \ldots, C_\varrho$ on F in such a manner that for every effective curve C on F a relation of the type $\lambda C \equiv \lambda_1 C_1 + \cdots + \lambda_\varrho C_\varrho$ or even of the type $\lambda C \parallel\!\parallel \lambda_1 C_1 + \cdots + \lambda_\varrho C_\varrho$ holds.

A base $C_1, C_2, \ldots, C_\varrho$ is called *intermediate*, if for every C the integer λ which occurs in the relation $\lambda C = \lambda_1 C_1 + \cdots + \lambda_\varrho C_\varrho$ is a divisor of each λ_i (SEVERI, 16, 17). The intermediate bases are those for which the determinant $|C_i \cdot C_j| = |n_{ij}|$ has a minimum value. If $\lambda C = \lambda(\mu_1 C_1 + \cdots + \mu_\varrho C_\varrho)$, then $(C^2) = \sum n_{ij} \mu_i \mu_j = \varphi(\mu)$. Thus for every intermediate base there is an associate *quadratic form* $\sum n_{ij} x_i x_j = \varphi(x)$ with integral coefficients. Two quadratic forms associated with two different intermediate bases are equivalent (i. e. are transformable into each other by a unimodular transformation). SEVERI (17) has applied the consideration of the quadratic form $\varphi(x)$ toward the study of surfaces (especially regular surfaces) admitting an infinite discontinuous group of birational transformations into themselves.

The following application of theorem 2 is due to SEVERI and answers a question raised by PICARD. On a regular surface F algebraic equivalence coincides with linear equivalence (section 3). Hence if F is regular and if J is the simple integral of the 3rd kind considered in theorem 2, then $\lambda_1 C_1 + \lambda_2 C_2 + \cdots + \lambda_l C_l \equiv 0$. There exists then on F a rational function R_1 whose zero and polar curves are among the curves $C_1, \ldots, C_l$. The integral $\int \frac{1}{R_1}\left(\frac{\partial R_1}{\partial x}\,dx + \frac{\partial R_1}{\partial y}\,dy\right) = \log R_1$ is of the 3rd kind and its logarithmic curves are among the curves C_i. Since the polar periods of $\log R_1$ do not all vanish, there exists a suitable linear combination $J' = aJ - a_1 \log R_1$, $a \neq 0$, such that one of the curves C_i, say C_1, is not a logarithmic curve of J'. If J' still has logarithmic curves, then $C_2, \ldots, C_l$ are algebraically, and hence also linearly, dependent, and we may find, as before, a rational function R_2 and constants a', a_1' ($a' \neq 0$), such that one of the curves $C_2, \ldots, C_l$, say C_2, is not a logarithmic curve of the integral $a'J' - a_1' \log R_2$. Proceeding in this manner we shall finally arrive at an integral of the type $I = cJ - c_1 \log R_1 - c_2 \log R_2 - \cdots - c_k \log R_k$ which has no logarithmic curves at all and hence is of the second kind (VII, 1). Since F is regular, I is necessarily a rational function on F (VII, 3), and therefore J is a logarithmo-rational function on F. Conversely, if every simple integral on F reduces to a logarithmo-rational function, then every simple integral of the 2nd kind is a rational function on F, and hence F is regular (VII, 3). Hence, *a necessary and sufficient condition that every simple integral on F be a logarithmo-rational function, is that F be a regular surface* (SEVERI, 13).

SEVERI's algebro-geometric theory of the base has been investigated by LEFSCHETZ from a topological point of view (see VI, 12). His results have thrown completely new light on the subject, by revealing wholly unsuspected topological connections.

7. The division group and the invariant σ of SEVERI (SEVERI, 16, 17). Let $\{E_1\}$ be a complete continuous system of curves E_1, satisfying the following conditions:

(a) A generic E_1 is arithmetically effective.

(b) The order of E_1 is greater than the order of the canonical curves K. From (a) it follows that $\{E_1\}$ consists of ∞^q distinct complete linear systems $|E_1|$, and from (b) it follows that these linear systems are all non-special. The curves E on F of the same order as the curves E_1 and satisfying the conditions

(5) $$(E^2) = (E \cdot E_1) = (E_1^2)$$

separate into a finite number of complete continuous systems. Let
$$\{E_1\}, \{E_2\}, \ldots, \{E_\sigma\}$$
be these systems, where σ is an integer $\geqq 1$. It is clear that each system $\{E_i\}$ satisfies the conditions (a) and (b) and therefore consists of ∞^q distinct linear systems. In view of the criterion 4 of section 5 we have $E_i \not\equiv E_1$, $\lambda_{i-1} E_i \equiv \lambda_{i-1} E_1$, where λ_{i-1} is a convenient integer >1 $(i = 2, 3, \ldots, \sigma)$. Since we are dealing with arithmetically effective curves we may write (section 3, p. 103) $E_i \neq E_1$, $\lambda_{i-1} E_i \equiv \lambda_{i-1} E_1$. We have thus found on F $\sigma - 1$ virtual curves $\Gamma_{i-1} = E_i - E_1$ $(i > 1)$, which are *divisors of zero* with respect to algebraic equivalence, i. e. such that $\quad \Gamma_i \neq 0, \quad \lambda_i \Gamma_i = 0, \quad \lambda_i > 1, \quad i = 1, 2, \ldots, \sigma - 1.$
Every divisor of zero on F coincides with (i. e. is algebraically equivalent to) *one of the above divisors Γ_i.* In fact, if $\overline{\Gamma}$ is a divisor of zero, then the virtual curve $E_1 + \overline{\Gamma}$ is of the same order as E_1, satisfies the conditions (5) and is arithmetically effective. Hence the system $\{E_1 + \overline{\Gamma}\}$ coincides with one of the systems $\{E_i\}, i \geqq 2$, and consequently $E_1 + \overline{\Gamma} = E_i$, or $\overline{\Gamma} = \Gamma_{i-1}$, q. e. d.

The $\sigma - 1$ divisors Γ_i are evidently all distinct. In fact from $\Gamma_i = \Gamma_j$ $(i \neq j)$ it would follow that $E_i = E_j$, and hence also $E_i \equiv E_j$, and this is impossible. The divisors Γ_i, together with a curve Γ_0 algebraically equivalent to zero, form a *finite abelian group G_σ of order σ* (since $\Gamma_i + \Gamma_j$ is also a divisor of zero), called the *division group of F*. We shall see in Chapter VII, 7 (p. 180) that G_σ coincides with the two-dimensional torsion group of the surface. From the manner in which the number σ has been obtained it clearly follows that *σ is an absolute invariant of F under birational transformations*. It is naturally independent of the system $\{E_1\}$ used throughout the discussion. If the curves $\{E_1\}$ do not satisfy the conditions (a) and (b), the number of systems $\{E_i\}$ may be different from σ (ALBANESE, 1). In that case, however, we would be dealing with curves for which effective equivalence ($\equiv$) is actually stronger than algebraic equivalence. In order to arrive at the same number σ it would then be necessary to deal with virtual complete system V_{E_i} (section 2) instead of with systems $\{E_i\}$.

If $C_1, C_2, \ldots, C_\varrho$ are ϱ curves constituting an intermediate base and if C is an arbitrary curve on F, then $\lambda C = \lambda (\lambda_1 C_1 + \cdots + \lambda_\varrho C_\varrho)$, $\lambda \neq 0$, and consequently $C = \lambda_1 C_1 + \lambda_2 C_2 + \cdots + \lambda_\varrho C_\varrho + \Gamma_i$, where Γ_i is an element of G_σ. The group G_σ is generated by a certain number ξ of elements $\Gamma_1, \Gamma_2, \ldots, \Gamma_\xi$ of orders $t_1, t_2, \ldots, t_\xi$, forming a base for G_σ. The integers t_i are the *coefficients of torsion of F*. It follows that every (effective or virtual) curve is algebraically equivalent to a sum of multiples of the ϱ curves C_i and of the curves $\Gamma_1, \Gamma_2, \ldots, \Gamma_\xi$. This property holds *a fortiriori* if the Γ_i's are replaced by the σ curves $E_1, E_2, \ldots, E_\sigma$

and if we suppose that $C_1 = E_1$. The set of $\varrho + \sigma - 1$ curves $C_i, E,$ (all effective) is said by SEVERI to constitute a *minimal base*.

GODEAUX (2, 3) has given interesting examples of surfaces for which $\sigma > 1$.

8. On the moduli of algebraic surfaces. In the theory of algebraic curves the following facts have been rigorously established (see VIII, 4). (1) An algebraic curve possesses only one numerical birational invariant, its genus. (2) The birationally distinct curves of a given genus p form one irreducible continuous system. (3) This continuous system is of dimension $3p - 3$ $(p > 1)$. It follows that the birationally distinct curves of a given genus p depend on $3p - 3$ continuous parameters or *moduli*, and that the birational equivalence of two curves of genus p is expressed by $3p - 3$ certain invariantive conditions: the equality of their moduli. Although the moduli of algebraic curves have been determined explicitly only for low values of the genus, nevertheless theoretically the problem of the moduli of algebraic curves is thus completely solved.

Much less is known on the similar problem for algebraic surfaces. In the first place an algebraic surface possesses several independent numerical birational invariants. Those known include p_g, p_a and $p^{(1)}$ (excluding the class of ruled surfaces), but there may be others, as yet undiscovered. In the second place, it is possible that the surfaces, whose numerical invariants have assigned values, form several irreducible continuous systems or classes, complete in the invariantive sense, and that these classes are of different dimensions. It is permissible to speak of the moduli of any one of these classes of surfaces, the number of the moduli being given by the dimension of the class.

The determination of the number M of moduli of a class of surfaces was first treated by NOETHER. By making certain assumptions on the postulation formulas, NOETHER shows that the regular surfaces for which $p_g = p_a = p > 3$ and $p^{(1)} > 5$ depend on

$$M = 10p - 2p^{(1)} + 12$$

moduli. ENRIQUES (14) treated the problem in all its generality and arrived at the following result:

The surfaces with given p_g', p_a, $p^{(1)}$ (excluding the class of ruled surfaces) depend on

(6) $$M = 10p_a - p_g - 2p^{(1)} + 12 + \Theta$$

moduli, where Θ—obviously an absolute birational invariant of the considered surfaces—is a non-negative integer.

We proceed to outline the proof. It seems to us that the proof makes use of an *assumption*, which raises an interesting question and which it would be highly important to establish with complete rigor.

Having excluded the class of ruled surface, we may assume that the surfaces of the considered class are free from exceptional curves (IV, 4). We consider on a generic surface of the class a regular system $|C|$, free from base points and of characters n, π, r $(r \geqq 3)$, and we transform birationally the surface into a surface F in S_3 by means of a generic web extracted from $|C|$. We have here several variable elements: the web varies in $|C|$; $|C|$ varies in a complete continuous system $\{C\}$, consisting of $\infty^{p_g - p_a}$ complete linear systems; the surface which carries $|C|$ varies in the class considered. As these elements vary, the surface F varies in an irreducible continuous system $\{F\}$ of surfaces of order $n = (C^2)$. The system $\{F\}$ is *complete* (in the sense that it is not contained in a larger continuous system of surfaces of order n, having the same characters p_g, p_a, $p^{(1)}$ as a generic F), and furnishes a projective realization in S_3 of our class of surfaces. We may assume also, by taking $|C|$ sufficiently general, that the generic F possesses ordinary singularities, i. e. a double nodal curve and a finite number of ordinary cuspidal points.

If D is the dimension of $\{F\}$ and if $\{F\}$ contains ∞^S surfaces birationally equivalent to a generic F, then obviously

(7) $$M = D - S.$$

The consideration of the variable elements of the construction, mentioned above, gives for S the following expression:

(8) $$S = 4n - 4\pi + 3p_a + p_g + 7 - \varrho,$$

where we assume that *a generic F admits ∞^ϱ birational transformations into itself*.

In order to evaluate D, we consider a generic but fixed surface F_0 in $\{F\}$ and we introduce the *characteristic linear system*, of dimension $D - 1$, cut out on F_0 by the surfaces F infinitely near F_0.[1]

The surfaces F, infinitely near F_0, pass through the double points of F_0, i.e. through the double curve and through the double points, outside the double curve, which are infinitely near the cuspidal points of F_0. It follows that these surfaces are *adjoint surfaces* of F_0, of order n, and that the characteristic curves are total curves of the system $|C' + 3C|$ (C a plane section of F), constrained to pass through the cuspidal points of F_0, i.e. total curves of the system $|C' + 3C|_H$, where H denotes the set of cuspidal points of F_0.

Using the Riemann-Roch theorem we find:

(9)   dimension of $|C' + 3C| = 4\pi + 6n + p_a - 4$,

---

[1] The notion of the characteristic linear system of a continuous system of surfaces in S_3 (or, more generally, on a V_3) is a straightforward generalization of the notion of the characteristic series of continuous system of curves on a surface, given in V, 1.

and a simple calculation, based on the fact that the cuspidal points are base points of the system of polar curves of F_0, leads to the following expression of the number t of cuspidal points:

(10) $$t = 2n + 8\pi + 2p^{(1)} - 12p_a - 22.$$

The t cuspidal points may not impose t independent conditions on the curves of the system $|C' + 3C|$. Let the number of independent conditions be $t - \Theta$, $\Theta \geqq 0$. Then we have: dimension of $|C' + 3C|_H$ = dimension of $|C' + 3C| - t + \Theta$, and hence, by (9) and (10),

(11)   dimension of $|C' + 3C|_H = 4n - 4\pi + 13p_a - 2p^{(1)} + 18 + \Theta.$

Here Θ appears as the superabundance of the system $|C' + 3C|_H$.

The characteristic curves on F_0 are total curves of $|C' + 3C|_H$, and hence the characteristic linear system on F_0, of dimension $D - 1$, either coincides with or is totally contained in $|C' + 3C|_H$. We have therefore: $D - 1 \leqq$ dimension of $|C' + 3C|_H$. At this stage of the proof, ENRIQUES assumes that *the characteristic system of $\{F\}$ on F_0 is complete*, and therefore coincides with $|C' + 3C|_H$. Under this assumption, the above inequality can be converted into an equality. Substituting into (7) the expressions of S and of $D - 1$, given by (8) and (11) respectively, we find

(12) $$M = 10p_a - p_g - 2p^{(1)} + 12 + \Theta + \varrho.$$

In general, the generic surface of the considered class will not possess infinitely many birational transformations into itself. Hence $\varrho = 0$ and we find (6).

The above assumption is in all respects similar to the assumption of the completeness of the characteristic series of a continuous system of curves, discussed in V, 3. A proof of the completeness of the characteristic system of a complete continuous system of surfaces in S_3 (with ordinary singularities) is not likely to be an easy undertaking.

We take an agnostic attitude and we assume that the characteristic system of $\{F\}$ has a certain deficiency $\omega \geqq 0$. The correct value of M is then given by the right-hand member of (12), *diminished by ω*. It is therefore not possible to affirm that M is not less then $10p_a - p_g - 2p^{(1)} + 12$.

For regular surfaces ($p_g = p_a = p > 3$) with an irreducible canonical system ($p^{(1)} > 5$), the formula (6) becomes: $M = 9p - 2p^{(1)} + 12 + \Theta$. ENRIQUES shows that in this case $\Theta = p + \Theta'$, $\Theta' \geqq 0$, and hence $M = 10p - 2p^{(1)} + 12 + \Theta'$. To within the additive term Θ', this expression of M coincides with the one given by NOETHER.

B. SEGRE (7) generalized the inequality of ENRIQUES, $\Theta \geqq p$, extending it to irregular surfaces. He proves that if $p_g > p_a$, then

(13) $$\Theta \geqq 2p_g - p_a - 1.$$

We conclude this section with a few remarks, which will illustrate from another point of view the question of the completeness of the characteristic system of $\{F\}$. We project the surfaces of $\{F\}$ from a generic fixed point O of the space on to a plane σ. The generic surface F is represented in this manner upon the n-fold covered plane σ. The branch curve $\varDelta$ in σ, section of σ with the tangent cone of F of vertex O, possesses a certain number of nodes and cusps. As F varies in the complete system $\{F\}$, $\varDelta$ varies in a continuous system $\{\varDelta\}$. *The system $\{\varDelta\}$ is complete*, i.e. it is not contained in a larger continuous system of curves of the same order and with the same number of nodes and cusps as a generic $\varDelta$. This statement, due to Enriques (16), *follows from the fact that the conditions of existence of an n-tuple plane, with a given branch curve, do not vary as the branch curve varies continuously without acquiring new singularities* (Enriques, 16). These conditions involve only the fundamental group of the residual space of the curve with respect to its carrying projective plane (VIII, 1), and this group is not altered by an isotopic deformation. If then $\{\varDelta\}$ were not complete, our system of n-tuple planes, i.e. the system $\{F\}$, could not be complete.

Let F_0 be a fixed generic surface in $\{F\}$ and let $\varDelta_0$ be the corresponding branch curve in $\{\varDelta\}$. Let moreover D_0 be the polar curve on F_0 with respect to the pole O. To a surface F, infinitely near F_0, there corresponds a branch curve $\varDelta$, infinitely near $\varDelta_0$, and it is seen immediately that *the intersections of F with D_0, outside the fixed points (i.e. outside the cuspidal points of F_0), are projected from O into the intersections of $\varDelta$ with $\varDelta_0$, outside the fixed intersections at the nodes and the cusps of $\varDelta_0$.* It follows that *the series cut out on D_0 by the characteristic system of $\{F\}$ on F_0, outside the cuspidal points of F_0, is projected into the characteristic series of $\{\varDelta\}$ on $\varDelta_0$.*

Now it can be easily shown that if $|C|$ is taken in a sufficiently general manner (for instance, if $|C|$, in addition to being regular, partially contains the canonical system), then the system $|C' + 3C|$ cuts out on D_0 a *complete non-special series* [see for instance, B. Segre (7)]. In this case, also $|C' + 3C|_H$ cuts out on D_0 a complete series, of which $\varTheta$ is the index of speciality. It follows that *if the characteristic system of $\{F\}$ on F_0 is complete (and hence coincides with $|C' + 3C|_H$), then the characteristic series of $\{\varDelta\}$ on $\varDelta_0$ is also complete.*

We prove that conversely, *if the characteristic series of $\{\varDelta\}$ on $\varDelta_0$ is complete, then also the characteristic system of $\{F\}$ on F_0 is complete.* From the hypothesis it follows that the characteristic system on F_0 cuts out on D_0 a complete series. In order to prove that the characteristic system on F_0 coincides with the complete system $|C' + 3C|_H$, it is therefore sufficient to prove that the residual systems of these two systems with respect to D_0 coincide. In other words, we have to prove that *the characteristic system on F_0 contains every reducible curve $D_0 + C_0$,* *where C_0 is any curve in* $|C|$. If C_0 is a plane section of F_0 the statement is immediate: the curve $D_0 + C_0$ is the section of F_0 with the surface F obtained by applying to F_0 an infinitesimal homology whose center and plane of invariant points coincide with the point O and with the plane of C_0 respectively. If C_0 is not a plane section of F_0, we consider the linear system Σ, ∞^4, determined by C_0 and by the web of plane sections of F_0, and we transform F_0 birationally by means of this system into a surface F_0^* in an S_4 containing the space S_3 of F_0. The surface F_0 is the projection of F_0^* from a certain point P_0 outside F_0^*, and to the hyperplane sections of F_0^* there correspond on F_0 the curves of Σ. We may assume that in these correspondence *the curve C_0 corresponds to and hence coincides with the section of F_0^* by the space S_3 of F_0*. Let p be the line joining the point O to the center of projection P_0. Projecting F_0^* on to the S_3 from a variable point P of p, we obtain in S_3 a continuous system, ∞^1, of surfaces F contained in $\{F\}$, and it is immediately seen that as P approaches P_0, the corresponding surface F, infinitely near F_0, cuts out on F_0 the curve $D_0 + C_0$, q.e.d.

We conclude that the assertion of the completeness of the characteristic system of a complete continuous system of surfaces with ordinary singularities in S_3 is equivalent to the assertion that the characteristic series of a continuous system of plane curves with nodes and cusps is complete. At present it is possible to prove the completeness of this series only if it is *non-special* (VIII, 5). However, in the case of the system $\{\Delta\}$, with which we were dealing, the *characteristic series is certainly special*. In fact, the index of speciality of this series equals Θ, and we have $\Theta \geq 2p_g - p_a - 1$, if $p_g > p_a$ (B. SEGRE), and $\Theta \geq p$, if $p_g = p_a = p > 3$, i.e. $\Theta > 0$ in all but a few insignificant cases.

Appendix to Chapter V

By David Mumford

Re: § 2. Instead of defining various types of strict algebraic equivalence, and of complete algebraic families of divisors, as in the text, it has proved more fruitful to construct, for each degree d, one all-inclusive family of all positive divisors of that degree, and to state results directly about the structure of that "universal" parameter space. Chow and van der Waerden introduced the Chow variety (more precisely, since it may be reducible, it should be called the Chow scheme) to parametrize cycles in P^n of fixed dimension and degree (Samuel, 3; Chow-van der Waerden, 1). In particular, if X is a normal variety, this constructs families of W-divisors[1], one for each degree, which are "universal" to a certain extent, although apparently this family is not actually characterized by a universal mapping property. Grothendieck (8) has constructed flat families of subschemes of an arbitrary projective scheme X, that are literally universal — any other flat family of subschemes of X, over any parameter scheme S, is induced from the given one by a unique morphism from S to his parameter space, which he calls the Hilbert scheme of X. (Cf. also Mumford, 5, lecture 15.) The Hilbert scheme has an open subset Div which is a universal parameter space for C-divisors on X.

One then says that two C-divisors D_1 and D_2 are *algebraically equivalent* if for some C-divisor E, $D_1 + E$ and $D_2 + E$ are positive and define points P_1, P_2 in Div which are in the same topological component of Div.

Re: § 4. The construction by purely algebraic means of the Picard variety has been one of the principal efforts in the 1950's. First, Weil constructed the Picard variety (or Jacobian variety) for curves (4; a much neater construction may be found in Chow, 4). The Picard variety in general has been treated by Matsusaka (3), (4), Lang (3), Chevalley (4), Seshadri (1), Grothendieck (9), Murre (4), Mumford (5), Artin (4). If X is any complete, or normal semi-complete variety, then one considers the group of C-divisors D on X algebraically equivalent to 0 modulo those linearly equivalent to 0. Then this group is always naturally isomorphic to the underlying point set of an algebraic group P, called the Picard variety of X. Moreover, if X is normal, then P is an

---

[1] W-divisors on a normal variety X are cycles of codimension 1 on X; C-divisors on any scheme X are Cartier divisors, i. e. elements of $\Gamma(X, K_X^*/\mathcal{O}_X^*)$, where $K_X^* = $ sheaf of units in the total quotient ring of $\mathcal{O}_X$.

abelian variety, i. e., it is complete. P can be characterized by universal mapping properties with respect to families of "C-divisors mod linear equivalence" over *reduced* parameter spaces. In case X is complete, then GROTHENDIECK has shown how to give P the structure of a group scheme which may not be reduced, so that the Universal mapping property holds for parameter spaces which may not be reduced.

Re: § 3. Now assume that X is a non-singular projective variety. If D is a positive C-divisor on X, then the "virtual characteristic series" is just the set of C-divisors on D cut out by sections:

$$s \in \Gamma(D, N_D)$$

where N_D is the normal sheaf to D in X, i. e., the dual of I/I^2, where $I = \mathcal{O}_X(-D)$, the sheaf of functions on X, zero on D. Given any positive C-divisor $\mathfrak{D} \subset X \times T$ such that $D = \mathfrak{D} \cdot X \times \{t\}$, there is a natural map:

$$\varrho : \left\{ \begin{array}{l} \text{Tangent space} \\ \text{to } T \text{ at } t \end{array} \right\} \to \Gamma(D, N_D)$$

(cf. KODAIRA-SPENCER, 3; MUMFORD, 5).

The image of ϱ is the "characteristic series" of the continuous system $\mathfrak{D}$ of divisors on X. To say that the characteristic series is complete is to assert that ϱ is surjective. But one warning: if T is allowed to be a scheme with only one underlying point, but non-zero nilpotent functions, then the surjectivity of ϱ is not very significant. The deep theorem, referred to by ZARISKI as the *Fundamental result*, should be interpreted to mean if D is sufficiently ample, then there exists a family $\mathfrak{D}$ over a *non-singular* parameter space T such that ϱ is surjective.

The irregularity q of a surface X, in cohomological terms, is nothing but $\dim H^1(X, \mathcal{O}_X)$. The second form of the fundamental result is that $q = \dim P$, where $P = $ the PICARD variety of X. (For the equivalence of these results, cf. MUMFORD; 5).

The attempted algebro-geometric proofs of the fundamental result by ENRIQUES and SEVERI have apparently never been completed. A very basic discovery is that the fundamental result is sometimes false in char p. IGUSA (6), SERRE (4).

A new and simple analytic proof in char O was given by KODAIRA and SPENCER (3). They showed that there is a non-singular T for which ϱ is surjective, whenever D is "semi-regular", meaning that the map $H^1(\mathcal{O}_X(D)) \to H^1(N_D)$ induced from the exact sequence:

$$0 \to \mathcal{O}_X \to \mathcal{O}_X(D) \to N_D \to 0$$

is O (note that the assumption that D is generic in the family $\mathfrak{D}/T$ has been eliminated here). Using a result of CARTIER, the first purely algebraic proof of the fundamental result in char O was given by GROTHENDIECK (9). This is based on showing that there is always a "PICARD" group

scheme P^*, representing the functor $S \to H^1(\mathcal{O}^*_{X \times S})/H^1(\mathcal{O}^*_S)$, such that
(i) Zariski tangent space to P^* at e is $H^1(\mathcal{O}_X)$, and (ii) P^*_{red} = usual
Picard variety. But Cartier showed that in char O all group schemes
are reduced, so $\dim(\text{Zar. tang. sp. } P^*) = \dim P^*_{\text{red}}$. An analysis of the
necessary and sufficient conditions for its validity in char p was then
given by Mumford (5): it turns out to be true if and only if certain
Bockstein operations

$$\beta_1\colon\ H^1(X, \mathcal{O}_X) \to H^2(X, \mathcal{O}_X)\,,$$

$$\beta_i\colon\ \mathrm{Ker}(\beta_{i-1}) \to \mathrm{Coker}(\beta_{i-1})\,, \quad i \geqq 2$$

are all zero. It seems clear that this is one instance where the higher-
order infinitesimal techniques made available by working with general
schemes instead of with varieties were exactly the right tools for solving
a classical problem.

Re: § 5. Weil (5) made a thorough study of the criteria of equivalence, strenghtening the results of § 5 very substantially. In fact, if
$X \subset \mathbf{P}^n$ is a non-singular surface, and $C \subset X$ is a hyperplane section of X,
generic over the field of definition of X: then in almost all cases, if D
is a divisor on X

$$D \cdot C \equiv 0 \quad \text{on} \quad C \Rightarrow D \equiv 0 \quad \text{on} \quad X\,.$$

The exception is when X itself is a ruled surface, and in this case the
result still holds if we assume that some multiple of D is algebraically
equivalent to 0. But given such an assumption on D, much stronger
results can be proven:

$$\left.\begin{cases} \text{Let } C \text{ be an irreducible} \\ \quad \text{curve on } X,\ (C^2) > 0. \\ \text{Let } D \text{ be a divisor on } X, \\ \quad mD \text{ alg. eq. to } 0 \\ D \cdot C \equiv 0 \text{ on } C. \end{cases}\right\} \Rightarrow \left.\begin{cases} D \equiv 0 \text{ on } X \text{ (char } O) \\ p^\nu \cdot D \equiv 0 \text{ on } X \text{ (char } p) \end{cases}\right\}$$

[*Proof.* Look at the kernel of the homomorphism $\mathrm{Pic}(X) \to \mathrm{Pic}(C)$.
If the assertion were false, this kernel would contain an element of
order l, $p \nmid l$. This would define an l-cyclic covering X' of X splitting
over C. This would give 2 disjoint curves C_1, C_2 in X' such that $(C_1^2) > 0$,
contradicting Hodge's Index Theorem.]

In treating many questions, it has proved very useful to introduce,
by the side of linear and algebraic equivalence, a third concept of equivalence between divisors: if X is a non-singular variety, 2 divisors D_1
and D_2 on X are numerically equivalent if for all irreducible curves C
on X, $(D_1 \cdot C) = (D_2 \cdot C)$[1]. A recent and systematic treatment of the
basic results on numerical equivalence is contained in Kleiman (1).

---

[1] $(D \cdot C)$ = degree of intersection of D and C.

Re: § 7. The first point is that the group of divisors numerically equivalent to 0, mod those algebraically equivalent to 0, is finite. This has been proven in all characteristics by MATSUSAKA (8), following SEVERI's method (cf. § 7). The proof on a surface X is as follows: by the theorem of RIEMANN-ROCH and duality on X, for all divisors D,

$$(*) \quad \dim H^0(\mathscr{O}_X(D)) + \dim H^0(\mathscr{O}_X(K - D)) \geqq \frac{(D \cdot D - K)}{2} + p_a(X) + 1.$$

Now choose a hypersurface section H of X of degree high enough so that

    a) degree $(H) >$ degree (K).

    b) $\dfrac{(H \cdot H - K)}{2} + p_a(X) + 1 > 0$.

Then if H' is any divisor on X numerically equivalent to H, it follows from (a) that degree $(K - H') < 0$, hence $K - H'$ is not linearly equivalent to any effective divisor, hence $\dim H^0(\mathscr{O}_X(K - H')) = 0$. Therefore by $(*)$ and (b), it follows that

$$\dim H^0(\mathscr{O}_X(H')) > 0.$$

In other words, every divisor numerically equivalent to H is linearly equivalent to an effective divisor. But the family of all effective divisors of the same degree as H is "of finite type", i. e., is parametrized by a finite set of components of the CHOW variety of cycles on X. It follows that every divisor D on X which is numerically equivalent to 0 is linearly equivalent to a difference $H' - H$, where H is fixed as above, and H' is taken from one of a finite set of algebraic families of divisors.

Re: § 6. The second point is that the group of *all* divisors, mod those numerically equivalent to 0, is a finitely generated free abelian group. This is much deeper, and all known proofs valid in all characteristics use arithmetic and GALOIS-theoretic considerations over function fields. The result was first established in char p by NERON (1); his proof was simplified in LANG-NERON (1). The most natural way to prove the theorem is to map the intersection *ring* of (algebraic) cycles on the nonsingular variety X to some cohomology ring with coefficients in a char O field k; then the rank ϱ of the group {divisors/numerical equivalence} is bounded by the dimension over k of the cohomology group into which the divisors are mapped:

Method 1. SEVERI's original method, using PICARD's Theorem on differentials of the 3rd kind, is essentially equivalent to mapping

$$\{\text{divisors}\} \to H^1(X, \Omega^1_X)$$

by taking the divisor D with local equations $f_i = 0$ to the ČECH 1-co-cycle $\{df_i/f_i - df_j/f_j\}$. Cup product makes the vector space

$$\sum_{i,j=0}^{\dim X} H^i(X, \Omega^j_X)$$

into a cohomology ring, and the above map can be extended so as to map cycles of codimension r to $H^r(X, \Omega_X^r)$. This cohomology theory has coefficients in the ground field k: so when the char is 0, we get the bound:

$$\varrho \leqq \dim H^1(X, \Omega_X^1) \, .$$

This bound is false in char$p$ (MUMFORD, 3). For a generalization of theorem 2, § 6, to arbitrary KÄHLER manifolds, cf. WEIL (2).

Method 2. LEFSCHETZ's method, using ordinary simplicial cohomology groups when $k = \mathbb{C}$, is based on mapping

$$\left\{ \begin{array}{c} \text{group of algebraic} \\ \text{cycles of} \\ \text{codimension } r \end{array} \right\} \to H^{2r}(X, \mathbf{Q}) \, .$$

In particular, we find that

$$\varrho \leqq B_2 \, ,$$

where $B_2 = $ the 2nd BETTI number of X
$= \dim H^2(X, \mathbf{Q})$.

Method 3. In char$p$, all known methods depend on either GALOIS cohomology, or on the stronger étale cohomology. NÉRON's original proof (1) has been successively reexamined and more deeply understood as this cohomology theory has progressed. After NÉRON, first IGUSA (9) applied GROTHENDIECK's results on π_1 (cf. Chapter 8) to refine NÉRON's result $\varrho < +\infty$ to

$$\varrho \leqq \deg(c_2) + 4q - 2$$

when X is a surface and $c_2 = $ 2nd CHERN class of tangent bundle to X; $q = \dim$ of PICARD variety of X. 2nd, OGG (1) and ŠAFAREVICH (1) gave more general formulations of IGUSA's techniques to arbitrary pencils of abelian varieties. 3rd ARTIN and GROTHENDIECK developed the étale cohomology groups $H^i(X, \mathbf{Q}_l)$ ARTIN (1), GROTHENDIECK (6), 1963–64 and 1965–66 and proved that they are finite dimensional. They proved further that

$$\sum_{i=0}^{2 \dim X} H^i(X, \mathbf{Q}_l)$$

is a ring and that each r-dimensional cycle has a fundamental class in H^{2r}. Then as in methods 1 and 2, one deduces:

$$\varrho \leqq \dim H^2(X, \mathbf{Q}_l) \, .$$

IGUSA's bound is reestablished when X is a surface by proving:

$$H^0(X, \mathbf{Q}_l) \cong H^4(X, \mathbf{Q}_l) \cong \mathbf{Q}_l \, ,$$
$$\dim H^1(X, \mathbf{Q}_l) = \dim H^3(X, \mathbf{Q}_l) = q \, ,$$
$$\sum (-1)^i \dim H^i(X, \mathbf{Q}_l) = \deg(c_2) \, .$$

A re-examination and strengthening of OGG's and ŠAFAREVICH's results using étale cohomology is found in RAYNAUD (1).

An outline of the essential idea behind Method (3.) is this:

(A) In view of the fact that $\left\{\dfrac{\text{divisors on } X}{\text{numerical equivalence}}\right\}$ admits a non-degenerate pairing into $\mathbb{Z}$, it is easy to check that this group is finitely generated if $\text{Pic}(X)/l \cdot \text{Pic}(X)$ is finite for some prime l [here $\text{Pic}(X)$ = divisors on X/linear equiv.].

(B) Blow up the base points of a generic pencil on X to obtain a surface X' and a proper morphism $f : X' \to \mathbf{P}^1$ such that i) there is a curve $E \subset X'$ mapped isomorphically to $\mathbf{P}^1$ by p, and ii) the curve $C_x = f^{-1}(x)$ is non-singular irreducible multiplicity 1 if $x \in U = \mathbf{P}^1 - \{x_1, \ldots, x_n\}$.

(C) If $J_x = $ the jacobian of C_x $(x \in U)$, then these abelian varieties fit together into a nice family $g : J \to U$ (a so-called abelian scheme). If $\Gamma(U, J)$ is the group of sections of J over U, define:

$$\text{Pic}(X) \xrightarrow{\ \phi\ } \Gamma(U, J)$$

as follows: if D is a divisor on X, let D' be its pullback on X', let $d = (D' \cdot C_x)$, and let A_x be the divisor $\text{Tr}_{C_x}(D' - dE)$ on C_x of degree 0 If $\alpha_x \in J_x$ is the point representing A_x, then set

$$\phi(D) = [\text{the morphism } x \mapsto \alpha_x] \,.$$

It is easy to check that the kernel and cokernel of ϕ are finitely generated abelian groups. Therefore, it is enough to prove $\Gamma(U, J)/l \cdot \Gamma(U, J)$ is finite.

(D) Let l be a prime, $l \neq \text{char}(k)$. Let J_l be the set of points of J which are points of order l or 1 in their fibres J_x over U. J_l is a finite union of non-singular curves, étale over U, and has the structure of a group scheme finite over U, i. e., there is a group law $(J_l) \times_U (J_l) \to J_l$ for adding any 2 points with the same image in U. A *principal homogeneous space* $p : K \to U$ over J_l is another finite union of non-singular curves, étale over U, on which J_l acts, i. e., with a morphism $(J_l) \times_U K \to K$, such that for any 2 points k_1, k_2 of K with the same image u in U there is a unique point of J_l over u taking k_1 to k_2. It is a standard fact that the set of all such principal homogeneous spaces over J_l forms an abelian group: $H^1(U, J_l)$, and that there is a canonical injection

$$\Gamma(U, J)/l \cdot \Gamma(U, J) \to H^1(U, J_l).$$

(E) Let $U' \to U$ be an étale GALOIS covering such that if V is a component of J_l, there is a diagram $U' \to V \to U$. Then the induced group scheme $J_l \times_U U'$ over U' "splits", i. e., is isomorphic to the product of $(\mathbb{Z}/l\,\mathbb{Z})^{2g}$, regarded as 0-dimensional algebraic group over k, and U'. Therefore, if $K \to U$ is any principal homogeneous space, $K \times_U U'$ is an étale GALOIS covering of U' with GALOIS groups $(\mathbb{Z}/l\,\mathbb{Z})^{2g}$. Now use the basic fact:

Lemma. *For any variety X over k, there is a maximal connected étale* GALOIS *covering $p: X' \to X$ with* GALOIS *group π abelian and $l \cdot \pi = (0)$.*

In particular, if $U'' \to U'$ is the maximal such covering of U', then if V is any component of any K, there is a diagram $U'' \to V \to U$. This shows that every principal homogeneous space K is a union of several of the finite set of curves V which factorize $U'' \to U$. From this it follows easily that there are only a finite number of such K's, i. e., $H^1(U, J_l)$ is finite.

The remark in ZARISKI's text that "the evaluation of ϱ for a given surface presents, in general, grave difficulties" is still very valid. The problem is to find a criterion for a cohomology class to be represented by a divisor. Some remarkable conjectures about this have been formulated in TATE (1) and partially proven in TATE (3): cf. Appendix to Chapter 7. NOETHER's theorem that divisors on a generic surface of degree d in $\mathbf{P}^3$ $(d \geqq 4)$ are cut out by other surfaces in $\mathbf{P}^3$ is proven in all characteristics by DELIGNE and GROTHENDIECK (GROTHENDIECK, 8, 1968).

When X is a surface, intersection of divisors defines a non-degenerate integral-valued quadratic form on the group of divisors mod numerical equivalence. HODGE (5) proved that this form has signature $(1, \varrho - 1)$. Equivalently, this means that if D is a divisor of degree 0 on a non-singular projective surface X, then $(D^2) \leqq 0$, and $(D^2) = 0$ holds only if D is numerically equivalent to 0. HODGE's method of proof is transcendental, but by far the simplest proof was given by GROTHENDIECK (2), using an extension of SEVERI's arguments (cf. Proof of criterion 4, §5 above).

Re: § 8. The problem of moduli of algebraic surfaces has been studied a great deal but it is still far from complete. To begin with, it was not even clear until recently what precisely a moduli space or the number of moduli were. In fact, in making these concepts precise, 2 important distinctions were discovered: (a) moduli of polarized vs. unpolarized varieties; (b) local vs. global moduli.

The concept of a polarized variety was introduced by MATSUSAKA (9). It is a pair $(X, \mathfrak{P})$ where X is a non-singular projective variety and for some ample divisor H on X, $\mathfrak{P}$ is the set of all divisors D on X such that $n D$ is equivalent to $m H$, where $n, m \geqq 1$ and if $\text{char}(k) = p > 0$ then $p \nmid n \cdot m$. This is an essential concept and one of the reasons that the Italian theory of moduli is so confusing is that they never made this concept explicit and distinguished which results applied to moduli of polarized varieties and which to unpolarized. Thus the theory in § 8 of ZARISKI's book basically is an analysis of the moduli of polarized varieties: if you have a class of pairs $(X, \mathfrak{P})$ with similar numerical characteristics, you can choose very ample divisors D simultaneously in all $\mathfrak{P}$'s and then map all the X's simultaneously into $\mathbf{P}^3$ and analyze the space of surfaces in $\mathbf{P}^3$ that you get. And conversely, when you have a family of surfaces in $\mathbb{P}^3$ with ordinary singularities with the same numerical characters, you can simultaneously normalize all of them, and the inverse image of their plane sections in $\mathbb{P}^3$ defines a polarization on each of them.

The idea of global moduli is the naive one: describe by some qualitative properties a big set $\mathfrak{M}$ of varieties or of polarized varieties, and then construct a variety M whose points are in a natural $1 - 1$ correspondence with the isomorphism classes of varieties in $\mathfrak{M}$. It is not hard to see what natural should mean (cf. MUMFORD, 4, p. 99). But what happens is that in *many* cases M does not exist and in *almost all* cases where M might exist, it is still not proven that M exists. More precisely, before M has a reasonable chance of existing, you should (i) define $\mathfrak{M}$ so that all varieties $X \in \mathfrak{M}$ have the same number of global vector fields, i. e., $\dim H^0(X, T_X)$ should be fixed, and (ii) either the varieties $X \in \mathfrak{M}$ should for some reason have natural projective embeddings (e. g., the canonical class K_X ample, or X a $\mathbb{P}^n$-bundle over a curve of genus $\neq 1$) or you should use instead a set $\mathfrak{M}$ of polarized varieties rather than varieties alone. (Cf. KODAIRA-SPENCER, 2, p. 413, for instance.) Even making these restrictions, the resulting M is known to exist as a variety only in the cases X a curve, X a $\mathbb{P}^n$-bundle over a curve, or X an abelian variety, or cases immediately reducible to one of these (BAILY, 2; MUMFORD, 4; SESHADRI, 2). However, in quite general cases M does exist as an analytic space (NARASIMHAN, SIMHA, 1) or as one of ARTIN's algebraic spaces (ARTIN, 3).

The difficulty in global moduli is that one is asking to do too much all at once. The problem breaks up quite naturally into local aspects and strictly global aspects. If one looks only at the local aspects, one can forget about the restrictions (i) and (ii) above and one can prove much more! The idea of a local moduli space is to fix a non-singular and complete variety X (or a polarized variety $(X, \mathfrak{P})$) and study all its "small" deformations. This was the basic idea of KODAIRA and SPENCER (2). They took an analytic point of view and posed the problem of constructing a universal analytic deformation of X: i. e., a proper smooth holomorphic map

$$p : \mathfrak{X} \to S$$

of analytic spaces, plus an isomorphism of X with the fibre $p^{-1}(s_0)$ over a fixed base point $s_0 \in S$, but where you are allowed to replace S by any open neighborhood $S_1 \subset S$ of s_0 (in other words, you look only at the *germ* of p, $\mathfrak{X}$ and S around the analytic subspace $X = p^{-1}(s_0) \subset \mathfrak{X}$). Their work was refined by KURANISHI (1) and DOUADY (1). It turns out that such a universal deformation *always* exists if universal is taken to mean: (i) every other deformation is induced from $p : \mathfrak{X} \to S$, and (ii) KODAIRA-SPENCER's map $\varrho : T_{s_0, S} \to H^1(X, T_X)$, T_X = sheaf of vector fields on X, is bijective. However, S must, in general, be singular and even non-reduced, in which case $T_{s_0, S}$ is the ZARISKI tangent space to S at s_0. Simultaneously, GROTHENDIECK (7) created a formal power series analog of this theory, valid in all characteristics. Here p is a proper, smooth morphism of formal schemes and $S = \mathrm{Spec}\,(A)$, where A is a complete local ring with residue field k, the ground field. Again a formal deformation in this sense, universal in senses (i) and (ii) *always* turns out to exist (cf. SCHLESSINGER, 1, 2).

Now here is what happens if you consider simultaneously the deformations of a variety X and all the associated polarized varieties $(X, \mathfrak{P}_\alpha)$. Notice that there are only a countable set of α's. Say the ground field is $\mathbb{C}$ first. Then let

$$p : \mathfrak{X} \to S$$

be KODAIRA-SPENCER's universal deformation. For each α, there will be a maximal closed analytic subspace $S_\alpha \subset S$ such that the polarization $\mathfrak{P}_\alpha$ lifts (necessarily uniquely) to $\mathfrak{X}_\alpha = p^{-1}(S_\alpha)$ and $\mathfrak{X}_\alpha \to S_\alpha$ will be the universal deformation of the pair $(X, \mathfrak{P}_\alpha)$. If $X_t = p^{-1}(t)$, $t \in S$, then if $t \in S - \bigcup_\alpha S_\alpha$, X_t will not, in general, be a variety at all but merely a compact analytic manifold. For instance, if X is an abelian variety of dimension n, $\dim S = n^2$ while $\dim S_\alpha = n(n+1)/2$ for all α. All the X_t's are complex tori, but they are abelian varieties only if $t \in S_\alpha$, some α. Or, if X is a KUMMER surface (i. e., X has first BETTI number 0 and $K_X \equiv 0$), then $\dim S = 20$ but $\dim S_\alpha = 19$ for all α. Exactly the same phenomenon happens with GROTHENDIECK's universal formal deformation over any ground field. Again we get closed subschemes $S_\alpha \subset S$. Now if $S_\alpha \subsetneq S$ for all α, then usually $\mathfrak{X}$ is a non-algebraizable formal scheme; but $\mathfrak{X}_\alpha = p^{-1}(S_\alpha)$ is algebraizable to a scheme $\mathfrak{X}'_\alpha$, projective over S_α.

Concerning the dimension of S and S_α, the following is known in general: for each α, form the "ATIYAH extension":

(*) $0 \to \mathcal{O}_X \to E_\alpha \to T_X \to 0$

where E_α is a locally free sheaf of rank $= \dim X + 1$, and the extension is given by:

fundamental class $(H) \in H^1(X, \Omega^1_X) \cong \mathrm{Ext}^1_{\mathcal{O}_X}(T_X, \mathcal{O}_X)$, $H \in \mathfrak{P}_\alpha$.

(Up to isomorphism, this extension is independent of the choice of H in $\mathfrak{P}_\alpha$.) Then

(**)   ZARISKI tang. sp. to $S \cong H^1(X, T_X)$,

ZARISKI tang. sp. to $S_\alpha \cong \mathrm{Im}\,[H^1(X, E_\alpha) \to H^1(X, T_X)]$.

Moreover, if $h^{(i)} = \dim H^i(T_X)$, S lies locally in $h^{(1)}$-dimensional affine space and is defined by *at most* $h^{(2)}$ equations. (KURANISHI, 1; in GROTHENDIECK's theory this has been proven by MUMFORD.) And if $h^{0,2} = \dim H^2(\mathcal{O}_X)$, then S_α, in S, is defined by at most $h^{0,2}$ additional equations. Finally, in case of surfaces, the RIEMANN-ROCH formula tells us that if $p_a = \chi(\mathcal{O}_X) - 1$ as usual, then:

(***) $\dim H^0(T_X) - \dim H^1(T_X) + \dim H^2(T_X) = 2(K^2) - 10(p_a + 1)$.

Now let's tie up these results with those in § 8 and see to what extent they have been recaptured and clarified by modern work. We assume that $\operatorname{char}(k) = 0$. What is really being done in § 8 is to describe all 1st-*order* deformations of a pair $(F_0, \mathfrak{P})$, where F_0 is a non-singular complete surface, and $\mathfrak{P}$ is the polarization defined by a regular very ample curve C. Let $f_0 = 0$ be the equation of a generic projection of F_0, embedded by $|C|$, into $\mathbb{P}^3$. Then all 1st order deformations of $(F_0, \mathfrak{P})$ are obtained by normalizing the surface $f_0 + \in f_1 = 0$ in $\mathbb{P}^3$, where $f_1 = 0$ is an adjoint surface to $f_0 = 0$ of the same degree, "passing through the double points infinitely near the cuspidal points". Thus the number M which ENRIQUES computes is nothing else but the dimension of the ZARISKI tangent space to S_α. Now it can be checked that the invariants introduced in the text have the following values:

$$\varrho = \dim H^0(T_F) \, ,$$
$$\theta = \dim H^2(E_\alpha)$$

[and $p^{(1)} = (K^2) + 1$, $p_g = \dim H^2(\mathcal{O}_F)$ as usual] ;

hence formula (12) reads

$$M = 10(p_a + 1) - 2(K^2) - \dim H^2(\mathcal{O}_X) + \dim H^0(T_F) + \dim H^2(E_\alpha)$$

and using the exact cohomology sequence of (*) and formula (***) this comes out as:

$$M = \dim [\operatorname{Im} H^1(E_\alpha) \to H^1(T_F)]$$

which is result (**). Moreover, SEGRE's result $\theta \geq 2p_g - p_a - 1$ if F is irregular and $p_g > 0$ can be proved by observing:

$$\begin{aligned}
\theta &= \dim H^2(E_\alpha) \\
&\geq \dim H^2(T_F) \\
&= \dim H^0(\Omega_F^1 \otimes \Omega_F^2) \text{ by SERRE duality} \\
&\geq \dim H^0(\Omega_F^1) + \dim H^0(\Omega_F^2) - 1 \\
&= 2p_g - p_a - 1
\end{aligned}$$

in view of the easy lemma:

Lemma. Let E, F be 2 locally free sheaves on a non-singular variety X. Then if $\Gamma(E)$, $\Gamma(F) \neq (0)$,

$$\dim \Gamma(E \otimes F) \geq \dim \Gamma(E) + \dim \Gamma(F) - 1 \, .$$

ENRIQUES's result $\theta \geq p_g$ applies actually in case K is ample on F *and* $\mathfrak{P}_\alpha$ is the polarization defined by K. Since the line bundle $\mathcal{O}_F(K) = \Omega_F^2$ lifts automatically to all deformations of F, it follows that in the cohomology sequence of (*)

$$d : H^1(T_X) \to H^2(\mathcal{O}_X)$$

is 0, hence

$$\theta = \dim H^2(E_\alpha)$$
$$\geqq \dim H^2(\mathcal{O}_X)$$
$$= p_g.$$

Thus the classical and modern theory tie up perfectly in regards to 1st order deformations. But as ZARISKI points out in § 3 and § 8 of Chapter 5 and § 5 of Chapter 8, the Italians never proved that 1st order deformations either of non-singular curves on an irregular surface or of surfaces in $\mathbb{P}^3$ with ordinary singularities, or of plane curves with nodes and cusps, could be realized as the 1st order terms of actual deformations, i. e., that in each case "the characteristic system is complete". In GROTHEN-DIECK's language, this is the question of whether the deformation problem is *unobstructed*, or whether the universal deformation space (our S_α) is *non-singular*. The invariant ω introduced by ZARISKI is, therefore:

$$\omega = \dim (\text{ZARISKI tang. sp. to } S_\alpha) - \dim S_\alpha.$$

We saw above that in the case of deformations of sufficiently ample curves on a surface, there are indeed no obstructions when $\operatorname{char}(k) = 0$, and the obstructions in $\operatorname{char} p$ can be analyzed. Unfortunately, the deformations of polarized surfaces are worse: even in $\operatorname{char} = 0$, obstructions *can exist*, and ω may be positive. MUMFORD (3) found an obstructed space curve γ, and combining this with the results in KODAIRA (11), p. 246, it follows that surfaces of high order with double curve γ are obstructed; moreover KAS (1) found that certain classes of elliptic surfaces are highly obstructed and presumably the same should apply to polarized elliptic surfaces. In fact, it appears very difficult to analyze the obstructions. The modern theory, however, has contributed at least an upper bound to the number of obstructions: since S_α is defined by $h^{0,2}$ equations in S, of which $\dim \operatorname{Im}[H^1(T_\lambda) \to H^2(\mathcal{O}_X)]$ have independent linear terms (since the ZARISKI tangent space to S_α lies in the subspace of $H^1(T_X)$ which is the image of $H^1(E_\alpha)$), it follows that:

$$\omega \leqq \dim (H^{\cdot}(\mathcal{O}_X)/\operatorname{Im} H^1(T_X)) + \dim (\text{Zar. tang. sp. to } S) - \dim S$$
$$\leqq \dim (\operatorname{Im} H^2(\mathcal{O}_X) \to H^2(E_\alpha)) + \dim H^2(T_X)$$
$$= \dim H^2(E_\alpha)$$
$$= \theta.$$

For a sheaf-theoretic treatment of the classical theory, and many interesting examples, see KODAIRA (11).

Chapter VI

Topological Properties of Algebraic Surfaces

1. Terminology and notations. In the following exposition the fundamental notions of Analysis Situs will be assumed known. The terms and notations are as in LEFSCHETZ, e. For the convenience of the reader we give a list of those most frequently used in the sequel.

E_n, *an oriented n-cell.*

K_n, *an oriented n-complex* (a set of distinct cells of dimensions $0, 1, \ldots, n$).

C_p, *a p-chain*; $C_p = \sum_{i=1}^{\alpha_p} \lambda_i E_p^i$, where the λ_i's are integers $\gtreqless 0$, and $E_p^1, E_p^2, \ldots, E_p^{\alpha_p}$ are the *p*-cells of K_n.

$F(C_p)$, *the oriented boundary of* C_p; in symbols: $C_p \to F(C_p)$.

Γ_p, *a p-cycle* $(\Gamma_p \to 0)$.

$\Gamma_p \sim 0$ $(\Gamma_p$ *is homologous to zero)* means that there exists a C_{p+1} such that $C_{p+1} \to \Gamma_p$.

$\Gamma_p \approx 0$ *(homology with division allowed)* means that $t \, \Gamma_p \sim 0$, where t is an integer. If $\Gamma_p \approx 0$ and $\Gamma_p \not\sim 0$, Γ_p is called a *divisor of zero.*

$C_p \cdot C_q$, the *intersection* of two chains C_p, C_q on K_n $(p + q \geqq n)$. If $q = n - p$, then $(C_p \cdot C_{n-p})$ is the KRONECKER *index*, or *the intersection number.*

Circuit: a complex K_n in which every $(n-1)$-cell is incident with an even number of *n*-cells and whose *n*-th Betti number mod 2 is 1.

M_n, *an n-dimensional manifold.*

2. An algebraic surface as a manifold M_4. The 4-dimensional real variety in $(1, 1)$ continuous correspondence (without exceptions) with the points of an algebraic surface F, free from singularities, is topologically determined and is called the *Riemannian variety of F*. We shall denote this variety by the same letter F. F, and more generally, the Riemannian variety of any algebraic variety can be triangulated and can therefore be covered by a complex (VAN DER WAERDEN, 2; VAN DER WAERDEN, 5, where the author points out and corrects an error in his preceding paper; for analytical manifolds, see LEFSCHETZ, e; KOOPMAN-BROWN, 1; LEFSCHETZ-WHITEHEAD, 1).

If F is free from singularities in S_r or with ordinary singularities in S_3, then it is possible to represent the complete neighborhood of any point P

of F by means of regular analytic functions of two complex parameters u and v, $x_i = f_i(u, v)$ $(i = 1, 2, \ldots, r)$, the Jacobian matrix of the r functions f_i being of rank 2 at the initial values u^0, v^0. Consequently the neighborhood of P is an E_4, and since F has no boundary, it follows that F *is an absolute* M_4.

If F possesses singularities, the singular locus L of F is two-dimensional at most. Hence the preceding consideration shows that in all cases F is a 4-circuit (LEFSCHETZ, e, Chapter 1), and $F - L$ is an open M_4. The point P being simple (i. e. not on L), and u, v, as above, let $u = u_1 + i u_2$, $v = v_1 + i v_2$. The 4 real variables taken in the order u_1, u_2, v_1, v_2 determine an orientation in the neighborhood of P. *This orientation is invariant under regular analytic transformations of the local coordinates u, v*. In fact, if $\bar{u} = \bar{u}_1 + i\bar{u}_2, \bar{v} = \bar{v}_1 + i\bar{v}_2$ are new coordinates at P, then

$$\frac{d(\bar{u}_1, \bar{u}_2, \bar{v}_1, \bar{v}_2)}{d(u_1, u_2, v_1, v_2)} = \left| \frac{d(\bar{u}, \bar{v})}{d(u, v)} \right|^2 > 0,$$

and the statement follows from the known fact that transformations with positive Jacobian preserve orientation. Thus, the analytic character of V_4 fixes at each point of V_4 an *intrinsic orientation*, and the intrinsic orientations at two distinct points of V_4 are analytical prolongations of each other. *Hence F (or $F - L$) is orientable.* In a similar manner it is possible to define the intrinsic orientation of any analytic manifold (LEFSCHETZ, e, Chapter 8; VAN DER WAERDEN, 2).

3. Algebraic cycles on F and their intersections (LEFSCHETZ, a, 9). Among the cycles Γ_p on F of particular interest are the *algebraic cycles Γ_2*, i. e. the cycles represented by or homologous to effective or virtual algebraic curves. If Γ_2 is algebraic, then $\Gamma_2 \sim \sum \lambda_i \Gamma_2^i - \sum \mu_j \bar{\Gamma}_2^j$, where the Γ_2's and the $\bar{\Gamma}_2$'s are effective algebraic curves traced on F. The algebraic cycles on F form a *modulus* (LEFSCHETZ, e, Chapter 1).

Given two effective algebraic curves Γ_2, Γ_2', the KRONECKER index $(\Gamma_2 \cdot \Gamma_2')$ is the algebraic sum of the intersection numbers (or of the indices) relative to their distinct common points. Let us suppose that Γ_2 and Γ_2' have at each common point a simple point with distinct tangents. Then the contribution of each intersection is ± 1. It is not difficult to show that the various indices are either all $+1$ or all -1, and that consequently, for proper orientations of Γ_2 and Γ_2', $(\Gamma_2 \cdot \Gamma_2')$ *is equal to the number of intersections of Γ_2 and Γ_2'*. The proof of this remarkable theorem (LEFSCHETZ, a, 5, 9) to the effect that *the algebraic number of intersections of two effective algebraic curves on F is equal to the arithmetic number of their intersections*, is extremely simple (LEFSCHETZ, 9). Let $u = u_1 + i u_2$, $v = v_1 + i v_2$ be the local coördinates at a common point P of the two curves, and let $\varphi(u, v) = \varphi_1(u_1, u_2, v_1, v_2) + i\varphi_2(u_1, u_2, v_1, v_2) = 0$ and $\psi(u, v) = \psi_1(u_1, u_2, v_1, v_2) + i\psi_2(u_1, u_2, v_1, v_2) = 0$ be the equations of Γ_2 and Γ_2' respectively. Since P is a simple intersection it follows that $\dfrac{d(\varphi, \psi)}{d(u, v)} \neq 0$ at P and that φ and ψ are analytically independent. We take φ and ψ as new local coördinates at P, φ as an analytic parameter on Γ_2' and ψ as an analytic parameter on Γ_2. The intrinsic orientations of F, Γ_2, Γ_2' are then given by the real coördinates $\varphi_1, \varphi_2, \psi_1, \psi_2$; ψ_1, ψ_2; φ_1, φ_2 respectively, in the orders written. The contribution of P to $(\Gamma_2 \cdot \Gamma_2')$ is seen to be $+1$, q. e. d.

The extension to the case of a multiple intersection P can be derived without difficulty by replacing Γ_2 and Γ_2' in the neighborhood of P by approximating complexes presenting only simple intersections (VAN DER WAERDEN, 2; LEFSCHETZ, d, Chapter VIII). It is not difficult to see that in our case we may choose a set of linear algebroid branches to represent these complexes.

Corollary. *If Γ_2 is an effective algebraic cycle* (i. e. $\sim$ to an effective algebraic curve), *then* $\Gamma_2 \not\approx 0$. In fact, if $\Gamma_2 \approx 0$, then $(\Gamma_2 \cdot \bar{\Gamma}_2) = 0$ for any cycle $\bar{\Gamma}_2$ on F, and this is impossible, since Γ_2 is met in a certain number of points by a hyperplane section of F.

4. The representation of F upon a multiple plane. We take as a projective model of F a surface

(1) $$f(x, y, z) = 0$$

of order n with ordinary singularities in S_3. We suppose that the polynomial f is of degree n in z (the point at infinity O_z on the z-axis is not on F). The surface F is mapped by (1) upon the n-tuply covered plane (x, y). The branch curve D of the n-valued function z is obtained by eliminating z between (1) and $\partial f / \partial z = 0$ and by neglecting in the resultant $R(x, y)$ the factor corresponding to the *apparent* branch curve, i. e. the projection of the double curve of F. We shall denote the order of D by m.

The lines of the space which are either bitangents of F or have with F at least a 3-point contact (stationary tangents) make up a finite number of line congruences (of dimension 2), and hence there is a finite number of such lines on a generic point of the space. Hence we may suppose that at all but a finite number of points of D two and only two values of z coincide, i. e. that D *is a simple branch curve*. Incidentally, the monodromy group of z, being primitive and containing transpositions, is the symmetric group of degree n.

The consideration of the surface dual to F shows that the generic tangent cone of F is irreducible. Hence we may suppose (although it is not essential) that D is irreducible. D necessarily possesses singularities. In general, i. e. if F is the generic projection of a surface free from singularities in S_r, $r \geqq 5$, D possesses only ordinary double points and cusps, corresponding to the bitangents and the stationary tangents of F on O_z. The number of cusps can be evaluated and shown to be always >0, if $n > 2$.[1] This also follows from a theorem on the fundamental group G of D with respect to its carrying complex projective plane to the effect that if D is irreducible and possesses only ordinary double points then G is cyclic (Zariski, 3; see VIII, 2).

5. The deformation of a variable plane section of F. Let p be the genus of a generic plane section of F and let $\{C_x\}$ denote the pencil of curves C_x cut out on F by the planes $x = \mathrm{const}$. We assume that: (1) the point at infinity on the y-axis is not on D; (2) the tangents of D parallel to the y-axis are all simple tangents; (3) none of the principal tangents at any multiple point of D is parallel to the y-axis. Let $x = \alpha_i$ be the tangents of D parallel to the y-axis, $i = 1, 2, \ldots, \mu$, where μ is the class of D, and let $\beta_1, \beta_2, \ldots$ be the abscissas of the multiple points of D. If

(2) $$\varphi(x, y) = 0$$

is the equation of D, we shall have then that φ is of degree m in y and that $x = \alpha_i$ is a simple branch point of the function y defined by (2). The branch points of the function z of y, defined by the equation of a $C_{\bar{x}}$,

$$f(\bar{x}, y, z) = 0,$$

are the roots $y_1, y_2, \ldots, y_m$ of $\varphi(\bar{x}, y) = 0$. For $x = \alpha_i$ two and only two branch points y_i coincide and C_{α_i} is of genus $p - 1$, having acquired a new double point O_i (at the point of contact of F with its tangent plane $x = \alpha_i$). On the contrary it is immediately seen that the curves C_{β_j} are of the same genus p as the generic plane section of F, provided that the singular tangent at each cuspidal point of F is not parallel to the z-axis.

We fix a generic curve C_a in the pencil $\{C_x\}$ and we cut the plane of the complex variable x open into a 2-cell E_2 along a set of simple arcs l_i, l_j', joining a to the points α_i and β_j and having only the point a in common. We suppose that as we turn about a in the counterclockwise sense we first encounter the arcs $l_1, l_2, \ldots, l_\mu$ in the order written and then the arcs l_j'. As x varies on E_2, the boundary included (but excluding the vertices α_i), C_x preserves its genus and therefore remains constantly homeomorphic to C_a. We now suppose that we have fixed a homeomorphism between C_a and any C_x so as to have a definite isotopic deformation of C_a into C_x. This will allow us to follow continuously and in a definite manner the variation of any locus on C_x, in particular of any cycle of C_x, as x varies on the closed E_2 (excluding the vertices α_i). Let R_x be the n-sheeted Riemann surface of C_x.

---

[1] The number of cusps of D is $r + \pi - 1 + m/2$, where π is the genus of D and r is the number of cuspidal points of F; $r = 2\mu(n - \mu - 1) + 4d + 6t$, where μ is the order of the double curve, d is the number of its apparent double points and t is the number of its triple points (Enriques-Chisini, a₂. See also Severi, 22, for a functional interpretation of the above relation between the number of cusps of D and the number of cuspidal points of F).

We fix a typical n-sheeted RIEMANN surface R_0 with branch points $t_1, t_2, \ldots, t_m$, homeomorphic to R_a and similarly connected and we denote by T_x the homeomorphism between R_0 and R_x which is the product of the homeomorphisms between R_0 and R_a and between R_a and R_x. If P_0 is any point on R_0 the point $T_x(P_0)$ varies continuously as x varies. There is however a discontinuity along the cuts l_i, l_j'. If x is on one of these cuts it corresponds to *two* points on the boundary of E_2 and consequently there are two distinct homeomorphisms between R_0 and R_x, say T_x and T_x'. The point $x = a$ corresponds to the $\mu + \mu'$ vertices of E_2 (μ' is the number of points β_j) and we have $\mu + \mu'$ distinct homeomorphisms between R_0 and R_a.

The homeomorphism T_x degenerates as $x \to \alpha_i$, since R_0 and R_{α_i} are not homeomorphic. We may, however, assume that T_x degenerates in the following manner. Let y_{i_1} and y_{i_2} be the two branch points of R_x which tend to coincide as $x \to \alpha_i$, and let t_{i_1}, t_{i_2} be the corresponding branch points of R_0. Then we may assume that T_{α_i} is $(1, 1)$ and continuous, except that to the double point O_i of C_{α_i} there corresponds on R_0 a certain cycle δ_i^0 relative to an open arc in the t-plane joining t_{i_1} and t_{i_2}, while to the points of R_{α_i} which overlap O_{α_i} on the other $n - 2$ sheets there correspond on R_0 $n - 2$ non-singular 1-cells having no points in common.

In a similar manner T_{β_j} will be $(1, 1)$ and continuous, except that to a certain number $n' < n$ of distinct overlapping points of R_{β_j} there will correspond on R_0 n' non-singular 1-cells having no points in common.

Given any chain C_p^0 on R_0 (assuming a certain subdivision of R_0 and hence also, in virtue of T_x, of R_x), we shall denote the corresponding chain $T_x(C_p)$ on R_x by C_p^x or simply by C_p. We observe that if C_p^0 is a one-cycle γ^0, then even without T_x the cycle γ is known to within a homotopy as a definite sum of loops g_i ($i = 1, 2, \ldots, m$), where g_i is a loop in the plane of the variable y issued from the point at infinity and surrounding the branch point y_i.

6. The vanishing cycles δ_i and the invariant cycles (PICARD, a$_1$, Chapter IV; LEFSCHETZ, a, 4, 5). We consider the cycles δ_i^0 ($i = 1, 2, \ldots, \mu$) introduced in the preceding section and the corresponding cycles $\delta_i = T_x(\delta_i^0)$ on C_x. As $x \to \alpha_i$ the cycle δ_i shrinks to the double point O_i of C_{α_i}. *Consequently* $\delta_i \sim 0$ *on* F. *On the contrary* $\delta_i \not\sim 0$ *on* C_x if $p > 0$, since δ_i is a cycle surrounding two branch points of R_x at which the same two sheets are interchanged, and it is well known that such a cycle is $\not\sim 0$ if R_x is of genus $p > 0$ (it is possible to find a cycle on R_x meeting δ_i in only one point).

If γ_0 is an arbitrary 1-cycle on R_0 and if x is on the cut l_i, we have on R_x two corresponding cycles $\gamma = T_x(\gamma_0)$ and $\gamma' = T_{x'}(\gamma_0)$. As x turns about α_i in the counterclockwise sense, the cycle γ is deformed into γ'. During the deformation the branch points y_{i_1} and y_{i_2} are interchanged and from this it is easily seen that *γ' differs from γ by a multiple of δ_i, $\gamma' \infty \gamma + k\delta_i$ on C_x* (PICARD), *and that the integer k is the Kronecker index $(\gamma \cdot \delta_i)$* (LEFSCHETZ). Hence

(3) $$\gamma' \infty \gamma + (\gamma \cdot \delta_i)\,\delta_i \text{ on } C_x.$$

In particular, δ_i is *invariant* (to within a homotopy) in the neighborhood of α_i, i. e. $\delta_i' \infty \delta_i$ on C_x. This follows immediately from the definition of δ_i and is in agreement with the well known fact that for any 1-cycle γ on R_x we have $(\gamma^2) = 0$.

If x is on the cut l_j', it is seen immediately that $\gamma' \infty \gamma$ on C_x. Hence *in regard to homologies on C_x there is no essential discontinuity along the cuts l_j'.*

Finally, on C_a there correspond to γ_0 μ cycles $\gamma = \gamma^a$, γ', γ'', ..., $\gamma^{(\mu-1)}$, where
(4) $$\gamma^{(i)} \infty \gamma^{(i-1)} + (\gamma^{(i-1)} \cdot \delta_i^a)\,\delta_i^a \text{ on } C_x.$$

A path in the x-plane which turns about all the branch points α_i leaves invariant each 1-cycle γ. Hence

(5) $$(\gamma \cdot \delta_1^a)\,\delta_1^a + (\gamma' \cdot \delta_2^a)\,\delta_2^a + \cdots + (\gamma^{(\mu-1)} \cdot \delta_\mu^a)\,\delta_\mu^a \infty 0 \text{ on } C_a,$$

for any cycle γ on C_a.

From the preceding considerations it follows that *a necessary and sufficient condition that a cycle γ on R_x be invariant in the whole x-plane is that all the* KRONECKER *indices $(\gamma \cdot \delta_i)$ $(i = 1, 2, \ldots, \mu)$ vanish.*

Given an arbitrary chain C_p^0 on R_0, the corresponding chain C_p on R_x generates, as x varies on the oriented E_2, a chain C_{p+2}. The boundary of C_{p+2} consists of the following chains: (1) the locus of C_p as x varies on the boundary of E_2; (2) the chain C_{p+1} generated by the boundary Γ_{p-1} of C_p as x varies on E_2. Let C_p and C_p' denote the chains which correspond to C_p^0 when x is on a cut l_i or l_j'. Then the part (1) of the boundary of C_{p+2} is a sum of chains $C_{p+1}^{(i)}$, $\overline{C}_{p+1}^{(j)}$, where $C_{p+1}^{(i)}$, for instance, is the locus of $C_p' - C_p$ as x describes the cut l_i from a to α_i.

Let us consider in particular the case of a 1-cycle $C_1 = \gamma$. Then the part (2) of the boundary of C_3 is not present. The part (1), the sum of the chains $C_2^{(i)}$, $\overline{C}_2^{(j)}$, is now a cycle Γ_2 on F. By a deformation it is possible to replace this cycle Γ_2 by a homologous cycle of simpler nature. Let in fact $C_{2,\,x}^{(i)}$ be the 2-chain on C_x which, in view of (3), is bounded by the cycle $\gamma' - \gamma - (\gamma\delta_i)\delta_i$, when x is on l_i. Similarly, let $\overline{C}_{2,\,x}^{(j)}$ be the chain on C_x bounded by $\gamma' - \gamma$, when x is on l_j'. Let moreover Δ_i denote the 2-chain generated by δ_i as x describes the cut l_i. The consideration of the boundary of the 3-chain generated by $C_{2,\,x}^{(i)}$ and $\overline{C}_{2,\,x}^{(j)}$, as x describes the cuts l_i and l_j' respectively, shows immediately that $\sum\limits_{i=1}^{\mu}(\gamma \cdot \delta_i)\Delta_i +$ a certain 2-chain on C_a is a 2-cycle homologous on F to the above cycle Γ_2. Indicating always by (C_a) a chain on C_a, we have therefore the following important homology:

$$(6) \qquad F(C_3) \sim \sum_{i=1}^{\mu} (\gamma \cdot \delta_i) \varDelta_i + (C_a) \text{ on } F,$$

where C_3 is the locus of γ as x varies on E_2.

Corollary. *If γ is an invariant cycle then C_3 is a cycle Γ_3.* In fact, the indices $(\gamma \cdot \delta_i)$ are then all zero and hence (6) becomes $F(C_3) \sim (C_a)$. Since C_a is irreducible, the only 2-cycles on C_a are the multiples of C_a. Hence $(C_a) = t C_a$ and since $C_a \not\sim 0$ on F (section 3, Corollary) it follows that $t = 0$, q. e. d.

7. The fundamental homologies for the 1-cycles on F. Let Γ_1 be an arbitrary 1-cycle on F, P a variable point on Γ_1 and C_x the curve of the pencil $\{C_x\}$ passing through P.[1] Let us first assume that Γ_1 is a simple circuit. Then we may suppose, deforming Γ_1 if necessary, that it is on a base point A_1 of $\{C_x\}$ and that it touches there C_a. We join A_1 to P by an arbitrary simple arc h on C_x which varies continously as P describes Γ_1. Let h' and h'' be the initial and the final position respectively of the arc h aus P describes Γ_1, startig from the point A_1 and returning to it. Choosing properly the arc h, we may assume that h' reduces to the point A. Then the two-dimensional locus of h is bounded by $\Gamma_1 - h''$, and hence $\Gamma_1 \sim h''$ on F. Since any 1-cycle consists of a finite number of simple circuits, we have therefore the following theorem:

Any 1-cycle of F is homologous on F to a 1-cycle lying on a generic C_x (PICARD, a_1; LEFSCHETZ, a; SEVERI, 12).

It clearly appears from the above proof that in the last theorem the pencil $\{C_x\}$ may be replaced by an arbitrary irreducible linear pencil of curves, of positive degree. In this general form the theorem is due to SEVERI (12), whose proof is transcendental. We shall prove later (VII, 7, e) that *if the pencil is of genus π there exist exactly 2π independent cycles of F which are not homologous to cycles belonging to a generic curve of the pencil.*

Let $\gamma_1, \gamma_2, \ldots, \gamma_{2p}$ be a minimal base of $2p$ 1-cycles on C_x. By the preceding theorem they also form a base for the 1-cycles on F. As cycles on F they are connected by the μ homologies, $\delta_i \sim 0$ on F, where δ_i should be expressed in terms of the γ_i's.

The following theorems hold:

a. *Any homology between the $2p$ cycles γ_i is a consequence of the homologies $\delta_i \sim 0$, which therefore constitute a fundamental set of homologies for the 1-cycles of F. Consequently, if among the cycles δ_i there are $2p - r$*

---

[1] If Γ_1 is on a base point A_i of $\{C_x\}$ the curve C_x passing through A_i is defined as the limiting position of C_x as P $\to A_1$.

which are independent on C_x, *then* $R_1 = r$, *where* R_1 *is the one-dimensional* BETTI *number of* F.

    b. *There are exactly r invariant cycles independent on* C_x.

    c. *These r invariant cycles are independent also on* F.

    d. *r is even.*

The theorems a. and b. have been proved by PICARD (a$_1$, p. 96; a$_2$, p. 398) and LEFSCHETZ (a, Chapter II) by transcendental and topological methods respectively. The theorem c. is stated in PICARD (a$_2$, p. 397) under an equivalent form, but, contrary to common belief, is given without proof. The only existing proof of c. is therefore the topological one due to LEFSCHETZ (a, Chapter II). The brief transcendental proof given by LEFSCHETZ (a, p. 62) cannot be accepted because it is based on the unproved statement of PICARD (see VII, 2, p. 159). PICARD's proof of a. will be given in VII, 2.

In the topological treatment the derivation of theorem b. comes first and is immediate. Let $2p - r$ be the rank of the matrix $\|(\gamma_k \cdot \delta_i)\|$, $(k = 1, 2, \ldots, 2p, i = 1, 2, \ldots, \mu)$, and let, for instance, $|(\gamma_k \cdot \delta_i)|$, $k, i = 1, 2, \ldots, 2p - r$, be the $(2p - r)$-row minor of the matrix which is different from zero. It is well known that if a given cycle γ on a RIEMANN surface R is such that $(\gamma \cdot \gamma') = 0$ for any γ' on R, then $\gamma \sim 0$ on R. From this it follows that *the cycles* $\gamma_1, \gamma_2, \ldots, \gamma_{2p-r}$ and also *the cycles* $\delta_1, \delta_2, \ldots, \delta_{2p-r}$, *are independent on* C_x, and that *the cycles* $\delta_{2p-r+1}, \ldots, \delta_\mu$ *depend on the cycles* $\delta_1, \delta_2, \ldots, \delta_{2p-r}$. In order that a cycle $\gamma = \sum_{j=1}^{2p} t_j \gamma_j$ be invariant it is therefore necessary and sufficient that $(\gamma \cdot \delta_i) = 0$ for $i = 1, 2, \ldots, 2p - r$. This gives $2p - r$ *independent* equations $\sum_{j=1}^{2p} (\gamma_j \cdot \delta_i) t_j = 0$, $i = 1, 2, \ldots, 2p - r$, in the t_j's, admitting r independent solutions, q. e. d.

The starting and most difficult point of the proof of the theorems a., c., d. is the following

    Lemma. *If* $\gamma \sim \lambda_1 \delta_1 + \lambda_2 \delta_2 + \cdots + \lambda_{2p-r} \delta_{2p-r}$ *on* C_x *and if γ is an invariant cycle, then* $\gamma \sim 0$ *on* C_x (LEFSCHETZ, a, p. 25).

The proof of this Lemma is rather involved and consists in showing that the locus of the invariant cycle γ as C_x varies is a cycle Γ_3 homologous to zero on F. From $\Gamma_3 \sim 0$ on F it follows immediately that $\gamma \sim 0$ on C_x, since $\gamma = \Gamma_3 \cdot C_x$.

Let now $\gamma_1', \gamma_2', \ldots, \gamma_r'$ be the r independent invariant cycles of C_x (theorem b.). By the above Lemma the $2p$ cycles $\delta_1, \delta_2, \ldots, \delta_{2p-r}$, $\gamma_1', \ldots, \gamma_r'$ are independent on C_x. Hence the determinant of their intersection numbers does not vanish. Since each γ_i' is invariant, $(\gamma_i' \cdot \delta_j) = 0$ for any i and j, and the above determinant is the product of the two determinants $|(\delta_i \cdot \delta_j)|$, $|(\gamma'_i \cdot \gamma'_j)|$. Consequently

(7) $$|(\delta_i \cdot \delta_j)| \neq 0, \qquad i, j = 1, 2, \ldots, 2p - r,$$

(7') $$|(\gamma'_i \cdot \gamma'_j)| \neq 0, \qquad i, j = 1, 2, \ldots, r.$$

Since we are dealing with skew-symmetric determinants $[(\delta_i \cdot \delta_j) = -(\delta_j \cdot \delta_i)]$, it follows that r is even, and this proves theorem d.

Let Γ_3^i be the 3-cycle generated by the invariant cycle γ'_i (section 6, Corollary). The intersections of Γ_3^j and γ'_i, where γ'_i is on a fixed C_x, are the intersections of γ'_i with $\gamma'_j = \Gamma_3^j \cdot C_x$. Hence $(\gamma'_i \cdot \Gamma_3^j) = (\gamma'_i \cdot \gamma'_j)$ and consequently, by (7'), $|\gamma'_i \cdot \Gamma_3^j| \neq 0$.

Suppose that $t_1 \gamma'_1 + \cdots + t_r \gamma'_r \backsim 0$ on F. Then it would follow that $\sum_{i=1}^{r} t_i (\gamma'_i \cdot \Gamma_3^j) = 0, j = 1, 2, \ldots, r$, and consequently $t_1 = t_2 = \cdots = t_r = 0$. Hence the cycles $\gamma'_1, \ldots, \gamma'_r$ are independent on F, and this proves theorems c. and a.

Remark. The theorem a. also follows directly from the theorem proved by VAN KAMPEN (1) (see also ZARISKI, 3) to the effect that a complete set of generating relations for the fundamental group of the residual space of the branch curve D (section 4) is given by the generating relations relative to the multiple points of D and the tangent lines $x = \alpha_i$ (see VIII, 1).

8. The reduction of F to a cell. (LEFSCHETZ, a). As a preparation for what follows we undertake the cutting of F along certain 3-dimensional loci so as to reduce F to a 4-cell. For this purpose we cut open C_x along $2p$ independent cycles (retrosections) $\gamma_h^x = \gamma_h$, all issued from a base point A_1 of $\{C_x\}$. The result is a 2-cell E_2^x. However, to avoid the other base points $A_2, \ldots, A_{n-1}$ being interior points of E_2^x we add to the cuts γ_h other cuts c_s $(s = 2, 3, \ldots, n)$, where c_s is an arc joining A_1 to A_s. As x varies on the cell E_2 (section 5), E_2^x generates a 4-cell E_4.

We use the following

Lemma. *Given a complex K and a subcomplex L of K, such that $K - L$ is homeomorphic to an open cell, every chain on K, of dimension lower than $K - L$, can be deformed continuously on K into a chain on L. During the deformation the points on L may be supposed to be fixed.* This lemma has been used frequently in various instances under one form or another, but in the above general form is due to EHRESMANN (1).

In our own case K is F and the subcomplex L of K is made up of the following loci:

1) The loci D_i, D'_j of C_x as x describes the cuts l_i and l'_j respectively.
2) The loci G_h, H_s of γ_h and c_s respectively, as x varies on E_2.

We add the following remark concerning the deformation of the arcs c_s. Let

$$l = \sum_{1}^{2p} \lambda_h \gamma_h + \sum_{2}^{n} \mu_j c_j.$$

Then $F(l) = \sum_{2}^{n} \mu_j A_j - (\mu_2 + \cdots + \mu_n) A_1 \neq 0$, unless $\mu_2 = \cdots = \mu_n$
$= 0$. Reasoning as in section 6 we see that as x turns about a branch
point α_i, l is increased by $(l \cdot \delta_i)\delta_i$. Let us suppose that l is an invariant
1-chain. If not all the μ_j's are zero, its locus C_3 as x varies cannot be
a cycle, because its intersection with a generic C_x is l, which is not
a cycle. On the other hand the formula (6), applied to l and to its locus
C_3, shows that the boundary of C_3 is homologous to a chain (C_a) and
hence to a multiple of C_a, and this is impossible since $C_a \neq 0$ on F.
Consequently, *if l is invariant, then all the coefficients μ_i are zero and l is
a cycle*. It also follows that *the matrix*

$$\left\| \begin{matrix} (\gamma_h \cdot \delta_i) \\ (c_j \cdot \delta_i) \end{matrix} \right\|, \quad h = 1, 2, \ldots, 2p; \, j = 2, 3, \ldots, n; \, i = 1, 2, \ldots, \mu,$$

is of rank $2p - r + n - 1$ (LEFSCHETZ, a, p. 31).

   9. The three-dimensional cycles. The loci H_s are 3-cells and so
are the loci G_h of the cycles γ_h cut open at A_1. Since we have assumed
a certain subdivision of C_x, we also have a subdivision of D_i and D'_j into
3-cells, each 3-cell being the locus of a 2-cell of C_x as x describes a cut
l_i or l'_j. Consequently any 3-chain C_3 on L is of the following form:

$$(8) \qquad C_3 = \sum_{1}^{2p} \lambda_h G_h + \sum_{2}^{n} \mu_s H_s - \sum_{1}^{\mu} C_3^i - \sum_{1}^{\mu'} C_3^{'j},$$

where C_3^i (or $C_3^{'j}$) is the locus of a chain C_2^i $(C_2^{'j})$ on C_x as x describes l_i
(or l'_j). We have $C_3 \cdot C_x = \sum_{1}^{2p} \lambda_h \gamma_h + \sum_{2}^{n} \mu_s c_s$, and if C_3 is a cycle Γ_3
this intersection must be a 1-cycle, i.e. $\mu_2 = \cdots = \mu_n = 0$. Hence
every Γ_3 on L is of the following form:

$$(8') \qquad \Gamma_3 = \sum_{1}^{2p} \lambda_h G_h - \sum_{1}^{\mu} C_3^i - \sum_{1}^{\mu'} C_3^{'j},$$

and

$$(9) \qquad \Gamma_3 \cdot C_x = \gamma = \sum_{1}^{2p} \lambda_h \gamma_h.$$

The boundary of $\sum_{1}^{2p} \lambda_h G_h$ is the locus of γ as x describes the boundary of
E_2. That portion of it which corresponds to the cut l_i is the locus of
$\gamma' - \gamma$ as x describes this cut (section 6). This locus must coincide with
the boundary of C_3^i, i. e. with the locus of $F(C_2^i)$ as x describes l_i. Consequently $C_2^i \to \gamma' - \gamma$, $\gamma' \sim \gamma$ on C_x, *and hence γ is an invariant cycle*.
Conversely, if the cycle γ given by (9) is invariant and if we define C_2^i
as the chain on C_x (x on l_i) which is bounded by the zero-cycle $\gamma' - \gamma$
(a similar definition holds for $C_2^{'j}$), then (8') yields a Γ_3.

   If $\gamma \sim 0$ on C_x then all the coefficients λ_h are zero, the right-hand
member in (8') disappears and consequently $\Gamma_3 \sim 0$. Conversely, if

$\Gamma_3 \sim 0$ then Γ_3 bounds a chain C_4 on F. Intersecting with a generic C_x we find $\gamma \sim 0$ on C_x. Consequently *there are as many independent cycles Γ_3 on F as there are invariant cycles γ on C_x*, i. e. $R_3 = R_1 \, (= r)$. This is one of the duality relations of POINCARÉ.

10. The two-dimensional cycles Every chain C_2 on L (section 8) is clearly of the form

$$C_2 = \sum_1^\mu C_2^i + \sum_1^{\mu'} C_2'^j + (C_a) - \sum_1^\mu (C_{\alpha_i}) - \sum_1^{\mu'} (C_{\beta_j}),$$

where C_2^i (or $C_2'^j$) is the locus of a 1-chain C_1^i (or $C_1'^j$) on C_x as x describes the cut l_i (or l'_j) and where (C_x) denotes a 2-chain on C_x. In order that C_2 be a cycle Γ_2 two things are necessary. In the first place the boundary of C_1^i must consist at most of a sum of multiples of the base points $A_1, A_2, \ldots, A_n$. In the second place C_1^i must reduce on C_{α_i} to the boundary of (C_{α_i}), i. e. at any rate to a 1-cycle ~ 0 on C_{α_i}. From this it follows first of all that C_1^i is a 1-cycle on any C_x. The only 1-cycles on C_x which reduce to zero-cycles on C_{α_i} as $x \to \alpha_i$ are the cycles which are homologous to a multiple of δ_i. Hence $C_1^i \sim t_i \delta_i$ on C_x. In a similar manner it can be shown that $C_1'^j$ is a 1-cycle ~ 0 on C_x. The consideration at the end of section 6 can now be applied and it follows that

$$(10) \qquad\qquad \Gamma_2 \sim \sum_1^\mu t_i \Delta_i + (C_a) \text{ on } F.$$

The condition that the right-hand member should represent a 2-cycle can now be derived immediately. Intersecting Γ_2 with C_a we see that the chain (C_a) must be bounded by the cycle $\gamma = \sum_1^\mu t_i \delta_i^a$, and hence $\gamma \sim 0$ on C_a. Conversely, if $\gamma \sim 0$ on C_a then (10) represents a 2-cycle provided we take for (C_a) the 2-chain on C_a which is bounded by γ. Hence *every cycle Γ_2 on F is of the form* (10), *with the necessary and sufficient condition that*[1]

$$(11) \qquad\qquad \sum_1^\mu t_i \delta_i^a \sim 0 \text{ on } C_a.$$

If a cycle Γ_2 on L is ~ 0 on F, it bounds a 3-chain C_3. By the Lemma of section 8 this chain can be deformed into the complex L, its boundary Γ_2 remaining fixed. Hence $\Gamma_2 \sim 0$ also on L. From the expression (8) of an arbitrary 3-chain on L it follows immediately that

$$(12) \qquad \Gamma_2^h = \sum_1^\mu (\gamma_h \cdot \delta_i) \Delta_i + (C_a) \sim 0 \text{ on } F, \qquad h = 1, 2, \ldots, 2p,$$

$$(12') \qquad \Gamma_2^s = \sum_1^\mu (c_s \cdot \delta_i) \Delta_i + (C_a) \sim 0 \text{ on } F, \qquad s = 2, \ldots, n,$$

---

[1] It is not permissible to replace in (11) C_a by C_x and to write $\sum t_i \delta_i \sim 0$ on C_x, because δ_i^a is obtained from δ_i by a deformation along a path in the x-plane *which depends on i*. This becomes clear if we recall that there are μ *distinct* homeomorphisms T_a^i between R_0 and R_a (section 5) and that $\delta_i^a = T_a^i(\delta_i^0)$.

and also, by the preceding remark, that *these homologies constitute a fundamental set.* Hence *a necessary and sufficient condition that the cycle* (10) *be* ~ 0 *on F is that the coefficients* t_i *be the intersection numbers of a 1-chain l with the cycles* δ_i^a, where

$$l = \sum_1^{2p} \lambda_h \gamma_h + \sum_2^n \mu_s c_s.$$

As a check on (11) we point out that in this case (11) becomes

$$\sum (l^{(i-1)} \cdot \delta_i^a)\, \delta_i^a \sim 0 \text{ on } C_a,$$

where $l^0 = l^a$ and $l', l'', \ldots, l^{(\mu-1)}$ are the transformed chains on C_a, and that if l is a cycle γ, this is nothing but the homology (5) of section 6. Obviously the proof of (5) can be repeated without modifications if γ is replaced by any chain l.

We may now evaluate the BETTI number R_2 of F. Since the number of independent cycles δ_i is $2p - r$, the number of independent sets of integers $t_1, t_2, \ldots, t_\mu$ for which (11) is satisfied is $\mu - 2p + r = \mu - 2p + R_1$. The number of independent homologies in (12) and (12') is $2p + n - 1 - \xi$, where ξ is the number of independent invariant 1-chains l (i. e. such that $(l \cdot \delta_i) = 0$ for all i's). Hence

$$R_2 = (\mu - 2p + R_1 + 1) - (2p + n - 1) + \xi,$$

where we add 1 to $\mu - 2p + R_1$ to account for the 2-cycle C_a, corresponding to $t_1 = \cdots = t_\mu = 0$. Now from the remark at the end of section 8 it follows that $\xi = r$. Substituting we find

$$(13) \qquad R_2 = \mu - 4p + 2R_1 - n + 2.$$

There follows an expression of R_2 in terms of the invariant $I = \mu - n - 4p$ of ZEUTHEN-SEGRE (III, 8), due to PICARD and ALEXANDER (1):

$$(13') \qquad R_2 = I + 2R_1 + 2.$$

This expression of I in terms of topological invariants of F yields the proof of the relative invariance of I under birational transformations.

We observe that $I + 4 = R_0 - R_1 + R_2 - R_3 + R_4 (R_0 = R_4 = 1)$ is the EULER-POINCARÉ characteristic of F (ALEXANDER, 1).

11. The group of torsion. We suppose that the subdivision of C_a satisfies the following conditions: 1) each base point A_s of $\{C_x\}$ is an interior point of a 2-cell of C_a; 2) no two base points are on the same 2-cell; 3) all 2-cells of C_a are oriented alike (i. e. as C_a). Let E_2^s $(s = 1, 2, \ldots, n)$ be the 2-cell containing A_s. If Γ_2 is any 2-cycle on F, we can write

$$(14) \qquad \Gamma_2 \sim \sum t_i \Delta_i + ((C_a)) + \sum_{s=1}^n \mu_s E_2^s,$$

where $((C_a))$ denotes a 2-chain on C_a in which the cells E_2^s do not occur. The cell E_2^s has the point A_s in common with a generic C_x, the intersection number $(E_2^s \cdot C_x)$ being $+1$. Hence

$$(\Gamma_2 \cdot C_x) = \mu' = \sum_{s=1}^{n} \mu_s.$$

Let us consider the zero-cycles $\Gamma_2^h, \overline{\Gamma}_2^s$, given by (12) and (12'). They are obtained by deformation from the cycles $F(G_h), F(L_s)$. Using a fundamental boundary relation (LEFSCHETZ, e, p. 169) we find:

$$F(G_h \cdot C_x) = G_h \cdot F(C_x) + F(G_h) \cdot C_x,$$
$$F(L_s \cdot C_x) = L_s \cdot F(C_x) + F(L_s) \cdot C_x,$$

or (since $G_h \cdot C_x = \gamma_h$, $L_s \cdot C_x = c_s$ and $F(C_x) = 0$) $F(\gamma_h) = F(G_h) \cdot C_x$, $F(c_s) = F(L_s) \cdot C_x$, or finally, since $F(\gamma_h) = 0, F(c_s) = A_s - A_1$,

$$\Gamma_2^h \cdot C_x = 0, \qquad \overline{\Gamma}_2^s \cdot C_x = A_s - A_1.$$

It follows that

(15) $$\Gamma_2^h = \sum_1^{\mu} (\gamma_h \cdot \delta_i) \Delta_i + ((C_a)),$$

(15') $$\overline{\Gamma}_2^s = \sum_1^{\mu} (c_s \cdot \delta_i) \Delta_i + ((C_a)) + E_2^s - E_2^1.$$

Using (15') and the homology $\overline{\Gamma}_2^s \infty 0$ $(s = 2, \ldots, n)$ we may eliminate from (14) the cells E_2^s $(s = 2, \ldots, n)$ and we find for any cycle Γ_2 a reduced form:

$$\Gamma_2 \infty \sum \lambda_i' \Delta_i + ((C_a)) + \mu' E_2^1.$$

If $\Gamma_2 \approx 0$, then $\mu' = (\Gamma_2 \cdot C_x) = 0$. Moreover, if $\Gamma_2 \infty 0$, then we must have

$$\sum_1^{\mu} \lambda_i' \Delta_i + ((C_a)) = \sum_1^{2p} \nu_h \Gamma_2^h + \sum_2^{n} \overline{\nu}_s \overline{\Gamma}_2^s,$$

and this must be an identity in the Δ_i's and in the cells of C_a. Consequently all the coefficients $\overline{\nu}_s$ are zero, so that if $\Gamma_2 \infty 0$, the reduced form of Γ_2 must be a linear combination of the cycles Γ_2^h alone. It follows that in order to obtain all two-dimensional divisors of zero it is sufficient to consider the homologies

$$\Gamma_2^h = \sum_1^{\mu} (\gamma_h \cdot \delta_i) \Delta_i + ((C_a)) \infty 0, \qquad h = 1, 2, \ldots, 2p.$$

The two-dimensional torsion coefficients of F are therefore the invariant factors $t_1, t_2, \ldots$ of the matrix $\| (\gamma_h \cdot \delta_i) \|$ which are greater than 1. The product $\sigma = t_1 t_2 \ldots$ is the order of the group of torsion.

We now pass to the one-dimensional zero-divisors. The fundamental homologies are $\delta_i \infty 0$ $(i = 1, 2, \ldots, \mu)$. Let $\delta_i \infty \sum_1^{2p} \beta_{ih} \gamma_h$. The retrosections γ_h satisfy the well known relations $(\gamma_i \cdot \gamma_{i+p}) = +1$ $(i = 1, 2, \ldots, p)$, $(\gamma_i \cdot \gamma_j) = 0$, if $j - i \neq \pm p$. Hence $\beta_{ih} = -(\gamma_{h+p} \cdot \delta_i)$, $\beta_{i, h+p} = (\gamma_h \cdot \delta_i)$ $(h = 1, 2, \ldots, p)$. If we put for simplicity $\gamma_{h+p} = \overline{\gamma}_h$, $\gamma_h = -\overline{\gamma}_{h+p}$

$(h = 1, 2, \ldots, p)$, the fundamental homologies become

$$\sum_1^{2p} (\gamma_h \cdot \delta_i)\, \bar{\gamma}_h \sim 0.$$

Consequently, *also the one-dimensional torsion coefficients are the invariant factors of the matrix* $\|(\gamma_h \cdot \delta_i)\|$ *which are greater than 1 and therefore coincide with the two-dimensional torsion coefficients.*

There are no 3-dimensional zero-divisors. This follows directly from section 9 and from the fact that C_x does not carry one-dimensional zero-divisors. We have already proved that $R_1 = R_3$ (section 9). *We have thus established for algebraic surfaces all the duality theorems of* Poincaré.

12. Homologies between algebraic cycles and algebraic equivalence. The invariant ϱ_0. Two algebraic cycles D_1 and D_2 which are algebraically equivalent $(D_1 \equiv D_2)$ (V, 2) are clearly also homologous, $D_1 \sim D_2$. In fact, if $D_1 \equiv D_2$ there exists a curve D such that $D_1 + D \parallel D_2 + D$ and hence $D_1 + D$ can be deformed into $D_2 + D$. Lefschetz (5, a) has proved by transcendental methods that conversely, *if $D_1 \sim D_2$ then $D_1 \equiv D_2$* (p. 180). Albanese (4) pointed out that the weaker form of Lefschetz's theorem, corresponding to homology $\approx$ with division allowed, follows from Severi's arithmetic criterion of algebraic equivalence (V, 5, criterion 4). In fact, if $D_1 \approx D_2$, then $(D_1 \cdot C) = (D_2 \cdot C)$, where C is an arbitrary curve. Hence D_1 and D_2 have the same order and moreover $(D_1^2) = (D_1 \cdot D_2) = (D_2^2)$, and from this it follows that $\lambda D_1 \equiv \lambda D_2$. From Lefschetz's theorem it follows that the number ϱ of Picard gives the maximum number of algebraic cycles which are independent with respect to homologies, and hence ϱ *appears as the* Betti *number of the algebraic modulus on* F (section 3).

Let $C_1, C_2, \ldots, C_\varrho$ be ϱ algebraic cycles forming a base (V, 6). We complete these cycles by $R_2 - \varrho = \varrho_0$ cycles Γ_2^i so as to have R_2 independent 2-cycles on F. *It is clear that no combination of the cycles Γ_2^i is algebraic.* Since $|(C_i \cdot C_j)| \neq 0$ we may replace the cycles Γ_2^i by other independent cycles $\bar{\Gamma}_2^i$ such that $(\bar{\Gamma}_2^i \cdot C_j) = 0$ $(i = 1, 2, \ldots, \varrho_0;$ $j = 1, 2, \ldots, \varrho)$. We suppose that the original cycles Γ_2^i already satisfy this condition. It is then seen that any 2-cycle Γ_2 whose algebraic number of intersections with *any* algebraic cycle is zero (and for this it is sufficient that $(\Gamma\ C_i) = 0$, $i = 1, 2, \ldots, \varrho)$ necessarily depends on the cycles Γ_2^i, i. e. $\lambda\Gamma + \lambda_1\Gamma_2' + \cdots + \lambda_{\varrho_0}\Gamma_2^{\varrho_0} \sim 0$ on F, $\lambda \neq 0$.

In particular Γ_2 may be a zero-divisor $(\lambda > 1, \lambda_1 = \cdots = \lambda_{\varrho_0} = 0)$. A remarkable result concerning the 2-dimensional zero-divisors of F, due to Lefschetz, is to the effect that they are all algebraic (VII, 7, c).

The number ϱ_0 of independent non-algebraic cycles on F acquires a definite significance in the theory of double integrals of the second kind attached to F, and from this theory it readily follows that ϱ_0 *is an absolute invariant of F under birational transformations* (VII, 10). From an intuitive standpoint the absolute invariance of ϱ_0 is indicated by the fact that in a birational transformation of F into a surface F' the only new 2-cycles created on either F or F' are the fundamental curves of the transformation, and therefore the new 2-cycles are all algebraic. Hence R_2 and ϱ are both increased or diminished by the same amount, namely by $e' - e$, where e, e' denote the number of exceptional curves of F, F' which are transformed into points, and the difference $R_2 - \varrho = \varrho_0$ is invariant.

A rigorous topological proof of the absolute invariance of ϱ_0 has been given by LEFSCHETZ (10).

13. The topological theory of algebraic correspondences (LEF-SCHETZ, 9). Given two curves $f_1(x_1, y_1) = 0$ and $f_2(x_2, y_2) = 0$ of genera p_1 and p_2 respectively, the two equations $f_1 = 0$, $f_2 = 0$ define a surface F in S_4 (x_1, y_1, x_2, y_2), called *the surface of the* (non-ordered) *pairs of points of the curves f_1 and f_2*. The points of F are in $(1, 1)$ continuous correspondence with the pairs of points P_1, P_2, where P_1 is on f_1 and P_2 is on f_2, and by this property F is birationally determined. As a Riemannian variety F is *the topological product $R_1 \times R_2$ of the* RIEMANN surfaces R_1, R_2 of f_1 and f_2 respectively (LEFSCHETZ, e, Chapter V). The surface F is characterized by the property of possessing two pencils of curves K_1 and K_2, such that the curves of one pencil are unisecants of the curves of the other pencil. A curve K_1 (or K_2) is the image of the set of point-pairs (P_1, P_2) where P_1 (or P_2) is a fixed point on f_1 (or f_2). The pencils are of genera p_1 and p_2 respectively. An algebraic correspondence between f_1 and f_2, of indices m_1 and m_2, is represented on F by an algebraic curve C, whose points are the images of the pairs of homologous points of the correspondence. We have obviously $(C \cdot K_1) = m_2$, $(C \cdot K_2) = m_1$.

It is seen from these preliminary remarks that the consideration of F reduces the study of the algebraic correspondences between f_1 and f_2 to the study of the algebraic curves which can be traced on F. From this point of view the surfaces of the pairs of points of two algebraic curves have been studied by C. SEGRE (1), DE FRANCHIS (1), MARONI (1), PICARD (2, 4), HUMBERT (1), especially and systematically by SEVERI (2). It was precisely the theory of correspondences, especially the problem of constructing a base for the singular correspondences between two curves, that led SEVERI to his general theory of the base. SEVERI's paper goes as far as it is possible to go by means of strictly algebro-geometric methods. However, to obtain a base he had to fall back on the transcendental theory of correspondences developed by HURWITZ in his fundamental paper (1).

In (9) LEFSCHETZ derives again the main results of HURWITZ and SEVERI in a purely topological manner and as a mere application of his basic coincidence and fixed point formula for topological transformations of manifolds (LEFSCHETZ, 7, 8; e, Chapter VII). LEFSCHETZ's methods have proved very effective both in this specific domain and in later applications by HODGE toward the solution of a fundamental question of the theory of double integrals of the first kind (VII, 8).

We fix a fundamental set of $2p_1$ 1-cycles γ_i^1 and of $2p_2$ 1-cycles γ_j^2 on f_1 and f_2 respectively. It is a simple matter to show that every cycle Γ_n on F depends on the products $\Gamma_p^1 \times \Gamma_{n-p}^2$, where Γ_p^1 is a cycle on f_1 and Γ_{n-p}^2 is on f_2. Minimal bases for the cycles of different dimensions are therefore the following:

1. *1-cycles*: the $2p_1$ cycles $\delta_i^1 = \gamma_i^1 \times P_2$ and the $2p_2$ cycles $\delta_j^2 = P_1 \times \gamma_j^2$, where P_1 and P_2 are fixed points on f_1 and f_2 respectively. Hence $R_1 = 2p_1 + 2p_2$.

2. *2-cycles*: the $4p_1 p_2$ cycles $\Gamma_2^{ij} = \gamma_i^1 \times \gamma_j^2$ and the cycles $A_1 = f_1 \times P_2$, $A_2 = P_1 \times f_2$. Hence $R_2 = 4p_1 p_2 + 2$.

3. *3-cycles*: the cycles $\gamma_i^1 \times f_2$ and $f_1 \times \gamma_j^2$. Hence $R_3 = 2p_1 + 2p_2 = R_1$.

There are no zero-divisors (LEFSCHETZ, a, Chapter III): this is implied by the absence of zero-divisors for f_1 and f_2 (LEFSCHETZ, e, Chapter V). The orientation of δ_i^1 and δ_j^2 is determined by the orientation of γ_i^1 and γ_j^2. We define the orientation of Γ_2^{ij} by the condition $(\gamma_i^1 \times Q_2 \cdot Q_1 \times \gamma_j^2) = +1$, where Q_1 and Q_2 are any two points on γ_i^1 and γ_j^2 respectively. In a similar manner the orientation of the remaining cycles can be defined. The order in which the two curves f_1 and f_2 are considered affects only the orientation of the cycles Γ_2^{ij}.

Let γ^1, $\bar{\gamma}^1$ be any two 1-cycles on f_1, γ^2, $\bar{\gamma}^2$ any two 1-cycles on f_2, and let $\Gamma_2 = \gamma^1 \times \gamma^2$, $\bar{\Gamma}_2 = \bar{\gamma}^1 \times \bar{\gamma}^2$. The following intersection formulas can be easily derived:

$$(16) \qquad (\Gamma_2 \cdot \bar{\Gamma}_2) = -(\gamma^1 \cdot \bar{\gamma}^1)(\gamma^2 \cdot \bar{\gamma}^2),$$

$$(16') \qquad (I_2 \cdot A_1) = (\Gamma_2 \cdot A_2) = 0,$$

$$(16'') \qquad (A_1^2) = (A_2^2) = 0; \qquad (A_1 \cdot A_2) = +1.$$

From (16') and (16'') it follows that if Γ_2 is any 2-cycle on F and if $(\Gamma_2 \cdot A_1) = \alpha_2$, $(\Gamma_2 \cdot A_2) = \alpha_1$, then

$$(17) \qquad \Gamma_2 \sim \alpha_1 A_1 + \alpha_2 A_2 + \sum_{i,j} \varepsilon_{ij} \Gamma_2^{ij}.$$

Let, in particular, Γ_2 be the image of an algebraic correspondence T between f_1 and f_2. Every 1-cycle γ^1 on f_1 is transformed into a cycle $T\gamma^1$ on f_2, in the sense that as the point P_1 varies on γ^1 the corresponding α_2 points $T(P_1)$ on f_2 describe paths whose sum is a cycle. The effect of T on the 1-cycles on f_1 is fully described by its effect on the cycle γ_i^1,

i. e. by a matrix $c = \|c_{ij}\|$, where

(18) $$T\gamma_i^1 \sim \sum_{j=1}^{2p_2} c_{ij}\gamma_j^2, \qquad i = 1, 2, \ldots, 2p_1.$$

The following relation can be proved:

(19) $$(\Gamma_2 \cdot \gamma^1 \times \gamma^2) = -(T\gamma^1 \cdot \gamma^2).$$

Putting in this relation $\gamma^1 = \gamma_h^1$ and $\gamma^2 = \gamma_k^2$ and using (16), (16'), (18), we find

(20) $$\sum_{i,j} \varepsilon_{ij}(\gamma_i^1 \cdot \gamma_h^1)(\gamma_j^2 \cdot \gamma_k^2) = \sum_{j=1}^{2p_2} c_{hj}(\gamma_j^2 \cdot \gamma_k^2).$$

Let $\varepsilon = \|\varepsilon_{ij}\|$, $C_1 = \|(\gamma_i^1 \cdot \gamma_j^1)\|$, $C_2 = \|(\gamma_i^2 \cdot \gamma_j^2)\|$. It is well known that C_1' (the transposed of C_1) $= -C_1$, $C_2' = -C_2$, and that det. $C_1 = \pm$det. $C_2 = \pm 1$. Hence C_1^{-1} and C_2^{-1} exist and we may write (20) in matrix notations as follows: $C_1' \varepsilon C_2 = -C_1 \varepsilon C_2 = cC_2$, or

(21) $$-C_1\varepsilon = c, \qquad \varepsilon = -C_1^{-1}c.$$

This shows that the effect of T on the cycles γ_i^1 on f_1 (indicated by the matrix c) determines completely the cycle Γ_2 (i. e. the matrix ε) and conversely. If we consider the inverse correspondence T^{-1}, we will have a corresponding matrix $\tilde{c}$, where $T^{-1}\gamma_h^2 \sim \sum \tilde{c}_{hk}\gamma_k^1$. Replacing in (17) Γ_2^{ij} by $\Gamma_2^{ji} = \gamma_2^j \times \gamma_1^i = -\Gamma_2^{ij}$ and ε by $-\varepsilon'$ and repeating the same reasoning we find

(22) $$\tilde{c} = C_2 c' C_1^{-1}.$$

Let T_1 and T_2 be two correspondences between f_1 and f_2 and let $\Gamma_2^h = \alpha_1^h A_1 + \alpha_2^h A_2 + \sum_{i,j} \varepsilon_{ij}^h \Gamma_2^{ij}$ $(h = 1, 2)$ be the corresponding cycles on F. The *number of common pairs of homologous points of* T_1 *and* T_2 is given by the KRONECKER index $(\Gamma_2^1 \cdot \Gamma_2^2)$, and by a simple calculation it is found that

(23) $$(\Gamma_2^1 \cdot \Gamma_2^2) = \alpha_2^1 \alpha_1^2 + \alpha_1^1 \alpha_2^2 - \text{trace } C_2 c_1' C_1^{-1} c_2,$$

a formula due to HURWITZ. If T_1 and T_2 coincide, the index $(\Gamma_2 \cdot \Gamma_2)$ gives the *virtual degree of the correspondence*.

The preceding considerations may be applied without modification to the case in which the curves f_1 and f_2 coincide. F represents then the *ordered pairs of points* of one and the same curve $f = f_1 = f_2$. In this case the identical correspondence I is of particular interest. It is represented by a cycle $\Gamma_2^0 \sim A_1 + A_2 + \sum_{i,j} \varepsilon_{ij}^0 \Gamma_2^{ij}$, where $\varepsilon^0 = -C_1^{-1} = -C_2^{-1}$ (since c^0 is the unit matrix). *The number of coincidences* of any correspondence T is given by the KRONECKER index $(\Gamma_2 \cdot \Gamma_2^0)$, or in view of (23), by

$$\alpha_1 + \alpha_2 - \text{trace } c,$$

a well known formula due to HURWITZ. In particular, if T is a *valence correspondence*, i. e. if the matrix c is a multiple of the unit matrix

$(c_{ij} = 0$, if $i \neq j$, $c_{ii} = -g$, where g is the valence of T), then we obtain the formula of Cayley-Brill: *number of coincidences* $= \alpha_1 + \alpha_2 + 2g\,p$.

A non-valence correspondence is called *singular*. However, if we deal with distinct curve f_1, f_2, we call *singular* a correspondence for which $\varepsilon \neq 0$ and hence also $c \neq 0$. If $\varepsilon = 0$, T is called an *ordinary* correspondence. For two coincident curves this definition of an ordinary correspondence leads to correspondences with zero valence.

The problem of the existence of algebraic correspondences between f_1 and f_2 reduces from the topological point of view to the following question: *under what conditions is a cycle Γ_2, given by* (17), *algebraic?* The answer is provided by a general theorem of Lefschetz to the effect that *a necessary and sufficient condition for a cycle Γ_2 on a surface F to be algebraic is that the corresponding period of any double integral of the first kind vanish* (VII, 5). Let u_i^1, u_j^2 ($i = 1, 2, \ldots, p_1$; $j = 1, 2, \ldots, p_2$) be the abelian integrals of the first kind attached to f_1 and f_2 respectively, and let $\omega_1 = \| \omega_{ih}^1 \|$, $\omega_2 = \| \omega_{jk}^2 \|$ be the corresponding period matrices. It can be easily shown that any double integral of the first kind attached to F depends on the integrals $\int\int du_i^1 du_j^2$. Integrating along Γ_2 and noting that u_i^1 is constant on A_2 and u_j^2 is constant on A_1, we find

$$\iint_{\Gamma_2} du_i^1 du_j^2 = \sum_{\mu, \nu} \varepsilon_{\mu \nu} \, \omega_{i\mu}^1 \, \omega_{j\nu}^2 \,.$$

Hence the required necessary and sufficient condition is the following well known matrix relation (Hurwitz, Scorza, Rosati):

$$(24) \qquad\qquad \omega_1 \, \varepsilon \, \omega_2' = 0.$$

If (24) is satisfied, Γ_2 may be *a priori* a virtual curve and in order to obtain an effective curve, corresponding to a solution ε of (24), it may be necessary to add to Γ_2 a sum of multiples of A_1 and A_2.

The number λ of independent solutions ε of (24) is called by Scorza *the simultaneity index of ω_1 and ω_2*. If ε^1 and ε^2 are two dependent solutions of (24) $(\lambda_1 \varepsilon^1 + \lambda_2 \varepsilon^2 = 0)$ and if Γ_2^1 and Γ_2^2 are the cycles relative to the correspondences T_1, T_2, then $\lambda_1 \Gamma_2^1 + \lambda_2 \Gamma_2^2 \sim \beta_1 A_1 + \beta_2 A_2$, i. e. the correspondence $\lambda_1 T_1 + \lambda_2 T_2$ is ordinary, and consequently the correspondences T_1 and T_2 are *dependent* in the sense of Hurwitz. There exist then λ independent correspondences between f_1 and f_2. The corresponding cycles Γ_2^i and the cycles A_1 and A_2 constitute a minimal base for the algebraic cycles on F. Hence $\varrho = \lambda + 2$, where ϱ is the number of Picard.

Appendix to Chapter VI

By DAVID MUMFORD

Re: § 2. In LOJASIEWICZ (1), it is shown that all varieties can be triangulated with real-analytic simplices. MUMFORD (2) showed that the topological space of a normal algebraic surface is a topological manifold if and only if the surface is non-singular (in the algebraic sense). BRIES-KORN (1) showed that, on the contrary, higher dimensional singular varieties (in fact, hypersurfaces with isolated singularities) can be topological manifolds.

Re: § 3. The situation discussed in § 3 has been developed to the point where, for any non-singular projective variety X, one constructs a ring homomorphism:

$$A(X) \to H^*(X, \mathbf{Z}),$$

$A(X) =$ Ring of cycles on X mod rational equivalence, or CHOW ring of X (cf. Appendix A),

$H^*(X, \mathbf{Z}) =$ cohomology ring of X.

See ATIYAH-HIRZEBRUCH (1) and BOREL-HAEFLIGER (1). The Corollary to the effect that an effective algebraic cycle is never homologous to 0 was disproven by HIRONAKA (3), for *non-projective* complete non-singular varieties. He found an effective 1-cycle on a 3-dimensional X which is algebraically equivalent to 0.

Re: §§ 4—11. §§ 4—11 of the text present explicitly the topological decomposition of an algebraic surface via a generic pencil of hyperplane sections. This has remained one of the most efficient methods of under-standing the topological structure of non-singular, projective varieties, in the étale as well as the classical cohomology theory. The generalization of this decomposition to n-dimensions is presented with close attention to detail in WALLACE (1). In char p, IGUSA (7, 8, 9) was the first to try to carry out this analysis. He showed that if X is any non-singular projective surface, then after blowing up a finite number of distinct points on X, one gets a surface X^* such that there exists a map:

$$X^*$$
$$p \downarrow$$
$$\mathbf{P}^1$$

where all but a finite number of the fibres $p^{-1}(x)$ are non-singular, and where the singular fibres $p^{-1}(x)$ are irreducible with one ordinary double point. Denote $p^{-1}(x)$ by C_x.

The first thing to look at in this situation is the structure of the "fibre system" p near one singular fibre. In particular, moving around a singular fibre induces a monodromy homeomorphism of the type known as a *Schraubung* (DEHN, 1; translated variously as "screw-map" MUMFORD, 10, "DEHN twist" Bers), which acts on the cohomology of the non-singular fibres by formula (3), §6. In the classical case, generalizations of this formula have been sought for whenever one has an analytic family of projective varieties Y_t given for all $t \in D$, D a disc in the complex plane, and Y_t non-singular if $t \neq 0$. GRIFFITHS, LANDMAN, CLEMENS (1), GROTHENDIECK (6, 1968), BRIESKORN (4), KATZ (2), BOREL have all shown that in all such cases, the monodromy maps $\gamma \mapsto \gamma'$, $\gamma \in H^k(Y_{t_0}, Q)$, for each k, are linear maps all of whose eigenvalues are roots of 1. GROTHENDIECK's proof will apply in char$p$ with the étale cohomology groups if resolution of singularities is proven.

I want to sketch how the description of the topology of X given in §§ 4—11 carries over almost cycle by cycle to the étale cohomology theory, valid in all characteristics. To do this efficiently, let me first restate the results in the classical case. We start with the basic map p: $X^* \to \mathbf{P}^1$. Let $x_1, \ldots, x_\mu \in \mathbf{P}^1$ be the points $x \in \mathbf{P}_1$ such that C_x is singular, let $a \in \mathbf{P}^1$ be any other point, and cut $\mathbf{P}^1$ open via arcs l_i as usual:

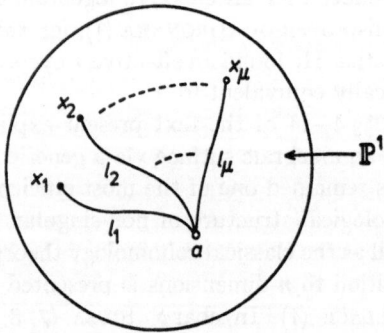

Let $\pi = \pi_1 (\mathbf{P}^1 - \{x_1, \ldots, x_\mu\})$. Moving out l_i, around x_i, and back on l_i, we get an element $\sigma_i \in \pi$, and π is just the free group on $\sigma_1, \ldots, \sigma_\mu$ modulo the relation $\sigma_\mu \ldots \sigma_1 = e$. Restricting the fibre space of curves X^* to the loop σ_i, we find a monodromy map

$$\tau_i \colon H_1(C_a, \mathbf{Z}) \to H_1(C_a, \mathbf{Z}) .$$

The results are these:

(i) If (,) is the intersection form $H_1(C_a, \mathbb{Z}) \times H_1(C_a, \mathbb{Z}) \to \mathbb{Z}$, then $\exists\ \delta_i \in H_1(C_a, \mathbb{Z})$ such that

$$\tau_i(\gamma) = \gamma + (\gamma, \delta_i) \cdot \delta_i, \quad \text{all} \quad \gamma \in H_1(C_a, \mathbb{Z}).$$

Define

$$\varphi : \mathbb{Z}^\mu \to H_1(C_a, \mathbb{Z})$$

by $\varphi(n_1, \ldots, n_\mu) = \sum n_i\, \delta_i$.

(ii) δ_i "vanishes" in C_{x_i} in the sense that $\exists$ a 2-chain $\varDelta_i$ in $p^{-1}(l_i)$ such that

a) $\partial \varDelta_i = \varDelta_i \cap C_a$,
b) $\varDelta_i \cap C_x$ is a 1-cycle representing δ_i for all $x \in l_i$, $x \neq x_i$,
c) $\varDelta_i \cap C_{x_i} = $ the double point of C_{x_i}.

In particular, this shows that $\delta_i \sim 0$ in X^*.

(iii) $H_1(X^*, \mathbb{Z}) \cong H_1(C_a, \mathbb{Z})/[\text{span of } \delta_i\text{'s, the vanishing cycles}]$ = coker (φ).

(iv) For all $\gamma \in H_1(C_a, \mathbb{Z})$, we can move γ around in all the fibres C_x, $x \in \mathbb{P}^1 - \overset{\mu}{\underset{1}{\cup}}\, l_i$, and sweep out a 3-chain. In fact, we cen get in this way a 3-chain Γ such that

$$\partial \Gamma = \sum t_i\, \varDelta_i + (\text{2-chain in } C_a)$$

where

$$t_k = (\tau_{k-1} \ldots \tau_1(\gamma), \delta_k).$$

Define $\psi : H_1(C_a, \mathbb{Z}) \to \mathbb{Z}^\mu$ by $\psi(\gamma) = (t_1, \ldots, t_\mu)$, t_k as above. Note that $\psi(\gamma) = 0$ if and only if $\tau_i(\gamma) = \gamma$, all i, i. e. γ is an "invariant" cycle.

(v) Let E be one of the exceptional curves on X^* — it is a section of the map p. Note that whenever $\sum n_i\, \delta_i \sim 0$ on C_a, there exists a 2-chain σ_2 on C_a such that $\partial \sigma_2 = -\sum n_i(\varDelta_i \cap C_a)$. Then $\sum n_i\, \varDelta_i + \sigma_2$ is a 2-cycle. The result is that $H_2(X^*, \mathbb{Z})$ is generated by the 2-cycles

$$E, C_a, \sum n_i\, \varDelta_i + \sigma_2$$

and the only relations are given by $\partial \Gamma \sim 0$, Γ as in (iv). In other words,

$$H_2(X^*, \mathbb{Z}) \cong \mathbb{Z} + \mathbb{Z} + \ker(\varphi)/\operatorname{Im}(\psi).$$

(vi) As for $H_3(X^*, \mathbb{Z})$, it is generated by the linear combinations $\sum n_i\, \Gamma_i$ whose boundaries are 0, i. e.,

$$H_3(X^*, \mathbb{Z}) \cong \operatorname{Ker}(\psi).$$

Now let's look at the étale theory of X^*. This theory is due to IGUSA (9), OGG (1), ŠAFAREVICH (1), ARTIN and GROTHENDIECK (GROTHENDIECK, 6, 1963—64). In this case, let

$$\pi = \pi_1^t(\mathbb{P}^1 - \{x_1, \ldots, x_\mu\})$$

be the algebraic fundamental group of $\mathbf{P}^1$, ramified only at $x_1, \ldots, x_n$, and tamely ramified in char p. Instead of choosing arcs l_i, we now choose embeddings of the algebraic closure of $k(\mathbf{P}^1)$ in the algebraic closure of the *completions* $k(\mathbf{P}^1)_{x_i}$ of $k(\mathbf{P}^1)$ in the x_i-adic topology. This choice defines a map:

$$I_i = \left\{ \begin{matrix} \text{tamely ramified} \\ \text{Galois group of} \\ k(\mathbf{P}^1)_{x_i} \end{matrix} \right\} \to \pi \ .$$

We also choose a collection ζ_n of primitive nth roots of 1, for all n prime to the characteristic, such that $(\zeta_{nm})^m = \zeta_n$, all n, m. Once this is done, I_i is a pro-cyclic group with canonical generator σ'_i. Let $\sigma_i = $ image of σ'_i in π. Grothendieck (6, 1960—61) showed that if the embeddings of the algebraic closures are suitably chosen, then $\sigma_1, \ldots, \sigma_\mu$ generate π and satisfy the relation $\sigma_1 \ldots \sigma_\mu = e$ as before.

Let $C_0 = $ generic fibre of p, and let $J_0 = $ jacobian variety of C_0. Choose a prime l different from the characteristic. The l-adic Tate group $T_l(J_0)$ replaces the group $H_1(C_a, \mathbf{Z})$ in the étale theory. Monodromy reappears since π acts in a natural way on $T_l(J_0)$[1]. The maps τ_i defined by σ_i generalize the previous monodromy maps. Igusa (8) proved that τ_i is again given by the formula:

$$\tau_i(\gamma) = \gamma + e(\gamma, \delta_i)\, \delta_i, \quad \text{all} \quad \gamma$$

for suitable δ_i, where the intersection form $(\ ,\ )$ is replaced by the skew-symmetric $e(\ ,\ )$ from the theory of abelian varieties (Lang, 3). For the étale cohomology groups of X^*, Artin showed

$$H^1(X^*, \mathbf{Z}_l) \cong H^0(\pi, T_l(J_0))$$

$$H^2(X^*, \mathbf{Z}_l) \cong \mathbf{Z}_l \cdot (\text{fund. class of } C_x)$$

$$+ \mathbf{Z}_l \cdot (\text{fund. class of } E)$$

$$+ \mathrm{Ker}\left[H^1(\pi, T_l(J_0)) \to \sum_{i=1}^{\mu} H^1(I_i, T_l(J_0)) \right]$$

$$H^3(X^*, \mathbf{Z}_l) \cong \mathrm{Coker}\left[H^1(\pi, T_l(J_0)) \to \sum_{i=1}^{\mu} H^1(I_i, T_l(J_0)) \right] .$$

But $H^1(I_i, T_l(J_0)) \cong T_l(J_0)/(\sigma_i - 1)$. $T_l(J_0)$, and Ogg (1) showed that the 1-cocycles $\{a_\sigma\}$ from π to $T_l(J_0)$ are given by any elements $a_{\sigma_i} \in T_l(J_0)$ such that $a_{\sigma_1} + \sigma_1(a_{\sigma_2}) + \cdots + \sigma_1 \cdot \cdots \cdot \sigma_{\mu-1}(a_{\sigma_\mu}) = 0$.

If this co-cycle splits over I_i, then $a_{\sigma_i} = \lambda_i \cdot \delta_i$, $\lambda_i \in \mathbf{Z}_l$, hence $\{a_\sigma\}$ is determined by a μ-tuple $(\lambda_1, \ldots, \lambda_\mu)$. It is now an easy verification that

$$H^1(X^*, \mathbf{Z}_l) \ ; \quad H^2(X^*, \mathbf{Z}_l)/\mathbf{Z}_l\langle C_a\rangle + \mathbf{Z}_l\langle E\rangle \ ; \quad H^3(X^*, \mathbf{Z}_l)$$

---

[1] This uses the fact proved by Igusa (8) that adjoining points of order l^n on J_0 to $k(\mathbf{P}^1)$ gives tamely ramified extensions.

are exactly the cohomology groups of the complex:

$$0 \to T_i(J_0) \xrightarrow{\alpha} \mathbf{Z}_i^\mu \xrightarrow{\beta} T_i(J_0) \to 0$$

where

$$\alpha(x) = (e(x, \delta_1), \ldots, e(x, \delta_\mu)),$$

$$\beta(\lambda_1, \ldots, \lambda_\mu) = \lambda_1 \delta_1 + \lambda_2 \sigma_1(\delta_2) + \cdots + \lambda_\mu \sigma_1 \ldots \sigma_{\mu-1}(\delta_\mu).$$

If we formally take the dual sequence to this, using e to identify $T_i(J_0)$ with its dual, we get

$$0 \leftarrow T_i(J_0) \xleftarrow{{}^t\alpha} \mathbf{Z}_i^\mu \xleftarrow{{}^t\beta} T_i(J_0) \leftarrow 0$$

where

$${}^t\alpha(\lambda_1, \ldots, \lambda_\mu) = \sum \lambda_i \delta_i,$$

$${}^t\beta(x) = (e(x, \delta_1), e(\sigma_1 x, \delta_2), \ldots, e(\sigma_{\mu-1}, \ldots, \sigma_1(x), \delta_\mu)).$$

This is formally identical with the sequence defining the classical homology groups!

In the above discussion, we have outlined the presentation of the homology of X and the results given in § 5—10 — with one exception. In § 7, 4 theorems a, b, c and d are discussed: we have used only a. and its corollary b. which can, following LEFSCHETZ, be proven by elementary topological means. However theorem c. and its corollary d. are much deeper. ZARISKI notes that c. is equivalent to LEFSCHETZ's lemma, stated on p. 136, that no linear combination of the vanishing cycles is invariant unless it is homologous to 0 on C_a. Unfortunately, LEFSCHETZ's topological proof of this lemma appears to be incomplete, and, to my knowledge, there is at present no topological proof of the lemma or of c. The first complete proof, therefore, must come from HODGE's theory of harmonic integrals. To see this, note that theorem c. is equivalent to the assertion:

(*) *Intersection with C_a defines an isomorphism from $H_3(X, \mathbf{Q})$ to $H_1(X, \mathbf{Q})$.*

In fact, consider the following subgroups of $H_1(C_a, \mathbf{Q})$:

Then $I \cong H_3(X, \boldsymbol{Q})$ under the map taking a γ to the 3-cycle Γ by "moving γ around in a pencil"; and $H_1(C_a, \boldsymbol{Q})/V \cong H_1(X, \boldsymbol{Q})$; and the composition:

$$H_3(X, \boldsymbol{Q}) \cong I \to H_1(C_a, \boldsymbol{Q}) \to H_1(C_a, \boldsymbol{Q})/V \cong H_1(X, \boldsymbol{Q})$$

is clearly intersection with C_a. It is an isomorphism if and only if $I \cap V = (0)$.

Finally, (*) is dual to the assertion that cup-product with the fundamental class ω of the hyperplane section C_a defines an isomorphism $H^1(X, \boldsymbol{Q}) \to H^3(X, \boldsymbol{Q})$. HODGE (5) proved, more generally, that if X is any projective variety and $\omega \in H^2(X, \boldsymbol{Q})$ is the fundamental class of a hyperplane section, then cup product with ω^{n-r} defines an isomorphism $H^r(X, \boldsymbol{Q}) \to H^{2n-r}(X, \boldsymbol{Q})$ (cf. WEIL, 6). It is not known whether the analogous result is true in char p, with the étale cohomology groups and l-adic coefficients. This conjecture is what GROTHENDIECK calls the "strong LEFSCHETZ theorem", cf. KLEIMAN (2).

A completely different method of analyzing the topology of a complex projective variety X of dimension n has been developed by BOTT (1) and ANDREOTTI-FRANKEL (1). This uses MORSE theory and is based on choosing

a) a hyperplane section $H \subset X$,

b) a "non-degenerate" real C^∞ function f on X such that $f(x) = 0$ if $x \in H$, $f(x) > 0$ if $x \notin H$.

BOTT uses f to build up X up to homotopy equivalence by attaching k-cells, where $n \leq k \leq 2n$, to H. Using this, he proves LEFSCHETZ's theorems:

$H_i(H, \boldsymbol{Z}) \to H_i(X, \boldsymbol{Z})$ bijective if $0 \leq i \leq n - 2$,

$H_{n-1}(H, \boldsymbol{Z}) \to H_{n-1}(X, \boldsymbol{Z})$ surjective.

The analogous results in char p for the étale cohomology groups and $\boldsymbol{Q}_l$-coefficients have been proven by GROTHENDIECK and ARTIN. This is what they call the "weak LEFSCHETZ theorem".

Re: § 12. In any characteristics $p \neq l$, torsion in the étale cohomology group $H^2(X, \boldsymbol{Z}_l)$ is "algebraic", i. e., equal to the fundamental class of a torsion divisor class. This follows immediately from the diagram

$$
\begin{array}{ccccc}
H^1(X, \boldsymbol{Z}/l^n) & \longrightarrow & H^2(X, \boldsymbol{Z}_l) & \xrightarrow{\ l^n\ } & H^2(X, \boldsymbol{Z}_l) \\
\uparrow \wr & & \text{fundamental} \uparrow \text{class} & & \text{fundamental} \uparrow \text{class} \\
H^1(X, \mu_{l^n}) & \longrightarrow & H^1(X, \mathbb{G}_m) & \xrightarrow{\ l^n\ } & H^1(X, \mathbb{G}_m) .
\end{array}
$$

The number ϱ_0 of transcendental 2-cycles has received a remarkable interpretation as the rank of the BRAUER group of X. This development started with AZUMAYA (1), AUSLANDER-GOLDMAN (1), and led to work of GROTHENDIECK (12), ARTIN (unpublished), TATE (2). Roughly the BRAUER group $\mathrm{Br}(X)$ is the set of locally free sheaves $\mathfrak{A}$ of $\mathcal{O}_X$-algebras, whose geometric fibres are algebras of $n \times n$-matrices, modulo a certain equivalence. It maps naturally into the cohomological BRAUER group which is the group $H^2(X, \mathbb{G}_m)$ in the étale topology. ARTIN showed that the two definitions agree when X is a non-singular curve or surface (cf. GROTHENDIECK, 12). Calling the cohomological BRAUER group $\mathrm{Br}'(X)$, we get exact sequences

$$0 \to \mathrm{Pic}(X)/n\,\mathrm{Pic}(X) \to H^2(X, \mu_n) \to \mathrm{Br}'(X)_n \to 0$$

for all n not divisible by the characteristic. For $l \neq \mathrm{char}$, this leads to

$$0 \to \begin{matrix}\text{Algebraic} \\ \text{2-cohomology}\end{matrix} \to H^2(X, \mathbb{Z}_l) \to T_l(\mathrm{Br}'(X)) \to 0$$

where T_l means the l-adic TATE group. GROTHENDIECK (12, p. 51) showed that in characteristic 0, or in dim 2, char p, the groups $\mathrm{Br}'(X)$ for all non-singular complete varieties X with the same function field are canonically isomorphic.

Re: § 13. The development of a completely algebraic theory of correspondences between 2 curves, valid in all characteristics, is due entirely to WEIL (3, 4) and is one of the great successes of modern algebraic geometry because of its application to counting solutions of congruences, and especially to the proof of the RIEMANN hypothesis for curves over finite fields. The extension of this theory to correspondences between higher dimensional varieties, and hopefully to a proof of the RIEMANN hypothesis for all varieties over finite fields, is one of the principal outstanding problems.

Let X_1 and X_2 be 2 non-singular complete curves. The first step in the classical theory of correspondences is to associate to every correspondence σ (i. e., σ is a divisor on $X_1 \times X_2$) a matrix by looking at the map induced by σ:

$$\sigma_* : H_1(X_1, \mathbb{Z}) \to H_1(X_2, \mathbb{Z}) .$$

If J_i is the Jacobian of X_i, then the first step in WEIL's approach is to notice that σ induces homomorphisms

$$\sigma_1 : J_1 \to J_2 ,$$
$$\sigma_2 : J_2 \to J_1 .$$

In fact, if $D(Z)$ denotes the group of divisors on Z, and $D_l(Z)$ are the divisors linearly equivalent to 0, then he proves that the map $\sigma \to \sigma_1$ is a surjective map $D(X_1 \times X_2) \to \mathrm{Hom}(J_1, J_2)$ with kernel $D(X_1) \times D(X_2) \times D_l(X_1 \times X_2)$. Thus we get:

$$\frac{D(X_1 \times X_2)}{D(X_1) \times D(X_2) \times D_l(X_1 \times X_2)} \cong \mathrm{Hom}(J_1, J_2)$$

$$\cong \mathrm{Hom}(J_2, J_1) .$$

Weil proves that these groups are free abelian of finite rank. Then using the l-adic Tate group to replace integral homology groups, he obtained induced linear maps:

$$\sigma_1^{(l)} : T_l(J_1) \to T_l(J_2) ,$$
$$\sigma_2^{(l)} : T_l(J_2) \to T_l(J_1) .$$

This gives matrix representations of σ. Via the canonical skew-symmetric forms e on $T_l(J_1)$ and $T_l(J_2)$, it turns out that $\sigma_1^{(l)}$, $\sigma_2^{(l)}$ are adjoint to each other. Since the development of the theory of étale cohomology by Artin, Grothendieck and Verdier, we can get directly at these matrices by looking at the map induced by σ from $H^1(X_1, \mathbb{Z}_l)$ to $H^1(X_2, \mathbb{Z}_l)$. In fact, using Poincaré duality and the Künneth formula, if σ is an n-dimensional cycle on $X_1 \times X_2$, X_i n-dimensional non-singular projective varieties, then σ induces a set of maps $\sigma(i)$ from $H^i(X_1, \mathbb{Z}_l)$ to $H^i(X_2, \mathbb{Z}_l)$ for each i (cf. Appendix B).

The 2nd step in the theory is Hurwitz' formula for the number of pairs of points related by 2 correspondences:

$$(\sigma \cdot \sigma') = \alpha_1 \cdot \alpha_2' + \alpha_1' \cdot \alpha_2 - \mathrm{Tr}(\sigma_2^{(l)} \cdot \sigma_1'^{(l)}) .$$

Here $(\sigma \cdot \sigma') =$ intersection number of σ, σ' on $X_1 \times X_2$; α_1, α_2 are the degrees of σ over X_1, X_2 resp.; α_1', α_2' are the degrees of σ' over X_1, X_2 resp. Weil proved this in the abstract case. A suitable generalization of this is now known in n-dimensions using the étale cohomology groups. Actually, the case of the formula which is most often used is the Lefschetz fixed point formula: if $X = X_1 = X_2$ is a non-singular projective n-dimensional variety, and σ is an n-dimensional cycle on $X \times X$, then

$$(\sigma \cdot \Delta) = \sum_{i=0}^{2n} (-1)^i \mathrm{Tr}(\sigma(i)) .$$

Here $\Delta =$ diagonal, and $\sigma(i) =$ induced endomorphism of $H^i(X, \mathbb{Z}_l)$.

The 3rd step in the theory is Castelnuovo's inequality:

$$\mathrm{Tr}(\sigma_1^{(l)} \cdot \sigma_2^{(l)}) \geqq 0$$

and >0 if $\sigma \notin D(X_1) \times D(X_2) \times D_l(X_1 \times X_2)$.

This is equivalent as Mattuck-Tate (1) observed to Hodge's Index theorem on the surface $X_1 \times X_2$ (cf. Appendix to Chapter 5), which has been proven in all characteristics by Grothendieck. Castelnuovo's inequality was first proven in all characteristics by Weil (3, 4); an arithmetic proof is due to Roquette (1). For expositions, see Lang (3) and Eichler (1). In characteristic p, no generalizations of Castelnuovo's inequalities in higher dimensions are known. Serre (7) gives generalizations to all Kähler varieties.

A 4th aspect of the theory is to give criteria for which matrices come from correspondences: more generally, given 2 abelian varieties A_1, A_2

and given

$$\sigma_* : H_1(A_1, \mathbb{Z}) \to H_1(A_2, \mathbb{Z}) \quad \text{(ground field } \mathbb{C})$$

or $\quad \sigma_l : T_l(A_1) \to T_l(A_2) \quad \text{(arbitrary ground field)}$

when is σ_* induced by a homomorphism $\sigma : A_1 \to A_2$ [resp. when is σ_l as $\mathbb{Z}_l$-linear combination of maps induced by homomorphisms $\sigma_i : A_1 \to$ $\to A_2$] ? When the ground field is $\mathbb{C}$, $H_1(A_i, \mathbb{Z})$ is just a lattice in $\tilde{A}_i$, the universal covering space of A_i, and the first question is answered by requiring that the unique real-linear extension of σ_* to a map $\sigma_* : \tilde{A}_1 \to \tilde{A}_2$ be complex-linear. This is formula (24), § 13. For ground fields k finitely generated of the prime field, TATE (1) conjectured that the answer to the second question is: if and only if σ_l commutes with the action of the GALOIS group of $\bar{k}/k$. He has proved this (3) for finite fields k with remarkable new methods.

Chapter VII

Simple and Double Integrals on an Algebraic Surface

1. Classification of integrals. a. *Simple integrals.* A simple integral, or integral of total differential, attached to an algebraic surface F, $f(x, y, z) = 0$, is an integral of the form

(1)
$$I = \int_{(x_0, y_0, z_0)}^{(x, y, z)} Q(x, y, z)\, dx + R(x, y, z)\, dy,$$

where Q and R are rational functions of x, y, z satisfying the integrability condition $\dfrac{\partial Q}{\partial y} = \dfrac{\partial R}{\partial x}$, the derivatives being evaluated by considering z as an implicit function of x and y. These integrals have been introduced by PICARD (a, 2, 3, 4). The integral I may possess on F either polar or logarithmic singularities, and their locus is referred to respectively as the *polar* or the *logarithmic curve* of I. The integral is of the *first, second or third kind* according as it possesses no singularities, polar singularities only, or logarithmic singularities. An integral I of the first or second kind is also characterized by the condition that $\int_{\Gamma_1} dI = 0$ for *any 1-cycle Γ_1 on F which is homologous to zero.* If I is any simple integral and if Γ_1 is any 1-cycle on F, the integral $\int_{\Gamma_1} dI$ is called *the period of I relative to Γ_1* if $\Gamma_1 \not\sim 0$ on F, *a logarithmic period* if $\Gamma_1 \sim 0$. A simple integral I without periods (i. e. whose periods all vanish) is a constant, a rational function or a logarithmo-rational function on F, according as I is of the first, second or third kind.

Given s simple integrals $I_1, I_2, \ldots, I_s$, they are said to be *dependent* or *independent* according as there does or does not exist a linear combination $\sum \lambda_i I_i$, with constant coefficients not all zero, which is without periods.

b. *Double integrals.* These integrals, which are of the form

(2)
$$J = \int\int R(x, y, z)\, dx\, dx, \quad (R, \text{ a rational function})$$

have been introduced by CLEBSCH and NOETHER. It is understood that the domain of integration is any analytical 2-chain C_2 on F not meeting the polar curve of R, including perhaps the curve at infinity of F.

If the domain of integration is a cycle Γ_2, then $\int_{\Gamma_2} dJ$ is called a *residue* or a *period* relative to Γ_2 according as Γ_2 is or is not $\infty 0$. If Γ_2 is deformed without meeting the curve of discontinuity of the integral, the residue or the period is unaltered.

The integral J is *of the first kind*, if $\int_{E_2} dJ$ is finite for any analytical 2-cell E_2. The integrals of the first kind are of the form $\int \frac{Q}{f_z'} dx\, dy$, where Q is an adjoint polynomial of order $n - 4$, n being the degree of f (CLEBSCH). It is clear that every residue of these integrals vanishes.

J is of the *second kind* if for any point O on F there exists an *improper* integral, i. e. an integral of the type

$$\iint \left(\frac{\partial U}{\partial x} + \frac{\partial V}{\partial y} \right) dx\, dy,$$

such that the difference of the two integrals is *finite at O*, i. e. finite for any analytical E_2 in a sufficiently small neighborhood of O. The double integrals of the second kind are also without residues and are characterized by this property (theorems 3 and 5, section 9).

This classification of simple and double integrals is invariant under birational transformations.

2. Simple integrals of the second kind. Given any $R_1 + 1$ simple integrals of the second kind, where R_1 is the one-dimensional BETTI number of F, there will exist a linear combination of these integrals, with coefficients not all zero, whose periods all vanish, and which reduces therefore to a rational function. Hence F *possess at most R_1 independent simple integrals of the second kind*.

For the proof that *there exist on F exactly R_1 independent simple integrals of the second kind* we show that there exists such an integral with R_1 *arbitrary periods* (PICARD, a_1, p. 99). From the fact that the adjoint surfaces of a sufficiently high order of F cut out on a generic plane $y = $ const. the complete system of the adjoint curves of the corresponding plane section C_y of F (III, 3), it follows that it is possible to find $2p$ adjoint polynomials $Q_h(x, y, z)$ (p is the genus of C_y) such that for y generic and fixed the $2p$ abelian integrals of the second kind

$$(3) \qquad\qquad I_h = \int \frac{Q_h(x, y, z)}{f_z'} dx, \qquad h = 1, 2, \ldots, 2p,$$

attached to the corresponding curve C_y are independent. We apply the considerations of VI, 5, 6 to the pencil $\{C_y\}$. Let $\delta_1, \ldots, \delta_\mu$ be the vanishing cycles of C_y, γ an arbitrary 1-cycle on C_y and let Ω_{hi}, ω_h be the period of I_h relative to δ_i and γ respectively. These periods are functions of y. As y turns around the critical point $y = \alpha_i$, the cycle γ is increased by $(\gamma \cdot \delta_i)\delta_i$ and hence ω_h is increased by $(\gamma \cdot \delta_i)\Omega_{hi}$. It can be proved that $(y - \alpha_i)\omega_h$ approaches zero as y approaches α_i. Consequently we have in the neighborhood of α_j

$$(4) \qquad \omega_h(y) = \omega_h^j(y) + \frac{(\gamma \cdot \delta_j)}{2\pi i} \Omega_{hj}(y) \cdot \log(y - \alpha_j),$$

where $\omega_h^j(y)$ is regular at α_j. At $y = \infty$ $\omega_h(y)$ possesses a pole of order $s_h - 1$, if the order of Q_h is $n - 3 + s_h$ (PICARD, a$_2$, p. 428). If $s_h \leqq 0$, then $\omega_h(y)$ vanishes at $y = \infty$.

Let $\delta_1, \ldots, \delta_{2p-r}$ be the $2p - r$ independent cycles δ_i and let us complete them by r other independent cycles $\gamma_1, \ldots, \gamma_r$ so as to have $2p$ independent cycles. Let B_{hj} be the period of I_h relative to γ_j and let $c_1, \ldots, c_r$ be r *arbitrary* constants. We consider the system of linear equations in the unknowns $a_1, \ldots, a_{2p}$:

$$(5) \qquad \begin{cases} \displaystyle\sum_{h=1}^{2p} a_h \Omega_{hi}(y) = 0, & i = 1, 2, \ldots, 2p - r, \\[2ex] \displaystyle\sum_{h=1}^{2p} a_h B_{hj}(y) = c_j, & j = 1, 2, \ldots, r. \end{cases}$$

The determinant of (5) is not identically zero since the $2p$ considered cycles δ_i, γ_j are independent, and consequently for a generic y there is a unique solution $a_1, \ldots, a_{2p}$. From the fact that in the first $2p - r$ equations the constants in the right hand members are all zero and also in view of the fact that the cycles $\delta_{2p-r+1}, \ldots, \delta_\mu$ depend on the cycles $\delta_1, \ldots, \delta_{2p-r}$, it follows immediately that the system (5) is invariant in the neighborhood of each critical point $y = \alpha_i$ ($i = 1, 2, \ldots, \mu$). Consequently the a_h's are uniform functions of y in the whole plane, and since they obviously do not possess essential singularities it follows that *each $a_h(y)$ is a rational function of y*.

The abelian integral

$$I = a_1(y) I_1 + \cdots + a_{2p}(y) I_{2p} = \int R \, dx,$$

where $R = \sum a_h Q_h / f_z'$, has all its periods relative to the vanishing cycles equal to zero, and its periods relative to cycles $\gamma_1, \ldots \gamma_r$ are the r arbitrary constants c_j. We obtain a simple integral u of a total differential of the second kind attached to F and having the same periods c_j by putting

$$u = \int R \, dx + S \, dy,$$

where

$$S = \frac{1}{n} \frac{\partial}{\partial y} \sum_{i=1}^{n} \int_{(x_0, z_i)}^{(x, z)} R(x, y, z) \, dx,$$

x_0 being a fixed value of x and $z_1, \ldots, z_n$ being the roots of the equation $f(x_0, y, z) = 0$. The rationality of the function S follows from the constancy of the periods of the integral I. Since $r = R_1$, our theorem is proved. We observe, however, that the above result yields also a transcendental proof of the relation $r = R_1$. In fact, since there exists a simple integral with r arbitrary periods, it follows that $r \leq R_1$ and consequently $r = R_1$ since by VI, 7 the cycles $\gamma_1, \ldots, \gamma_r$ form a fundamental set of 1-cycles on F.

The question as to the invariant cycles assumes from the transcendental point of view the following form. If γ is an invariant cycle, then the corresponding period $\omega(y)$ of any integral I of type (3) is a uniform function in the whole plane; hence, by (4) and by the remark on the behavior at $y = \infty$, $\omega(y)$ *is a polynomial of degree s_{h-1}*. *The number of independent invariant cycles is therefore equal to the number of independent periods of I which are polynomials in y.* We have seen in VI, 7 that this number is $r = R_1$, a result which PICARD (a_2, p. 389) also obtains by means of the differential equation of order $2p$ satisfied by the $2p$ independent periods of I (equation of FUCHS-PICARD). In view of the preceding considerations it is clear that for the proof of the theorem c. of VI, 7, to the effect that these invariant cycles are independent on F, it is sufficient to show that *there exists on F a simple integral with arbitrarily assigned periods relative to the invariant cycles,* or, what is the same, that *there exists an abelian integral I of type (3) attached to C_y, having zero periods relative to the vanishing cycles δ_i and arbitrarily assigned constant periods relative to the r invariant cycles.* Under this equivalent form theorem c. of VI, 7 is stated without proof in PICARD (a_2, p. 397). It would be interesting to find a transcendental proof of this theorem.

3. On the number of independent simple integrals of the first and of the second kind attached to a surface of irregularity q. The fundamental theorem. The existence of a possible connection between the irregularity q of a surface and the number of the integrals of the first and of the second kind attached to the surface was indicated by early results of HUMBERT and ENRIQUES, and thereafter the establishment of the exact nature of this connection was the object of investigations by ENRIQUES, CASTELNUOVO, SEVERI, PICARD and POINCARÉ. The combined efforts of these authors have led to the following

Fundamental theorem: *An algebraic surface of irregularity q possesses exactly q independent simple integrals of the first kind and $2q \, (= r)$ independent simple integrals of the second kind.*

A first result in this direction was obtained by HUMBERT (2): *if a surface F does not possess simple integrals of the first kind, then every continuous system of curves on F is contained in a linear system.* To see this we replace in the proof of theorem 2 of V, 6 the abelian integral $\bar{\omega}(\xi, \eta)$ of the third kind by an abelian integral $u(\xi, \eta)$ of the first kind. It is then seen immediately that the sum $\sum\limits_{i=1}^{\nu} u(\xi_i, \eta_i) = I(x, y, z)$ is a simple integral of the first kind attached to F. Hence, by hypothesis, $\sum\limits_{i=1}^{\nu} u(\xi_i, \eta_i) = \text{const.}$ and consequently, by ABEL's theorem, the ∞^2 sets of ν points (ξ_i, η_i) vary in a linear series g_ν^r. The rest of the proof is as in V, 1, p. 93 (the theorem of ENRIQUES on rational algebraic systems).

ENRIQUES proved (see V, 3) that in HUMBERT's theorem the concluding part follows also from the hypothesis that F is regular and thus pointed to a possible connection between the class of irregular surfaces and the class of surfaces possessing simple integrals of the first (and hence also of the second) kind.

Subsequently ENRIQUES inverted his last mentioned result (see V, 3), obtaining thus as a corollary of HUMBERT's theorem the following result: *an irregular surface necessarily possesses simple integrals of the first kind.* We have seen however (V, 3) that ENRIQUES' proof is based on an assumption (the completeness of the characteristic series), and that a complete proof of the existence on an irregular surface of continuous systems not contained in linear systems was given only in 1910 by POINCARÉ (see section 5).

In 1901 ENRIQUES (8) proved the following theorem: *if F possesses p' independent simple integrals $I_1, \ldots, I_{p'}$ of the first kind with $2p'$ distinct periods, then there exist on F continuous systems consisting of $\infty^{p'}$ complete linear systems.* The proof was simplified by SEVERI (4) and is as follows: The integrals I_h, considered in a plane section C_y, give rise to a complete regular system (section 6) of p' reducible integrals (this follows from the fact that every 1-cycle of F can be deformed into C_y, see VI, 7). By a well known theorem of PICARD (1) the equations $\sum\limits_{i=1}^{p'} I_h(x_i, y, z_i) = k_h$ $(h = 1, 2, \ldots, p')$ determine, for generic constants k_h, a finite number m of sets of p' points on C_y. As C_y varies in the pencil $\{C_y\}$ these mp' points describe a curve K on F. As the constants k_h vary, this curve K varies in a continuous system, $\infty^{p'}$, and it is immediately seen that there is only a finite number of curves in the system which are linearly equivalent to a generic K, q. e. d.

The value of ENRIQUES' theorem consists in the fact that it would provide a rigorous proof of the completeness of the characteristic series (of sufficiently general complete continuous systems) as soon as it is proved in some manner, *independent of the consideration of continuous systems of curves,* that F possess q independent simple integrals of the first kind with $2q$ periods. Such a proof, however, is not available. The proof given in section 7, a cannot be used advantageously in connection with the theorem of ENRIQUES, because this proof is based on a result of POINCARÉ (sections 4, 5), from which, without any further considerations, the existence of continuous systems consisting of ∞^q linear systems, and hence also the completeness of the characteristic series (V, 3), follows as an immediate corollary.

SEVERI (8) introduces the notion of *the residue function* of an integral $J = \int R \, dx + S \, dy$ of the second kind. We shall assume for simplicity that the polar curve D of J is of the first order, is irreducible and free from singularities, and that the functions R and S vanish to the 1st order on the curve at infinity of F and do not become infinite on any plane section $x =$ const. or $y =$ const. These conditions can always be satisfied by subtracting, if necessary, a suitable rational function from J (SEVERI, l. c., and PICARD, a_1, p. 114). We consider the integral $J(x, \bar{y}, z) = \int R(x, \bar{y}, z) dx$ attached to $C_{\bar{y}}$. Let $(x_0, \bar{y}, z_0)$ be an intersection of D with $C_{\bar{y}}$. Putting $R = A/B$ it is found that the expansion of $J(x, \bar{y}, z)$ on $C_{\bar{y}}$ in the neighborhood of $(x_0, \bar{y}, z_0)$ is of the form

$$J(x, \bar{y}, z) = \frac{c}{x - x_0} \cdot \frac{A(x_0, \bar{y}, z_0)}{B_{z^2}''(x_0, \bar{y}, z_0)} + \text{terms of degree} \geqq 0 \text{ in } x - x_0,$$

where c is a constant. The function $\varphi(x, y, z) = A(x, y, z)/B_{z^2}''(x, y, z)$, *considered on* D, is called by SEVERI *the residue function* determined by J on its polar curve D. It is found that: (1) the poles (of the first order) of φ are at the points at infinity of D and at the points of contact with D of the planes $y =$ const.; (2) the zeros (of the first order) of φ are at the points of D at which the tangent plane to F is parallel to the z-axis, and *at the points of a characteristic set G on D* (V, 1).

The consideration of the residue function φ of J has an immediate application. Suppose that J is a rational function. Then G is simply the set of base points of the pencil $J =$ const., i. e. G is a set of the characteristic series of the linear system $|D|$. Conversely, let G be a set of this series cut out on D by another curve $\overline{D}$ of $|D|$. If R is the rational function possessing D as a polar curve and $\overline{D}$ as a curve of zeros and if φ_1 is the residue function of R on D, it is immediately seen that φ and φ_1 have the same zeros and poles on D and that consequently $\varphi = \lambda \varphi_1$, where λ is a constant. It follows that $J' = J - \lambda R$ is an integral whose residue function is identically zero and which is therefore of the first kind. Consequently, *a necessary and sufficient condition that J be reducible to an integral of the first kind* (by subtraction of a rational function) *is that the set G should belong to the characteristic series of $|D|$.* Now it is clear that if F possesses simple integrals J of the second kind which are not rational functions, i. e. if the BETTI number R_1 of F is

>0 (see section 2), then F also possesses such integrals which are not reducible to integrals of the first kind. In fact, the R_1 periods of J can be assigned arbitrarily, in particular can be taken all real or all pure imaginary, and our statement follows from well known properties of abelian integrals. The above set G for such an integral J does not belong, by the preceding result, to the characteristic series of the complete linear system $|D|$. Consequently *this characteristic series is not complete, and therefore F is irregular* (IV, 3). This result to the effect that *a surface possessing simple integrals of the second or, in particular, of the first kind is necessarily irregular*, is the converse of the theorem of Enriques quoted above. This establishes the complete identity between the class of irregular surfaces and the class of surfaces possessing simple integrals of the first or of the second kind.

We know that the number of independent simple integrals of the second kind is $r = R_1$. Let r_0 be the number of integrals of the first kind. *A first quantitative result was obtained by* Severi (8), *to the effect that* $r - r_0 \leqq q$. This follows immediately from the above criterion for the reducibility of an integral of the second kind to one of the first kind and from the fact that the deficiency of the characteristic series of $|D|$ is $\leqq q$ (IV, 3), and that consequently there exist at most q independent characteristic sets G no combination of which is a set of the characteristic series of $|D|$.

In (9) Severi *derives the equality* $r - r_0 = q$, but he uses the theorem of the completeness of the characteristic series of the complete continuous system $\{D\}$. It is clear that assuming this theorem and taking as $|D|$ a regular system, the equality $r - r_0 = q$ follows, provided it is proved that *any* characteristic set G on D corresponds to some simple integral of the second kind having D as polar curve. Let Σ be a continuous system, ∞^1, containing D and determining on D the given characteristic set G, and let $\varphi(\xi, \eta) = 0$ be the representative curve of the system Σ, in $(1 - 1)$ correspondence with the curves of Σ. Humbert's procedure, repeatedly used before, shows immediately that an abelian integral of the second kind attached to φ and possessing as its only pole the point (ξ_0, η_0) corresponding to D furnishes the required simple integral of the second kind on F.

In the same paper (9) Severi *derives the inequalities* $r_0 \leqq q$ and $r \leqq 2q$. Let $\omega_{hj} = \alpha_{hj} + \beta_{hj}$, $j = 1, 2, \ldots, R_1 (= r)$, be the periods of the integral of the first kind I_h ($h = 1, 2, \ldots, r_0$). We can find an integral A_h of the second kind with periods α_{hj} and a similar integral B_h with periods β_{hj}. From the independence of the integrals I_h it follows immediately that the $2r_0$ integrals A_h and B_h are independent. Hence $r \geqq 2r_0$, and since $r - r_0 \leqq q$, it follows $r_0 \leqq q$ and $r \leqq 2q$.

The inequality $r \leqq 2q$ has also been derived by Picard (a_2, p. 421) from well known properties of reducible abelian integrals (see section 7, a).

The last step toward the proof of the fundamental theorem was made by CASTELNUOVO (9) through the consideration of the variety of PICARD of F (V, 4). CASTELNUOVO *proves that* $r_0 \geq q$, and from this inequality taken together with the inequalities of SEVERI it follows immediately that $r_0 = q$ and $r = 2q$. The following is an outline of CASTELNUOVO's proof. Assuming the existence on F of a continuous system consisting of ∞^q linear systems, let $\{C\}$ be a system ∞^q consisting of linearly isolated curves and let $\varphi(\xi_1, \xi_2, \ldots, \xi_{q+1}) = 0$ be the equation of a projective model in an S_{q+1} of the variety of PICARD V_q attached to F, in $(1-1)$ correspondence with the curves C of $\{C\}$. The variety V_q admits a continuous transitive group G_q of ∞^q birational transformations into itself (see V, 4) and therefore, by general theorems concerning varieties enjoying this property, due to PICARD (4, 5) and PAINLEVÉ (1), possesses q independent simple integrals of total differentials of the first kind:

$$u_i = \int P_1^{(i)} d\xi_1 + \cdots + P_q^{(i)} d\xi_q, \qquad i = 1, 2, \ldots, q,$$

where the P's are rational functions. The inversion of the integrals u_i leads to a parametric representation of V_q by means of $2q$-fold periodic functions ξ_j of the parameters u_i. A transformation of G_q is given by the equations: $u_i' = u_i + h_i$ $(i = 1, 2, \ldots, q)$, where $h_1, \ldots, h_q$ are arbitrary constants. Let γ be a fixed algebraic curve on V_q, the image of an algebraic system Σ, ∞^1, immersed in $\{C\}$, and let $(\xi^{(1)}), \ldots, (\xi^{(m)})$ be the points of γ which correspond to the curves of Σ passing through a point $P(x, y, z)$ on F (m is the index of Σ).

The sums

(6)
$$\sum_{j=1}^{m} u_h(\xi^{(j)}) = I_h(x, y, z), \qquad h = 1, 2, \ldots, q,$$

define simple integrals of the first kind attached to F. It these integrals are independent, the inequality $r_0 \geq q$ is proved. The real difficulty arises if these integrals are not independent. An essential point of the proof consists first of all in showing that *the extreme case in which all the integrals* I_h *reduce to constants does not arise*. The proof is based on the following lemma of SEVERI (11):

Lemma. *If an algebraic series* $s_{\nu, m}$, ∞^1, *of sets of* ν *points and of index* m *on an algebraic curve* Γ *is such that the collections of* m *sets of the series passing through a variable point of* Γ *vary in a linear series (of order* $m\nu$*), then the sets of* $s_{\nu, m}$ *are themselves contained in a linear series (of order* ν*).*

To make use of this lemma we observe that to each element T of the group G_q there corresponds an operation in $\{C\}$, of the type $C' \equiv C + C_T - C_0$, where C_0 is a fixed curve in $\{C\}$ and where C_T is fixed but depends on T (see V, 4). Consequently the congruences $u_i + u_i' \equiv \bar{u}_i + \bar{u}_i'$ (mod. periods) $(i = 1, 2, \ldots, q)$ imply the equivalence relation:

$C + C' \equiv \bar{C} + \bar{C}'$, and conversely. More generally, it is seen immediately that the congruences $u_i^{(1)} + u_i^{(2)} + \cdots + u_i^{(m)} \equiv \bar{u}_i^{(1)} + \bar{u}_i^{(2)} + \cdots + \bar{u}_i^{(m)}$ (mod. periods) $(i = 1, 2, \ldots, q)$ imply the equivalence relation $C_1 + C_2 + \cdots + C_m \equiv \bar{C}_1 + \bar{C}_2 + \cdots + \bar{C}_m$, where C_α $(\bar{C}_\alpha)$ is the curve of $\{C\}$ which corresponds to the point $u^{(\alpha)}$ $(\bar{u}^{(\alpha)})$ of V_q. Let us suppose that $I_h = \text{const.}$ $(h = 1, 2, \ldots, q)$. Then (6) shows that the ∞^2 sets of m curves of Σ which pass through a variable point of F vary in a linear system. We fix an irreducible curve A on F and we consider the series $s_{\nu, m}$ $(\nu = (A \cdot C))$ cut out on A by the curves of Σ. This series satisfies the condition of the preceding lemma and is consequently contained in a linear series on A. If we assume for A a generic curve of an irreducible linear pencil of positive degree, then it follows by Severi's criterion 1 of V, 5 that Σ is contained in a linear system. This is a contradiction, since Σ is an arbitrary algebraic system immersed in $\{C\}$ and since the generic curve C is isolated with respect to linear equivalence.

Let q' be the number of independent integrals I_h. We have $0 < q' \leqq q$. Castelnuovo proves that if $q' < q$, then *the group G_q is imprimitive*: V_q is ruled out by $\infty^{q-q'}$ algebraic varieties $V_{q'}$ and by $\infty^{q'}$ algebraic varieties $V_{q-q'}$, constituting two systems of imprimitivity of G_q. For a proper choice of the integrals u_i, $q - q'$ integrals $u_{q'+1}, \ldots, u_q$ are constant on $V_{q'}$, while the q' integrals $u_1, \ldots, u_{q'}$ are constant on $V_{q-q'}$ and their corresponding sums (6) give q' independent simple integrals I_h. Each $V_{q'}$ and $V_{q-q'}$ is again a variety of Picard. To complete the q' integrals I_h by $q - q'$ other independent integrals of the first kind it is sufficient to consider an algebraic curve γ' on a $V_{q-q'}$ and to apply to γ' the same method as was applied to γ. The simple integrals furnished by γ' are independent of the q' integrals I_h and are $q - q'$ in number or less. In the second case $V_{q-q'}$ presents the same features of imprimitivity as V_q and the above procedure can be repeated, until a set of q independent simple integrals of the first kind is obtained.

A more simple proof of the fundamental theorem was given subsequently by Severi (11). Using his lemma stated above, Severi proves the following *transcendental criterion of linear equivalence*:

Let Σ be an algebraic system, ∞^1, of curves C on F and let $|A|$ be an irreducible linear pencil of positive degree. A necessary and sufficient condition that Σ be contained in a linear system is that for any simple integral I of the first kind attached to F the sum of the values which I assumes at the intersections of C with a fixed A should remain constant as C varies in Σ.

Proof. Let $P_i(x_i, y_i, z_i)$ $(i = 1, 2, \ldots, n)$ be the intersections of a variable C with the fixed A. That the condition $\sum_{i=1}^{n} I(x_i, y_i, z_i) = \text{const.}$

is necessary, follows from the ordinary theorem of ABEL. To prove that it is sufficient, we consider the usual curve Γ, $\varphi\,(\xi,\eta)=0$, the image of Σ, and the usual sets of points (ξ_i,η_i), $i=1,2,\ldots,\nu$, where ν is the index of Σ. If $u(\xi,\eta)$ is an abelian integral of the first kind on Γ, then $\sum_{i=1}^{\nu} u(\xi_i,\eta_i)=I(x,y,z)$ is a simple integral of the first kind attached to F. From the hypothesis $\sum_{i=1}^{n} I(x_i,y_i,z_i)=\text{const.}$ it follows immediately, by the ordinary theorem of ABEL, that the series $s_{\nu,n}$ on Γ, of order ν and index n, whose sets correspond to the points of the fixed curve A, satisfies the condition of SEVERI's lemma. Hence the sets of this series are equivalent to each other. The (n,ν) correspondence between A and Γ is therefore of valence zero in one sense and consequently, by the classical theory of algebraic correspondences, is of valence zero in both senses. We have then that *the curves C cut out on the fixed A equivalent sets of points.* Let A_1 be *any* irreducible curve of $|A|$. The curves A and A_1 cut out on C equivalent sets and therefore the corresponding abelian sums relative to any integral I are congruent. It follows that these sums remain constant also for an arbitrary curve A_1 of $|A|$, and consequently *the curves C cut out equivalent sets on any curve of $|A|$*. The transcendental criterion now follows from the criterion 1 of V, 5.

We observe incidentally that from the criterion just proved follows the criterion 3 of V, 5.

Let $\{C\}$ be the continuous system, ∞^q, considered in the proof of CASTELNUOVO, and let C_0 be a generic curve of $\{C\}$, such that every curve C in a q-dimensional neighborhood N_q of C_0 is linearly isolated. From the above transcendental criterion it follows that the sums $\sum_{i=1}^{m} I_h(\xi_i,\eta_i)=c_h$ $(h=1,2,\ldots,r_0)$ relative to the intersections of a variable C of N_q with a fixed A assume ∞^q distinct sets of values $(c_1,\ldots,c_{r_0})$. Hence $r_0 \geqq q$, q. e. d.

4. The normal functions of POINCARÉ (POINCARÉ, 8, 9). We shall now assume that of the $2p$ abelian integrals (3) of section 2, p are abelian integrals of the first kind:

$$I_h = \frac{P_h(x,y,z)}{f_z'}\,dx, \qquad\qquad h=1,2,\ldots,p,$$

where P_h is a polynomial of degree $n-3$ in x and z and of a certain degree $n-3+s_h$ $(s_h \geqq 0)$ in x,y and z. To determine rationally the integrals I_h we may proceed as follows. We consider $p-1$ generic lines in S_3. Their $p-1$ intersections with the plane of a generic C_y determine uniquely in this plane an adjoint curve Γ_{n-3} of C_y, of order $n-3$. The locus of Γ_{n-3}, as y varies, is an adjoint surface $P(x,y,z)=0$ of a certain order $n-3+s$, passing s times through the axis of

the pencil $y = $ const., i. e. $P(x, y, z)$ is of degree $n - 3 + s$ in all the
variables and of degree $n - 3$ in x and z. In this manner we may
determine the p independent integrals I_h.

From Chapter IV, 1 it follows that there exist $p - \omega_{n-3} (\geqq p - q)$ polynomials P_h of degree $n - 3$ in all the variables. Although it is not essential for
what follows, it will to some extent simplify matters, if we suppose that
the adjoint system of the system $|C|$ of plane sections of F is regular[1].
There will be no loss of generality, since we are dealing with birational pro-
perties of F, and we may, if necessary, replace $|C|$ by a sufficiently high
multiple of $|C|$ (p. 78; see also section 7, d, of this chapter). Under this
hypothesis we shall have $\omega_{n-3} = q$, $\omega_{n-2} = 0$ (IV, 1) and hence *we may
choose $p - q$ of the polynomials, say $P_{q+1}, \ldots, P_p$, to be of total degree
$n - 3$ and the remaining q polynomials $P_1, \ldots, P_q$ of total degree $n - 2$.*

Let $\Omega_{hj}(y)$ and $\omega_h(y)$ be the periods of I_h relative to the vanishing
cycle δ_j and an arbitrary cycle γ respectively. The period $\Omega_{hi}(y)$ is
uniform and regular at $y = \alpha_i$ (section 2), while for $\omega_h(y)$ the representation (4) holds. From (4) it follows that the difference

$$(7) \qquad \omega_h(y) - \sum_{j=1}^{\mu} \frac{(\gamma \cdot \delta_j)}{2\pi i} \int_a^{\alpha_j} \frac{\Omega_{hj}(Y)}{Y - y} dY$$

is uniform and regular at each critical point α_i. The path of integration
from a to α_i is the cut l_i. As y turns about $y = a$, the second term of
the above difference increases by the period of I_h relative to the corresponding increment of the cycle γ, and this increment is ~ 0 on C_y
[see VI, 6, formula (5)]. Hence (7) is regular at $y = a$. As $y \to \infty$ the
integrals in the second term of (7) approach zero. From the behavior
of $\omega_h(y)$ at $y = \infty$ (section 2) we derive the following representation of
$\omega_h(y)$ in the whole plane (POINCARÉ, 8, 9):

$$(8) \qquad \omega_h(y) = \sum_{j=1}^{\mu} \frac{\lambda_j'}{2\pi i} \int_a^{\alpha_j} \frac{\Omega_{hj}(Y)}{Y - y} dY + d_h', \qquad h = 1, 2, \ldots, q,$$

$$(8') \qquad \omega_h(y) = \sum_{j=1}^{\mu} \frac{\lambda_j'}{2\pi i} \int_a^{\alpha_j} \frac{\Omega_{hj}(Y)}{Y - y}, \qquad h = q + 1, \ldots, p,$$

where $\lambda_j' = (\gamma \cdot \delta_j)$, and where $d_1', \ldots, d_q'$ are constants[2].

---

[1] We shall see later (section 7, d) that in point of fact this system is always
regular (theorem of PICARD).

[2] If the adjoint system of the system of plane sections were not assumed to
be regular, than instead of the constants d_h' we would have polynomials in y.
This remark applies also to the constants d_h in relations (10). In this last form
(i.e. with polynomials instead of constants) the relations (10) were given by
POINCARÉ. LEFSCHETZ (a, p. 73) proved directly that the polynomials necessarily
reduce to constants.

Let D be an algebraic curve traced on F and let $B_i(x_i, y, z_i)$, $i = 1$, $2, \ldots, k$, be the intersections of D with a generic C_y. POINCARÉ considers the abelian sums

$$(9) \qquad \nu_h(y) = \sum_{i=1}^{k} I_h(x_i, y, z_i) = \sum_{i=1}^{k} \int_{A_1}^{B_i} dI_h, \qquad h = 1, 2, \ldots, p,$$

where A_1 is one of the base points $A_1, \ldots, A_n$ of $\{C_y\}$ and where the paths of integration are arcs m_i on C_y joining A_1 to B_i, which are continuously deformed as y varies. A representation analogous to (8), (8') holds also for $\nu_h(y)$, the only difference being that the integration domain is now the 1-chain $m = m_1 + \cdots + m_k$ instead of the cycle γ. Hence

$$(10) \qquad \nu_h(y) = \sum_{j=1}^{\mu} \frac{\lambda_j}{2\pi i} \int_{a}^{\alpha_j} \frac{\Omega_{h,j}(Y)}{Y - y} dY + d_h, \qquad h = 1, \ldots, q,$$

$$(10') \qquad \nu_h(y) = \sum_{j=1}^{\mu} \frac{\lambda_j}{2\pi i} \int_{a}^{\alpha_j} \frac{\Omega_{h,j}(Y)}{Y - y} dY, \qquad h = q + 1, \ldots, p,$$

where $\lambda_j = (m \cdot \delta_j)$. For the curve D we have a homology of the form $D \sim \sum \mu_i \Delta_i + (C_a)$ [VI, 10, formula (10)]. It is seen immediately that the integers μ_i coincide with the λ_i's by considering the boundary of the 3-chain C_3 generated by m. In fact, we have (VI, 6) $F(C_3) \sim D - \sum \lambda_i \Delta_i - (C_a)$, and hence

$$(11) \qquad D \sim \sum_{i=1}^{\mu} \lambda_i \Delta_i + (C_a) \quad \text{on } F.$$

Given D, the integers λ_i are not uniquely determined, because they depend on the paths of integration m_i. If we change the m_i's into new paths $\bar{m}_i$, the difference $\bar{m} - m$, where $\bar{m} = \bar{m}_1 + \cdots + \bar{m}_k$, is a cycle γ; the abelian sum $\nu_h(y)$ is increased by the period $\omega_h(y)$ relative to γ and given by (8), (8'), and the coefficients λ_i are changed into $\bar{\lambda}_i = \lambda_i + \lambda_i' = \lambda_i + (\gamma \cdot \delta_i)$. The new homology $D \sim \sum \bar{\lambda}_i \Delta_i + (C_a)$ is a consequence of (11) in view of the homology $\sum (\gamma \cdot \delta_i) \Delta_i + (C_a) \sim 0$ on F, which holds for any cycle γ [VI, 10, formula (12)].

If the λ_i's are fixed, then we may vary the chain m only to within invariant cycles. The constants d_h will then be increased by a linear homogeneous combination with integral coefficients of the periods of I_h relative to the r invariant cycles. These periods are constants for $h = 1, 2, \ldots, q$ and are all zero for $h = q + 1, \ldots, p$ (section 2, p. 159).

If some of the intersections of D with C_y are at the base points A_i of $\{C_y\}$, these may be all or in part included among the points B_i or may all be disregarded. The effect of this on the abelian sums is the addition or subtraction of the abelian sum relative to a sum of multiples of the points A_i, $t_1 A_1 + t_2 A_2 + \cdots + t_n A_n$, i. e. of the quantity

$$
(12) \qquad \sum_{s=2}^{n} t_s \int_{A_1}^{A_s} d I_h = \sum_{j=1}^{\mu} \sum_{s=2}^{n} \frac{t_s(c_s \cdot \delta_j)}{2\pi i} \int_{\alpha}^{\alpha_j} \frac{\Omega_{hj}(Y)}{Y-y} \, dY + k_h ,
$$

where $c_2, \ldots, c_n$ are the open arcs from A_1 to A_s considered in VI, 8 and where $k_1, \ldots, k_p$ are fixed constants ($k_h = 0$, if $h > q$). It is important to observe that if the integers t_s, $s > 1$, are not all zero, then the abelian sums (12) never reduce to periods $\omega_h(y)$ or to $\omega_h(y) + \text{const.}$ (Lefschetz, a, p. 65). In fact, assuming the contrary, let γ be the cycle defining the period $\omega_h(y)$. Since the abelian sum (12) and $\omega_h(y)$ increase in the neighborhood of α_i by $\sum_{s=2}^{n} t_s (c_s \cdot \delta_i) \Omega_{hi}$ and by $(\gamma \cdot \delta_i) \Omega_{hi}$ respectively and since their difference is a constant, it follows that $(c \cdot \delta_i) = 0$, $i = 1, 2, \ldots, \mu$, where $c = \gamma - t_2 c_2 - \cdots - t_n c_n$. Hence c is invariant, and this is impossible if the t_i's are not all zero (VI, 8).

The abelian sums $\nu_h(y)$ were called by Poincaré (8) the *normal functions* belonging to the *algebraic* curve D. Given D and assuming a definite convention concerning the possible fixed intersection of D with C_y, the normal functions of D are determined to within periods of the integrals I_h.

We give below a few properties of the normal functions, which easily follow from their definition. In all the propositions below it is assumed, for simplicity, that the set of intersections of a D with a C_y includes all the variable *and* fixed intersections and that the normal functions of D are evaluated accordingly.

a. *A necessary and sufficient condition that $D_1 \equiv D_2$ is that their normal functions ν_h^1 and ν_h^2 coincide to within periods, i. e. that $\nu_h^1 \equiv \nu_h^2$ mod. (periods)* (Poincaré, 8). That the condition is necessary is obvious. As for its sufficiency, if the curves D_1 and D_2 are assumed to have the same order, the statement is nothing but a transcendental formulation of Severi's linear equivalence criterion 1 (V, 5). Lefschetz (a, p. 77) has observed that this assumption is superfluous. If $(D_1 \cdot C_y) = k_1$, $(D_2 \cdot C_y) = k_2$ and if $k_1 \leq k_2$, the two sets of k_2 points, $D_1 \cdot C_y + (k_2 - k_1) A_1$, $(D_2 \cdot C_y)$ are equivalent on C_y. Repeating the reasoning employed in the proof of Severi's criterion referred to above, we derive that $D_1 + t C_y \equiv D_2$, where t is an integer. Since the normal functions of D_1 and D_2 coincide, it follows that those of $t C_y$, i. e. the expressions (12), where now $t_2 = \cdots = t_n = t$, must be equal to periods. This implies $t = 0$, q. e. d.

b. If $D \equiv D_1 + D_2$, then $\nu_h \equiv \nu_h^1 + \nu_h^2$.

c. *A necessary and sufficient condition that D_1 and D_2 be algebraically equivalent* $(D_1 = D_2)$ *is that $\nu_h^1 \equiv \nu_h^2 + \text{const.}$ ($h = 1, 2, \ldots, q$), $\nu_h^1 \equiv \nu_h^2$ ($h = q + 1, \ldots, p$)* (Severi, 20). The necessary character of the condition follows from the fact that if $D_1 = D_2$, then there exists an integer t such that $D_1 + tC_y \equiv D_2 + tC_y$ (V, 2), and from the fact that if a curve D varies in a continuous system $\{D\}$, then the integers λ_i which occur in the expression of ν_h must remain fixed. Conversely, if the condition is satisfied then it follows from the homology (11) that $D_1 \sim D_2 + tC_y$, where t is an integer. From this it follows in the first place, as in a., that $t = 0$, and then from $D_1 \sim D_2$ it follows, by LEF-SCHETZ's theorem (section 7, b; see also VI, 12), that $D_1 = D_2$.

d. *A necessary and sufficient conditions that given curves* $D_1, D_2, \ldots, D_t$ *be algebraically dependent is that their normal functions* $\nu_h^1, \nu_h^2, \ldots, \nu_h^t$ *satisfy the congruences*

$$(13) \begin{cases} \mu_1 \nu_h^1 + \mu_2 \nu_h^2 + \cdots + \mu_t \nu_h^t \equiv \text{const. (mod. periods)}, & h = 1, \ldots, q, \\ \mu_1 \nu_h^1 + \mu_2 \nu_h^2 + \cdots + \mu_t \nu_h^t \equiv 0 \ \text{(mod. periods)}, & h = q + 1, \ldots, p, \end{cases}$$

where the μ's are integers, not all zero. This is a corollary of c.

By the subtraction of periods it is seen that the equations (13) are equivalent to the following equations:

$$\mu_1 \lambda_i^{(1)} + \mu_2 \lambda_i^{(2)} + \cdots + \mu_t \lambda_i^{(t)} = 0, \quad i = 1, 2, \ldots, \mu.$$

If t is sufficiently high (for instance if $t \geq \mu = $ class of F), these equations admit a solution different from the trivial solution $\mu_1 = \cdots = \mu_t = 0$. This furnishes a new proof of the existence of a base for the algebraic curves on F and of the invariant ϱ of PICARD (V, 6).

We add one remark concerning the normal functions of the curves tC. They are the abelian sums relative to the set of points $t(A_1 + A_2 + \cdots + A_n)$, i. e. the functions (12), where now $t_2 = t_3 = \cdots = t_n = t$. Let us impose on the system $|tC|$ the set G_t of base points $A_2, \ldots, A_m$, each taken with an assigned multiplicity t. Observing that $|C|_{G_1} (G_1 = A_2 + \cdots + A_n)$ is of virtual positive degree $(=1)$ and applying the RIEMANN-ROCH theorem, it is seen immediately that if t is sufficiently high the system $|tC|_{G_t}$ is irreducible and does not possess accidental base points. The curves of this system have t variable intersections with C_y, and the set of these intersections varies in the linear series $|tA_1|$. *The normal functions of the curves of* $|tC|_{G_t}$, *evaluated with reference to the variable intersections only, are therefore equal to periods.* For the curves of the continuous system $\{tC\}$ passing through the base set G_t, the first q normal functions are congruent to constants while the remaining $p - q$ normal functions are equal to periods. These particular normal functions constitute the point of departure of SEVERI's re-elaboration of POINCARÉ's method (SEVERI, 20).

5. The existence theorem of LEFSCHETZ-POINCARÉ (POINCARÉ, 8, 9; LEFSCHETZ, a, 5; SEVERI, 20). The consideration of the normal functions of POINCARÉ suggests *two* existence problems.

The first is the following. Let ν_h be the normal functions of an algebraic curve D. From the theorems a. and c. of the preceding section it follows that as D varies in a continuous *non-linear* system $\{D\}$ the q additive constants d_h in (10) vary. *We may ask whether there exist curves in $\{D\}$ corresponding to any preassigned values of the constants d_h.* The importance of this question appears immediately if it is observed that the affirmative answer would provide, at last, the proof of the fundamental theorem repeatedly used before (see especially V, 3 and VII, 3): *there exist on F continuous systems consisting of ∞^q distinct linear systems.* These linear systems of $\{D\}$ correspond to the ∞^q non-congruent sets of the constants d_h.

If the first question is answered in the affirmative, it would follow that given p function ν_h of type (10) and (10′), the conditions that there should exist an algebraic curve of which they are the normal functions involve only the integers λ_i. *What are these conditions?* In substance the question we are dealing with is the following: *under what conditions is the cycle $\sum \lambda_i \Delta_i + (C_a)$ algebraic?*

The proof that the first question is to be answered in the affirmative is due to POINCARÉ (8, 9). POINCARÉ's proof has been subsequently simplified by SEVERI (20). The second problem has been solved by LEF-SCHETZ (a, 5). The following treatment is as in LEFSCHETZ and settles simultaneously both existence problems.

It is sufficient to consider the theoretically important case $k = p$. We are, therefore, dealing with the following equations:

$$(14) \quad \sum_{s=1}^{p} I_h(x_s, y, z_s) = \sum_{j=1}^{\mu} \frac{\lambda_j}{2\pi i} \int_a^{\alpha_j} \frac{\Omega_{h,j}(Y)}{Y-y} dY + d_h = \nu_h(y), \quad h = 1, 2, \ldots, p,$$

where $d_h = 0$, for $h > q$. We have here the typical equations of the problem of inversion of the abelian integrals attached to C_y. Since the functions $\nu_h(y)$ are determined to within periods Ω_{hi}, the equations (14) *determine* for a generic y a set of p points $B_s(x_s, y, z_s)$ on C_y. It may happen that the functions $\nu_h(y)$ give for any y a set of p characteristic constants of a *special set of p points.* In this case there are infinitely many sets of p points B_s satisfying (14), and they constitute a linear series $g_p^{p'}$, $p' > 0$. It is then sufficient to fix rationally on C_y p' generic points, and the remaining $p - p'$ points will be uniquely determined. As y varies, the set $B_1, \ldots, B_p$ varies and describes a certain locus. The question is: under what conditions is this locus an algebraic curve?

Any rational symmetric function $R(x_1, z_1; x_2, z_2; \ldots; x_p, z_p)$ of the points $B_1, \ldots, B_p$ is a uniform function of y. In order that to (14) there should correspond an algebraic curve it is, therefore, necessary and sufficient that R be a meromorphic function in the whole plane of the variable y, or, what is the same, that x_s and z_s be algebroid functions of y. For a generic value y_0 of y each of the equations (14) is of the type $F(x_1, z_1, \ldots; x_p, z_p) = 0$, where F is a regular function of its variables. The only values of y which appear as possible singularities of R are the following:

1. $y = \alpha_i$, $i = 1, 2, \ldots, \mu$.

2. The values b of y such that the constants $v_h(b)$ do not determine a unique set of p points B_i.

3. $y = a$.

4. The so-called *singular values* y_0 of y, i. e. such that on C_{y_0} the integrals I_h cease to be independent.

Very simple considerations show that at the values (1) and (2) of y the functions x_i and z_i do not cease to be algebroid. The case $y = a$ leads to the trivial condition that *it be possible to form with the integers* λ_i *a 2-cycle* $\sum \lambda_i \Delta_i + (C_a)$. In fact, if this condition is satisfied, then $\sum \lambda_i \delta_i^a \sim 0$ on C_a (VI, 10), and consequently the functions $v_h(y)$ do not depend on a, since their derivatives with respect to a are $\sum \lambda_i \Omega_{h i}(a)$ and therefore vanish.

The consideration of the singular values (4) is of special interest and leads to significant conclusions. If $y = y_0$ is a singular value of y, the polynomials $P_i(x, y_0, z)$ $(i = 1, 2, \ldots, p)$ are dependent and consequently the following identity holds:

$$\sum_{i=1}^{q} c_i P_i(x, y, z) + \sum_{j=q+1}^{p} d_j P_j(x, y, z) = (y - y_0) \psi(x, y, z),$$

where $\psi(x, y, z)$ is an adjoint polynomial of degree $n - 3$ at most (and hence can be expressed as a linear combination of $P_{q+1}, \ldots, P_p$) and the c's and d's are constants. This identity is impossible, unless all the c's are zero, because in the contrary case the integrals I_h would be dependent for *each* value of y. Hence

$$\sum_{j=q+1}^{p} d_j P_j(x, y, z) = (y - y_0) \psi(x, y, z),$$

where now ψ is of degree $n - 4$ at most. We may replace one of the polynomials P_j (such that $d_j \neq 0$) by ψ, without creating new singular values of y for the new set of $p - q$ polynomials. Repeating this procedure of replacing adjoint polynomials of degree $n - 3$ by adjoint polynomials of degree $\leq n - 4$, we shall finally arrive at a set of $p - q$ polynomials for which the only singular value is $y = \infty$. We suppose that the polynomials P_i already satisfy this condition. It is immediately seen that $y = \infty$ is a singular value if the following identity holds: $\sum d_j P_j = \psi(x, y, z)$, where ψ is of degree $\leq n - 4$. The maximum number of independent adjoint polynomials of degree $n - 4$ being p_g, we may suppose that the polynomials P_{q+i}, $i = 1, 2, \ldots, p_g$, are of degree $n - 4$ (and not less) and that the polynomials P_{q+p_g+j}, $j = 1, 2, \ldots, p - q - p_g$, of degree $n - 3$, are such that no combination of these polynomials is of degree $\leq n - 4$. It is immediately seen that the integrals $I_{q+1}, \ldots, I_{q+p_g}$ are then identically zero for $y = \infty$ (while the integrals $I_{q+p_g+1}, \ldots, I_p$ are independent on C_∞).

To investigate the equations (14) in the neighborhood of $y = \infty$, we apply the homographic transformation: $x = \dfrac{x'}{y'}$, $y = \dfrac{1}{y'}$, $z = \dfrac{z'}{y'}$. The equations (14) become

$$(14') \qquad \sum_{j=1}^{p} \int_{A_1}^{B_j} \frac{\overline{P}_h(x', y', z')}{\overline{P}_{z'}} dx' = y'^{\varepsilon}\left(y' \sum_{j=1}^{\mu} \frac{\lambda_j}{2\pi i} \int_a^{\alpha_j} \frac{\Omega_{hj}(Y)}{Yy'-1} dY + d_h\right),$$

where $\overline{P}_h$ and $\overline{f}$ are the transforms of P_h and f, and $\varepsilon = 0$, -1 or -2 according as P_h is of degree $n-2$, $n-3$ or $n-4$. Since $d_h = 0$ for $h > q$, it follows that the right-hand member of (14') is regular at $y' = 0$, except when $h = q+1, \ldots, q+p_g$, in which case, expanding $\dfrac{1}{Yy'-1}$ in the neighborhood of $y' = 0$, we find a term $y'^{-1} \cdot \sum \dfrac{\lambda_j}{2\pi i} \int_a^{\alpha_j} \Omega_{hj}(Y)\,dY$. Hence the condition of regularity in this case is

$$(15) \qquad \sum_{j=1}^{\mu} \lambda_j \int_a^{\alpha_j} \Omega_{hj}(Y)\,dY = 0, \qquad h = q+1, \ldots, q+p_g.$$

We now interprete the equations (15). In a general manner, let us have a double integral $\iint R(x,y,z)\,dx\,dy$ and a cycle $\Gamma_2 \sim \sum \lambda_j \Delta_j + (C_a)$. If we calculate the period of the double integral relative to Γ_2 by integrating first with respect to x along the cycle $\sum \lambda_j \delta_j$ and then with respect to y, we find

$$(16) \qquad \iint_{\Gamma_2} R(x,y,z)\,dx\,dy = \sum \lambda_j \int_a^{\alpha_j} \Omega_j(Y)\,dY,$$

where $\Omega_j(y)$ is the period of $\int R(x,y,z)\,dx$ relative to δ_j. The equations (15) show that *the periods of the p_g double integrals $\iint \dfrac{P_{q+i}}{f_z'} dx\,dy$ $(i = 1, 2, \ldots, p_g)$ relative to the cycle $\sum_{j=1}^{\mu} \lambda_j \Delta_j + (C_a)$ are all zero.* Since these are the p_g independent double integrals of the first kind, we have the following

Existence theorem. *A necessary and sufficient condition that the normal functions $v_h(y)$ belong to an algebraic curve D is that $\sum_{j=1}^{\mu} \lambda_j \Delta_j + (C_a)$ be a cycle and that the period of any double integral of the first kind relative to this cycle vanish* (LEFSCHETZ, a; 5).

The second condition characterizes the algebraic cycles from a transcendental point of view.

Remark. The necessary character of the above condition can also be derived directly as follows. It obviously holds for any curve C_y, since $dy = 0$. For an arbitrary algebraic curve, whose projection on to the (x, y) plane is $\varphi(x, y) = 0$, it is sufficient to replace the variable x by $\varphi(x, y)$. This is possible except for the neighborhoods of the points (x', y') at which $\varphi_x' = 0$. The contribution of these neighborhoods to the value of any double integral may be supposed to be arbitrarily small.

6. Reducible integrals. Theorem of POINCARÉ. We first recall a few classical properties of RIEMANN matrices. Matrix notations shall be used throughout this section (LEFSCHETZ, c, Chapter XVII).

A matrix $\omega = \|\omega_{ij}\|$ with p rows and $2p$ columns of complex elements is called a RIEMANN matrix of genus p, if there exists a rational skew-symmetric matrix C of $2p$ rows, such that

$$(17) \qquad\qquad \omega C \omega' = 0,$$

$$(17') \qquad \sqrt{-1}\, \omega C \bar{\omega}' \text{ is a positive definite Hermitian matrix,}$$

where $\bar{\omega}'$ is the conjugate transposed of ω. C is called a *principal* matrix of ω. This definition implies that C is non-singular. Two RIEMANN matrices ω and ω^* of the same genus are isomorphic, if there exists a non-singular rational $2p$-rowed square matrix G and a non-singular complex p-rowed square matrix A, such that $\omega^* = A \omega G'$. If G is unimodular then ω and ω^* are *equivalent*. If A and G are given, then also ω^* is a RIEMANN matrix, and its principal matrix is

$$(18) \qquad\qquad C^* = (G^{-1})' C G^{-1},$$

and is independent of A. For a proper unimodular matrix G, C^* assumes the canonical form (FROBOENIUS):

$$(19) \qquad\qquad C^* = \begin{pmatrix} 0 & d \\ -d & 0 \end{pmatrix},$$

where

$$(19') \qquad\qquad d = \begin{pmatrix} d_1 & & 0 \\ & \ddots & \\ 0 & & d_p \end{pmatrix},$$

$d_1, d_2, \ldots, d_p$ being the invariant factors of C. Choosing properly the matrix A, it is possible to get for ω^* a matrix of the form: $\omega^* = (E_p, \omega_1^*)$, where E_p is the p-rowed unit matrix, and ω_1^* is necessarily non-singular. It follows incidentally that ω is of rank p.

The period matrix ω of p independent abelian integrals I_h of the first kind, attached to an algebraic curve of genus p, relative to a given set of $2p$ independent 1-cycles γ_i, is a RIEMANN matrix of genus p. The passage from ω to an equivalent matrix ω^* signifies in this case a transformation of the integrals I_h:

$$I_h^* = \sum_{k=1}^{p} a_{hk} I_k, \qquad\qquad h = 1, 2, \ldots, p,$$

where $\|a_{hk}\| = A$, accompanied by a transformation of the cycles γ_i:

$$g \gamma_i^* \sim \sum_{j=1}^{2p} g_{ij} \gamma_j, \qquad\qquad i = 1, 2, \ldots, 2p,$$

where $\left\|\dfrac{g_{ij}}{g}\right\| = G$. If the cycles γ_i and γ_i^* both form minimal bases, then $g = 1$ and G is unimodular. If the cycles are retrosections, then C has the canonical form (19), and $d_1 = d_2 = \cdots = d_p = 1$.

The most general RIEMANN matrix of genus p is the period matrix of a class of $2p$-ply periodic meromorphic functions of p independent variables.

Given $p - q$ $(q > 0)$ independent abelian integrals $I_1, \ldots, I_{p-q}$, they are said to form a set of *reducible integrals*, if they have zero periods along r independent 1-cycles $(r > 0)$, $\gamma_1, \gamma_2, \ldots, \gamma_r$. There always exists a minimal base of cycles $\bar{\gamma}_i$ $(i = 1, 2, \ldots, 2p)$, such that $\gamma_i = t_i \bar{\gamma}_i$ $(i = 1, 2, \ldots, r;\ t_i$ an integer). The given $p - q$ integrals I_h have zero periods also along the cycles $\bar{\gamma}_1, \ldots, \bar{\gamma}_r$; hence we may assume that the cycles $\gamma_1, \ldots, \gamma_r$ can be completed by $2p - r$ cycles $\gamma_{r+1}, \ldots, \gamma_{2p}$, so as to have $2p$ cycles forming a minimal base. The given integrals I_h have essentially only $2p - r$ distinct periods at most, called *reduced periods*: those relative to the cycles $\gamma_{r+1}, \ldots, \gamma_{2p}$. The non-reduced periods of I_h are its $2p$ periods relative to an arbitrary minimal base. With our choice of the cycles $\gamma_{r+1}, \ldots, \gamma_{2p}$, the reduced periods and the non-reduced periods are integral linear combinations of each other.

If the first $p - q$ of the p independent integrals I_h are the above reducible integrals, and if the minimal base is as above, then ω is of the form $\begin{pmatrix} 0 & \omega_1 \\ \omega_2 & \omega_3 \end{pmatrix}$, where ω_1 is the matrix of the reduced periods of $I_1, \ldots, I_{p-q}$. If we write C in the form $\begin{pmatrix} C_1 & C_2 \\ C_3 & C_4 \end{pmatrix}$, where C_4 has $2p - r$ rows and columns, we find from (17) and (17'):

$$(20) \qquad\qquad \omega_1 C_4 \omega_1' = 0;$$

$$(20') \qquad \sqrt{-1}\, \omega_1 C_4 \bar{\omega}_1' \text{ is a positive definite Hermitian matrix.}$$

It is seen immediately that if $r > 2q$, there would exist a linear combination of the integrals $I_1, \ldots, I_{p-q}$, whose periods are all real. The existence of such an integral would make (20) and (20') inconsistent. Hence necessarily $r \leqq 2q$, i.e. $p - q$ *independent reducible integral cannot have less than* $2(p - q)$ *reduced periods.*

If r has the maximum value $2q$, the integrals $I_1, \ldots, I_{p-q}$ are said to form a *regular system of reducible integrals*. From (20) and (20') it follows that *in this case the matrix* ω_1 *of the reduced periods is a* RIEMANN *matrix of genus* $p - q$, *and that* C_4 *is its principal matrix.*

We now prove the following lemma, due to ALBERT (3):

Lemma. *Let* ω *be an* RIEMANN *matrix of genus* p, *C a principal matrix of* ω. *If* φ *is a matrix satisfying the relation* $\varphi C \omega' = 0$, *then there exists a matrix* δ, *such that* $\varphi = \delta \omega$.

Proof. We first consider the case in which C has the canonical form (19) and $\omega = (E_p, \omega_1)$, where $\omega_1 = \|\omega_{hk}\|$, h, $k = 1, 2, \ldots, p$. Let $\varphi = \|\varphi_{ij}\|$, $i = 1, 2, \ldots, s$, $j = 1, 2, \ldots, 2p$. We have from (17):

$$(21) \qquad d_h\,\omega_{kh} = d_k\,\omega_{hk}, \qquad h, k = 1, 2, \ldots, p,$$

and from $\varphi C \omega' = 0$ we have the following relations:

$$\sum_{j=1}^{p} d_j\,\varphi_{ij}\,\omega_{hj} = d_h\,\varphi_{i, p+h}, \quad i = 1, 2, \ldots, s; h = 1, 2, \ldots, p.$$

Using (21) we can write these relations as follows:

$$(21') \qquad \sum_{j=1}^{p} \varphi_{ij}\,\omega_{jh} = \varphi_{i, p+h}.$$

If we put $\delta = \|\varphi_{ij}\|$, $i = 1, 2, \ldots, s$; $j = 1, 2, \ldots, p$, wo see immediately from (21') that $\delta\omega = \varphi$, q.e.d.

In the general case, we first reduce both ω and C to their respective canonical forms, assumed in the preceding proof. Let $\omega^* = A\,\omega\,G'$ and $C^* = (G^{-1})'\,C\,G^{-1}$ be the transformed matrices, having the canonical form. We have $\varphi C \omega' = \varphi G' C^* \omega^{*'} A'^{-1} = 0$, and hence $\varphi G' C^* \omega^* = 0$, since A is non-singular. By the preceding proof there exists a matrix δ^*, such that $\varphi G' = \delta^* \omega^* = \delta^* A \omega G'$, and hence $\varphi = \delta\omega$, where $\delta = \delta^* A$, q.e.d.

We are in position to give a very simple proof of the following:

Theorem of POINCARÉ. *If an algebraic curve of genus p possesses a regular system of $p - q$ reducible integrals, it also possesses a complementary regular system of q reducible integrals* (POINCARÉ, 1, 2; a special case in PICARD, 1).

Proof. We take on the curve p independent abelian integrals $I_1, \ldots, I_p$ and a minimal base $\gamma_1, \ldots, \gamma_{2p}$ for the 1-cycles, and we assume that the integrals $I_{q+1}, \ldots, I_p$ are the supposedly existing reducible integrals and that they have zero periods along the cycles $\gamma_1, \gamma_2, \ldots, \gamma_{2q}$. The matrix ω is then of the form

$$\omega = \begin{pmatrix} \omega_1, & \omega_3 \\ 0, & \omega_2 \end{pmatrix},$$

where, as we have seen before, ω_2 is a RIEMANN matrix of genus $p - q$.

Let

$$C = \begin{pmatrix} C_1, & C_3 \\ C_4, & C_2 \end{pmatrix},$$

where C_2 is a principal matrix of ω_2 and hence is non-singular. Let H be a rational matrix with $2p - 2q$ rows and $2q$ columns, satisfying the matrix equation $H'C_2 + C_3 = 0$, and let t be the smallest integer, such that the elements of the matrix $H_1 = tH$ are integers ($t = 1$, if C_2 happens to be unimodular). We define a transformation of the cycles γ_i into new cycles γ_i^* by means of the following matrix:

$$G = \begin{pmatrix} E_{2q} & 0 \\ -H_1 & t E_{2p-2q} \end{pmatrix}.$$

It is important to notice that *the cycles* $\gamma_1, \ldots, \gamma_{2q}$ *are not affected, and that the new cycles* γ_i^* *do not necessarily form a minimal base.* The transformed matrix ω^* has the following form:

$$\omega^* = \begin{pmatrix} \omega_1 & \omega_3^* \\ 0 & t\,\omega_2 \end{pmatrix},$$

while for C^* we find immediately the following expression:

$$C^* = \begin{pmatrix} t\,C_1^* & 0 \\ 0 & C_2 \end{pmatrix},$$

where $C_1^* = t C_1 + H_1' C_4$.

From $\omega^* C^* \omega^{*\prime} = 0$ we deduce the following matrix equations:

(22) $$\omega_2 C_2 \omega_2' = 0;$$

(22') $$\omega_3^* C_2 \omega_2' = 0;$$

(22'') $$\omega_3^* C_2 \omega_3^{*\prime} + t\omega_1 C_1^* \omega_1' = 0.$$

The equation (22) is in agreement with the fact that C_2 is a principal matrix of the Riemann matrix ω_2. From (22') we deduce, in view of Albert's lemma, that $\omega_3^* = \delta \omega_2$. From this it follows that $\omega_3^* C_2 \omega_3^{*\prime} = 0$ and hence, by (22''),

(23) $$\omega_1 C_1^* \omega_1' = 0.$$

The consideration of the matrix $\sqrt{-1}\,\omega^* C^* \overline{\omega}^{*\prime}$ shows in a similar manner that $\sqrt{-1}\,\omega_1 C_1^* \overline{\omega}_1'$ is a positive definite Hermitian matrix. *Hence* ω_1 *is a Riemann matrix of genus* q, *and* C_1^* *is a principal matrix of* ω_1.

We now transform the integrals I_h into new integrals I_h^*, the matrix of the transformation being:

(24) $$A = \begin{pmatrix} E_q & -\delta/t \\ 0 & E_{p-q} \end{pmatrix}.$$

This transformation affects only the integrals $I_1, \ldots, I_q$. The principal matrix is not affected, while the new period matrix ω^{**} is now the matrix

$$\omega^{**} = \begin{pmatrix} \omega_1 & 0 \\ 0 & t\,\omega_2 \end{pmatrix}, \quad \text{q.e.d.}$$

We have started with $p - q$ reducible integrals $I_{q+1}, \ldots, I_p$, which have zero periods along the cycles $\gamma_1, \ldots, \gamma_{2q}$. We have found $2p - 2q$ cycles $\gamma_{2q+1}^*, \ldots, \gamma_{2p}^*$ and q integrals $I_1^*, \ldots, I_q^*$, whose periods along these cycles vanish and whose reduced periods, relative to the cycles $\gamma_1, \ldots, \gamma_{2q}$, are the elements of the Riemann matrix ω_1. The proof shows clearly the following facts: (1) *the cycles* $\gamma_{2q+1}^*, \ldots, \gamma_{2p}^*$ *are uniquely determined* (to within a linear combination of the same cycles); (2) *the cycles* $\gamma_1, \ldots, \gamma_{2q}, \gamma_{2q+1}^*, \ldots, \gamma_{2p}^*$ *do not necessarily form a minimal base*; (3) *the integrals* $I_1^*, \ldots, I_q^*$ *are uniquely determined* (to within a linear combination of the same integrals).

7. Miscellaneous applications of the existence theorem.

a. The most important application, the existence on F of continuous systems constituted of ∞^q linear systems, has already been discussed in (V, 3). From this we have derived in section 3 the fundamental theorem on the number of simple integrals of the first and of the second kind.

We give here a direct proof of this theorem, using the results of the preceding section (LEFSCHETZ, a, p. 74).

The integrands P_h of the integrals $I_{q+1}, \ldots, I_p$ of section 4 are of degree $n-3$ in all the variables, and hence these integrals have zero periods along the invariant cycles $\gamma_1, \ldots, \gamma_r$. It follows, by the preceding section, that $r \leq 2q$ (see also SEVERI's proof of this inequality in section 3). The existence theorem of LEFSCHETZ-POINCARÉ shows, however, that r cannot be less than $2q$. In fact, as the q arbitrary constants $d_1, \ldots, d_q$ in the equations (14) vary continuously, the algebraic curve, say D—solution of the equations (14)—varies in an algebraic system Σ, ∞^q. The integers λ_j being fixed, the constants d_h which give rise to one and the same curve D, are determined to within periods of the integrals $I_1, \ldots, I_q$ along the r invariant cycles (see section 4). In other words, *the variable element D of the variety Σ is an r-ply periodic function of the q parameters d_h.* On the other hand, to each curve D of Σ there corresponds a set of finite constants d_h. The consideration of the period-network in the space of $2q$ real dimensions of the parameters d_h shows immediately that this would be impossible if r was less than $2q$.[1]

We now have that the integrals $I_{q+1}, \ldots, I_p$ form, on a generic C_y, a regular system of $p-q$ reducible integrals, with $2q$ vanishing periods along the invariant cycles $\gamma_1, \ldots, \gamma_{2q}$. By the theorem of POINCARÉ, there exists, on each C_y, a complementary set of q reducible integrals $I_1^*, \ldots, I_q^*$, with $2(p-q)$ vanishing periods along certain cycles $\delta_1^*, \ldots, \delta_{2(p-q)}^*$, and these form, together with the invariant cycles, a base (not necessarily minimal) for the 1-cycles on C_y. Here the cycles δ_i^* play the role of the cycles $\gamma_{2q+1}^*, \ldots, \gamma_{2p}^*$ of the preceding section. Since the cycles $\gamma_1, \ldots, \gamma_{2q}$ are invariant, it follows that the cycles δ_i^* are uniquely determined on each C_y (to within a linear combination of the same cycles). From the preceding section it follows that the integrals I_h^* (determined to within a linear combination) can be so chosen that the matrix of their reduced periods coincide with the matrix ω_1 of the periods of the integrals $I_1, \ldots, I_q$ relative to the invariant cycles (we keep the notations of the preceding section except that the cycles $\gamma_{2q+1}, \ldots, \gamma_{2p}$ and the cycles $\gamma_{2q+1}^*, \ldots, \gamma_{2p}^*$ are now replaced by the cycles δ_i and δ_i^* respectively). Since these periods are

---

[1] The fundamental theorem could now be derived from SEVERI's inequalities (section 3) $r_0 \leq q$, $r - r_0 \leq q$. The proof of these inequalities does not make use of the completeness of the characteristic series.

constants (i.e. independent of y), it follows in the first place, that the integrals I_h^* are *rationally* determined on each C_y, i.e. the integrand of each I_h^* is rational in x, z and y. In fact, as y describes a closed path, we come back with an integral which has the same periods as, and which therefore coincides with, I_h^*, the lower limit of both integrals being always the base point A_1.

In the second place, since all the periods of the integrals I_h^* are constant, and since there exists on C_y at least one cycle which is increased by a multiple of a vanishing cycle δ_i as y turns about α_i, it follows that *the periods of the integrals I_h^* along the cycles δ_i all vanish, and that consequently the cycles δ_i coincide with the cycles δ_i^*.*

We may derive from this another proof of the preceding statement, to the effect that the integrals I_h^* are rationally determined. In fact, the matrix δ, which appears in the matrix A (24) of the transformation from the integrals I_h to the integrals I_h^*, is defined by the matrix equation $\omega_3^* = \delta \omega_2$, where $\omega_3^* (= \omega_3)$ and ω_2 are now the period matrices of $I_1, \ldots, I_q$ and of $I_{q+1}, \ldots, I_p$, respectively, along the $2p - 2q$ vanishing cycles δ_i. It is then seen immediately that this equation is invariant in the neighborhood of each critical point α_i, and hence the elements of A are rational functions of y.

From the vanishing of the periods of the integrals I_h^* along the vanishing cycles δ_i, we conclude, as in section 2, that these integrals give rise to q independent simple integrals u_h of total differential, attached to the surface F. These integrals are of the first kind, because the integrals I_h^* ($h = 1, 2, \ldots, q$) are of the first kind on *each C_y*. This is obvious if $y \neq \alpha_i$ and $y \neq \infty$. For $y = \alpha_i$, this follows from the fact that the period of I_h^* along δ_i vanishes, and that consequently, as $y \to \alpha_i$, I_h^* does not acquire a logarithmic period. The value $y = \infty$ is in no way exceptional, as can be seen by making a generic homographic transformation.

We have thus proved the inequality $r_0 \geqq q$, where r_0 denotes, as in section 2, the number of independent simple integrals of total differential of the first kind attached to the surface. We have proved already in section 2, by elementary considerations, that $r_0 \leqq q$. Hence $r_0 = q$, q e.d.

PICARD (a_1, p. 116) showed that every simple integral u of the first kind, attached to the surface $f(x, y, z) = 0$, has necessarily the following form:

$$u = \int \frac{A \, dx + B \, dy}{f_z'},$$

where A and B are adjoint polynomials of order $n - 2$, A being of degree $n - 3$ in x, z and B being of degree $n - 3$ in y, z. The surface $A = 0$ cuts out on a generic plane $y = $ const., outside the line at infinity, an adjoint curve of order $n - 3$ of the corresponding plane

section C_y. This adjoint curve cannot be cut out by an adjoint surface of order $n-3$, because otherwise the periods of the abelian integrals $\int \frac{A\,dx}{f_z'}$ along the invariant cycles, and hence also all the periods
of the integral u, would vanish. Consequently, the number of independent simple integrals of the first kind cannot be greater than the deficiency of the linear system cut out in the plane $y = $ const. by the adjoint
surfaces of order $n-3$, and this shows again that necessarily $r_0 \leqq q$.

We make an important remark concerning the period matrix ω_1 of
the integrals u_h (the same as that of the integrals I_h^*). It is well known
that the period matrix of the abelian integrals, attached to an algebraic
curve of genus $p \geqq 4$, is not the most general RIEMANN matrix of
genus p. The $\frac{p(p+1)}{2}$ elements of the normalized period matrix must
satisfy certain $\frac{p(p+1)}{2} - (3p-3)$ relations[1], leaving a degree of
freedom $3p-3$, equal to the number of the moduli of the curve.
On the contrary, the period matrix of the q simple integrals of the first
kind, attached to an algebraic surface of irregularity q, is the most general
RIEMANN matrix of genus q. This theorem is due to CASTELNUOVO and
ENRIQUES (4), and results from the following topological property of
an algebraic variety V_d: If $d > 2$, then the one-dimensional BETTI number
R_1 of V_d is the same as the one-dimensional BETTI number of a sufficiently
general V_{d-1} on V_d (for instance, of a V_{d-1}, which is a member of an
infinite linear system on V_d). A more general topological property,
concerning the BETTI number R_k, $k \leqq d-2$, was proved by LEF-
SCHETZ (5). From this topological property of V_d, it follows that any
sufficiently general surface on V_d is necessarily of irregularity $q = R_1$.
Since the most general RIEMANN matrix of genus q is the period-matrix
of the simple integrals of the first kind, attached to a suitable abelian
variety of dimension q, the above theorem of CASTELNUOVO-ENRIQUES
follows immediately.

The elements of the period matrix ω_1 are birational invariants of
the surface F. A change of basis of the integrals and of the 1 cycles
transforms ω_1 into an isomorphic RIEMANN matrix. Thus, with every
algebraic surface there is associated an invariant class of isomorphic
matrices. RIEMANN matrices constitute therefore a fundamental class,
not only for analysis in general, but especially for the birational theory
of algebraic surfaces. Their theory has been created in a large measure
by SCORZA. Its most salient feature is the existence of a linear algebra A
associated with ω_1 whose elements are the *complex multiplications* of ω_1
and which is invariant under isomorphisms of ω_1. The problem of
classifying all the algebras A thus arising, and hence all RIEMANN

---

[1] These relations are known only in the case $p = 4$ (SCHOTTKY).

matrices, initiated by SCORZA, LEFSCHETZ, ROSATI, has just been completely solved by ALBERT (2).

We make a last remark. The simple integrals $u_1, \ldots, u_q$ attached to F may be supposed to stand for the integrals $I_1, \ldots, I_q$ of section 4. *For this particular choice of the integrals I_h the first q normal function v_h of any algebraic curve D reduce to constants.* In fact, the periods Ω_{hi} all vanish for $h = 1, 2, \ldots, q$, since $\delta_i \sim 0$ on F. The theorems a. and c. of section 4 now assume a simpler form. Thus in c. it is sufficient to require that the $p - q$ normal functions $v_{q+1}, \ldots, v_p$ coincide (to within periods). In a. it is sufficient to add to the conditions just mentioned the condition that the first q normal functions (constants) $v_1, \ldots, v_q$ coincide (to within periods) on *one* curve of the pencil $\{C_y\}$. This gives SEVERI's transcendental criterion of VII, 3, p. 164.

b. *If $D_1 = D_2$, then $D_1 \sim D_2$, and conversely* (LEFSCHETZ, a, 5; compare with VI, 12). The direct theorem is trivial. To prove the converse we consider a system $\{K\}$ of ∞^q linear systems. From $D_1 \sim D_2$ it follows that $D_1 + K_1 \sim D_2 + K_2$, where K_1 and K_2 are any two curves of $\{K\}$, and hence the normal functions $v_{q+1}, \ldots, v_p$ of $D_1 + K_1$ and $D_2 + K_2$ are congruent (mod. periods). We fix K_1 and we let K_2 vary in $\{K\}$. The constants $v_1, \ldots, v_q$ giving the first q normal functions of $D_2 + K_2$ assume ∞^q sets of values and hence, for some curve $\overline{K}_2 = K_2$ in $\{K\}$, will coincide with the analogous constants relative to the fixed curve $D_1 + K_1$. By theorem a. of section 4 it follows that then $D_1 + K_1 \equiv D_2 + \overline{K}_2$, and hence $D_1 = D_2$, since $K_1 \equiv \overline{K}_2$.

c. *Every two-dimensional divisor of zero on F is algebraic* (LEFSCHETZ, a, 5). This is an immediate corrollary of the existence theorem, since the period of any double integral of the first kind relative to a zero divisor vanishes. As a consequence, the group of torsion of F coincides with the group of torsion of the algebraic modulus (VI, 3) and the order of this group coincides with the number σ of SEVERI (V, 7).

d. **The theorem of PICARD-SEVERI on the regularity of ad-joint systems.** From the equality $r = 2q$ proved in section 3, PICARD (a_2, p. 437) has derived an extremely interesting result concerning the adjoint system of the system of plane sections of F. We have indicated by ω_r the deficiency of the linear system cut on a generic plane by the adjoint surfaces of order r of F (IV, 1). The deficiencies $\omega_{n-3}, \omega_{n-2}, \ldots$ are connected to the irregularity q of F by the formula of ENRIQUES: $\omega_{n-3} + \omega_{n-2} + \cdots = q$. Hence $\omega_{n-3} \leqq q$. On the other hand we may form with $p - \omega_{n-3}$ independent adjoint polynomials $P_h(x, y, z)$ of order $n - 3$ the independent abelian integrals $I_h = \int P_h / f'_z \, dx$ of the first kind attached to C_y. Of the $2p$ periods of I_h $2q$ vanish, namely the periods relative to the $r (= 2q)$ invariant cycles of C_y. The $p - \omega_{n-3}$

integrals I_h being independent, the number of their arithmetically distinct periods cannot be less than $2(p - \omega_{n-3})$ (section 6). Hence $2(p - \omega_{n-3})$ $\leqq 2p - 2q$, i. e. $\omega_{n-3} \geqq q$. Consequently, $\omega_{n-3} = q$ and $\omega_{n-2} = \omega_{n-1}$ $= \cdots = 0$. We have thus the following

Theorem of PICARD. *The adjoint system of the system of plane sections of F is always regular. Or, in projective language, the adjoint surfaces of order $n - 3$ cut out on a generic plane a system of maximum deficiency q, while the adjoint surfaces of order $r \geqq n - 2$ cut out on a generic plane the complete system of adjoint curves of order r of the corresponding plane section of F.*

SEVERI (15) gave an algebro-geometric proof of PICARD's theorem and arrived at a more general result:

If an irreducible curve D is a member of a continuous system Σ which is not an irrational pencil, then the adjoint system $|D'|$ is regular.

e. *Under the hypothesis of the preceding theorem of* SEVERI, *D contains $2q$ independent 1-cycles of F* (SEVERI, 12). For the proof we suppose that the theorem is not true and we observe that then the q simple integrals of the first kind attached to F *are dependent on D*, since they possess on D less than $2q$ distinct periods. There exists therefore a simple integral u of the first kind which reduces on D to a constant. The integral u reduces to a constant also on every other curve D_1 of Σ, since the 1-cycles on D_1 are obtained from the 1-cycles on D by a deformation. Two cases are possible: either the curves of Σ meet in variable points, or Σ is a pencil, necessarily linear, by hypothesis. In the first case, the constant value of u on a variable D_1 of Σ must coincide with the value of u on D, since D_1 meets D in variable points. Hence u is constant on F, and this is impossible. In the second case u is a function of the parameter t of the pencil, defined to within periods and *finite for every value of t*. Hence again u is a constant, q. e. d.

For irrational pencils the following theorem holds: *if Σ is an irrational pencil of genus π, there exist exactly $2q - 2\pi$ 1-cycles on D which are independent on F*. In fact, the π independent abelian integrals of the first kind attached to the pencil, considered as a one-dimensional algebraic variety, give rise to π independent simple integrals of the first kind which reduce to constants on any curve D of Σ. Conversely, any effective simple integral of the first kind (i. e. not a constant), which is constant on a generic D of Σ, is an abelian integral of the first kind attached to the pencil Σ. Hence there exist π and only π independent simple integrals of the first kind which reduce to constants on a generic D, and consequently the number of independent 1-cycles of F homologous to 1-cycles of D is $\geqq 2(q - \pi)$. On the other hand this number cannot be greater than $2q - 2\pi$, because the real parts of the $2q$ periods of a simple integral of the first kind can be assigned arbitrarily, q. e. d.

8. Double integrals of the first kind. Theorem of Hodge. We have seen (section 5) that if U is any double integral of the first kind attached to a surface F, its periods relative to the algebraic cycles all vanish. It follows that the number of arithmetically distinct periods of U is at most ϱ_0, where $\varrho_0 (= R_2 - \varrho)$ is the maximum number of independent non-algebraic cycles no combination of which is algebraic (VI, 12). *Is it possible to have an integral U whose periods all vanish?* This important question has been solved only recently by Hodge (2) not only for algebraic surfaces but more generally for any n-dimensional algebraic variety V_n and for the n-fold integrals of the first kind attached to V_n. An n-fold integral U of the first kind on V_n is defined, as in the case of surfaces, by the condition that $\int_{C_n} dU$ be finite for every analytical n-chain C_n on V_n.

Theorem of Hodge. *An n-fold integral of the first kind attached to an algebraic variety V_n cannot have all its periods equal to zero.*

Hodge's proof is based on the consideration of the topological product of V_n by itself and makes use of ideas used by Lefschetz in his topological theory of algebraic correspondences (VI, 13). Subsequently another proof was given by G. de Rham (2) as a consequence of general properties of the integrals attached to a topological manifold established by the same author in his interesting dissertation (1). See also Kähler (2) and the simplified proof by Kneser (1). The original proof of Hodge, given below, is, however, simple enough in its essential aspects.

Let V_n be in S_{n+1}, given by an equation $f(x_0, x_1, \ldots, x_n) = 0$, and let $f(x_0', x_1', \ldots, x_n')$ be another copy of the same variety in S_{n+1}'. We shall denote by A and B the corresponding (homeomorphic) real Riemannian $2n$-dimensional varieties. The homeomorphism between A and B is given by the identical correspondence $x_i' = x_i$ on V_n. We shall denote by F the topological product $A \times B$ as well as the algebraic $2n$-dimensional variety $f = 0$, $f' = 0$ in $S_{2n+2} (x_0, \ldots, x_n, x_0', \ldots, x_n')$ of which $A \times B$ is the Riemannian variety. This variety is in $(1-1)$ correspondence with the ordered pairs of points of A and B. Let a_p^h, $h = 1, 2, \ldots, r_p$, be a minimal base of r_p independent p-cycles on A, and let b_p^h be the corresponding cycles on B $(p = 0, 1, \ldots, 2n)$. In order not to interrupt the proof we first make a few remarks of which we shall make use in the sequel. They refer to well known facts.

a. *A minimal base for the q-cycles Γ_p on F is given by the cycles $a_p^i \times b_{q-p}^j$, $p = 0, 1, \ldots, q$* (see, for instance, Lefschetz, e, Chapter V).

b. Let $z_1, \ldots, z_m$ denote real variables and let $\omega = A_{i_1 \ldots i_p} dz_{i_1} \ldots dz_{i_p}$ (the i's are summation indices) be a differential form of degree p, where the functions $A_{i_1 \ldots i_p}$ are the components of a covariant skew-symmetric tensor satisfying the usual conditions of continuity and differentiability up to second order in a given domain T. According to Poincaré (see PICARD, a_1, Chapter 1; DE RHAM, 1) the condition that the p-fold integral $\int_{C_p} \omega$ be an *invariant integral*, in the sense that its value remains the same if the chain C_p is deformed over T in such a manner that its boundary $F(C_p)$ remains fixed, is that $B_{i_1 i_2 \ldots i_{p+1}} = \delta^{i_1 \ldots i_{p+1}}_{\alpha_1 \ldots \alpha_{p+1}} \dfrac{\partial A_{\alpha_1 \ldots \alpha_{p+1}}}{\partial z_{\alpha_1}} = 0$, where the δ's are the well known KRONECKER symbols. The property of which we shall make use is the following and is an immediate consequence of POINCARÉ's condition. Let us have two differential forms $\omega_1 = A_{i_1 \ldots i_p} dz_{i_1} \ldots dz_{i_p}$, $\omega_2 = B_{j_1 j_2 \ldots j_q} dz_{j_1} \ldots dz_{j_q}$, $p + q \leqq m$, and let us consider the differential form of degree $p + q$ *(the direct product of ω_1 and ω_2)*: $\omega_1 \times \omega_2 = \delta^{\alpha_1 \alpha_2 \ldots \alpha_{p+q}}_{i_1 i_2 \ldots i_{p+q}} A_{\alpha_1 \ldots \alpha_p} B_{\alpha_{p+1} \ldots \alpha_{p+q}} dz_{i_1} \ldots dz_{i_{p+q}}$. *If the integrals $\int \omega_1$ and $\int \omega_2$ are invariant, then also the integral $\int \omega_1 \times \omega_2$ is invariant* It should be noted that if $\int \omega$ is invariant, then $\int_{\Gamma_p} \omega = \int_{\bar{\Gamma}_p} \omega$, if $\Gamma_p \sim \bar{\Gamma}_p$.

c. Let $\sigma_j (\sigma'_j)$ denote a cell on $A (B)$ and let $z_1, z_2, \ldots, z_{2n} (z'_1, \ldots, z'_{2n})$ be real coördinates defined in a region of $A (B)$ containing $\sigma_j (\sigma'_j)$. Let $\omega_1 = A_{i_1 \ldots i_n} dz_{i_1} \ldots dz_{i_n}$, $\omega_2 = B_{j_1 \ldots j_n} dz'_{j_1} \ldots dz'_{j_n}$. Then the *integral* $\int \omega_1 \times \omega_2$ over $\sigma_j \times \sigma'_{2n-j}$ is zero, if $j \neq n$, and is equal to $\int_{\sigma_n} \omega_1 \cdot \int_{\sigma'_n} \omega_2$, if $j = n$.

With these preliminaries, let $U = \int \varphi(x_0, \ldots, x_n) dx_1 \ldots dx_n$ be an n-fold integral of the first kind attached to A, and let $V = \int \varphi(x'_0, \ldots, x'_n) dx'_1 \ldots dx'_n$ be the corresponding integral on B. Let $x_j = z_j + i z_{j+n}$, $x'_j = z'_j + i z'_{j+n}$, $U = U' + i U''$, $V = V' + i V''$, $\varphi = R + i S$. Let

$$U = \int (a'_{j_1 \ldots j_n} + i a''_{j_1 \ldots j_n}) dz_{j_1} \ldots dz_{j_n},$$

$$V = \int (b'_{j_1 \ldots j_n} + i b''_{j_1 \ldots j_n}) dz'_{j_1} \ldots dz'_{j_n},$$

all the j's being summation indices. Since the real and imaginary parts of U or of V are invariant integrals, it follows, by b., that also

$$(25) \qquad U' \times V'' = \int a'_{i_1 \ldots i_n} b''_{j_1 \ldots j_n} dz_{i_1} \ldots dz_{i_n} dz'_{j_1} \ldots dz'_{j_n}$$

is an invariant integral.

Let Γ_{2n} be the cycle on F representing the pairs of homologous points $x' = x$ of A and B. This cycle is defined by the identical correspondence on V_n and plays an important rôle in the proof. By a., we have

$$(26) \qquad t \Gamma_{2n} \sim \sum_{h,k,j} \lambda^{(j)}_{h,k} a^h_j \times b^{(k)}_{2n-j},$$

t an integer, and consequently, by c.,

$$(27) \qquad \frac{1}{t} \int_{\Gamma_{2n}} d(U' \times V'') = \sum \lambda^{(n)}_{h,k} \omega'_h \omega''_k,$$

where $\omega_h = \omega_h' + i\omega_h''$ is the period of U relative to a_n^h. On the other hand, if we evaluate this integral directly from (25), noting that on Γ_{2n} we have $z_i' = z_i$, we find that its value is zero, if n is even, and is equal to

$$(28) \qquad (-1)^{\frac{n-1}{2}} 2^{n-1} \int_A (R^2 + S^2)\, dz_1 \ldots dz_{2n} > 0,$$

if n is odd. *Comparing* (28) *and* (27) *we immediately conclude that our theorem is true if n is odd.* Suppose that the theorem is not true if n is even, and let U be without periods. We consider an algebraic curve C of genus $p > 0$ and an abelian integral $\int \psi(x, y)\, dx$ of the first kind attached to C. Then it is immediately seen that $\int \varphi(x_0, \ldots, x_n)$ $\psi(x, y)\, dx_1 \ldots dx_n\, dx$ is an $(n+1)$-fold integral of the first kind without periods attached to $A \times C$, and this is impossible, q. e. d.

We consider any two integrals U_1 and U_2 of the first kind attached to A, and we denote by V_1 and V_2 the corresponding integrals on B. Let

$$U_1 \times V_2 = \int \varphi_1(x_0 \ldots x_n)\, \varphi_2(x_0', \ldots, x_n')\, dx_1 \ldots dx_n\, dx_1' \ldots dx_n'.$$

It is immediately seen that the value of this integral over $\Gamma_{2n}(x_i' = x_i)$ is zero. This also follows from a general remark of Lefschetz to the effect that the value of a multiple integral of the first kind over an algebraic cycle is zero (compare with section 5). Using (26), we find that *the periods ω_h^1 and ω_h^2 ($h = 1, 2, \ldots, R_n$) of any two (distinct or coincident) n-fold integrals of the first kind satisfy the bilinear relation* (Hodge, 2)

$$(29) \qquad \sum_{h,k} A_{h,k}\, \omega_h^1 \omega_k^2 = 0, \qquad\qquad A_{h,k} = \lambda_{h,k}^{(n)}.$$

The topological significance of the coefficients $\lambda_{h,k}^{(n)}$ can be derived from the intersection formulas (Lefschetz, e, p. 241):

$$(a_n \times b_n \cdot a_p \times b_{2n-p}) = 0, \qquad p \neq n,$$

$$(a_n \times b_n \cdot a_n' \times b_n') = (-1)^n (a_n \cdot a_n')(b_n \cdot b_n'),$$

$$(\Gamma_{2n} \cdot a_p^i \times b_q^j) = (-1)^p (a_p^i \cdot a_q^j).$$

These formulas generalize the formulas (16), (16'), (16'') and (19) of (VI, 13). Let $a = \|(a_n^h \cdot a_n^k)\|$ *be the matrix of the intersection numbers of the n-cycles a_n^h.* Then it follows from the above relations that

$$\lambda = \|\lambda_{hk}^{(n)}\| = \pm t a'^{-1}.$$

From the inequality (28) it follows that if n is odd *the periods of U cannot be all real or pure imaginary.* This can be proved for any value of n (Kähler, 2) by considering, instead of the integral $U' \times V''$, the integral

$$U \times \overline{V} = \int \varphi(x_0, \ldots, x_n)\, \overline{\varphi}(\bar{x}_0', \ldots, \bar{x}_n')\, dx_1 \ldots dx_n\, d\bar{x}_1' \ldots d\bar{x}_n',$$

where $\overline{\varphi}(\bar{x}') = \overline{\varphi(x')}$ denotes the conjugate of $\varphi(x')$. The value of this integral over $\varGamma_{2n}$ is easily found to be equal to $i^n \varrho$, where $i = \sqrt{-1}$ and ϱ is a real number $\neq 0$. Hence

$$(30) \qquad\qquad \sum A_{hk}\, \omega_h\, \overline{\omega}_k = i^n \varrho\,, \qquad\qquad \varrho \neq 0\,,$$

and this proves that the periods ω_h cannot be all real or all pure imaginary, because in the contrary case the relation (30) would contradict the relation (29).

An immediate corollary of this theorem, for algebraic surfaces, is *the inequality* $\varrho_0 \geqq 2p_g$. In general $\varrho_0 > 2p_g$, so that in general *the real (or imaginary) parts of the ϱ_0 periods of a double integral of the first kind cannot be assigned arbitrarily* (only $2p_g$ of them can be assigned arbitrarily).

We point out two interesting applications of HODGE's theorem. Necessary and sufficient conditions that $\dfrac{A\,dy - B\,dx}{f_z'}$ be a total differential of the first kind have been given by PICARD (a_1, p. 117) and are the following: A and B are adjoint polynomials of degree $n - 2$, and of degree $n - 3$ in y, z and x, z respectively; moreover, there exists an adjoint polynomial C of total degree $n - 2$ and of degree $n - 3$ in x, y, such that the polynomials A, B, C satisfy the following two identities:

$$(31) \qquad\qquad A f_x' + B f_y' + C f_z' = N f\,,$$

$$(31') \qquad\qquad A_x' + B_y' + C_z' - N = 0\,,$$

where N is an adjoint polynomial of degree $n - 3$.

LEFSCHETZ (5) has proved that a necessary and sufficient condition that a double integral $\displaystyle\iint \dfrac{Q(x, y, z)\, dx\, dy}{f_z'}$ of the first kind be without periods is that there exist three adjoint polynomials A, B, C of the same degrees as above, such that

$$(32) \qquad\qquad A f_x' + B f_y' + C f_z' = N f\,,$$

$$(32') \qquad\qquad A_x' + B_y' + C_z' - N = Q\,.$$

It is easily seen that if A, B, C satisfy an identity of type (32), then $A_x' + B_y' + C_z' - N$ is necessarily an adjoint polynomial of degree $\leqq n - 4$. From this it follows immediately, in view of HODGE's theorem and in view of LEFSCHETZ's result, that *the identity (31') is a consequence of the identity* (31). We have here a purely algebraic statement which is equivalent to HODGE's theorem. A direct proof of this algebraic result does not seem however an easy undertaking.

SEVERI (22) considers differential forms $\dfrac{A\,dy - B\,dx}{f_z'}$ enjoying the property that, *without being necessarily exact differentials, they give rise to an integral* $\displaystyle\int \dfrac{A\,dy - B\,dx}{f_z'}$ *which is of the first kind on every irreducible curve of the surface.* He calls such differential forms *semiexact differentials of the first kind.*

Severi proves that the identities (32) and (32') of Lefschetz give also the necessary and sufficient conditions that $\dfrac{A\,dy - B\,dy}{f_z}$ be a semiexact differential of the first kind. In view of Hodge's theorem *the semiexact differentials of* Severi *coincide necessarily with* Picard's *exact differentials of the first kind.*

In (4) Hodge introduces the notion of a harmonic integral attached to an analytic manifold which has attached to it a Riemannian (positive definite) metric, and establishes important properties of these integrals. The connection with algebraic varieties lies in the fact that the multiple integrals attached to them are harmonic integrals whose integrands are algebraic. Applying his results to algebraic surfaces Hodge deduces the following *topological characterization of the geometric genus p_g of an algebraic surface: if the matrix $a = \| a_2^h \cdot a_2^k \|$ of the intersection numbers of the 2-cycles is transformed into a diagonal matrix $a' = \mu a \mu'$, then the number of positive terms in a' equals $2p_g + 1$* (Hodge, 3).

9. Residues of double integrals and the reduction of the double integrals of the second kind (Picard, a_2; Lefschetz, a, Note I; Lefschetz, 2, 3, 4; Poincaré, 7). The theory of the double integrals of the second kind has been created by Picard. Emphasizing its topological aspects Lefschetz has considerably simplified the theory and has enriched it by new results. Following in the main Lefschetz's treatment we shall give in this section a few general properties of double integrals and we shall undertake a preliminary reduction of the double integrals of the second kind.

Let $D_1, D_2, \ldots, D_k$ be the irreducible curves of discontinuity of the double integral

$$(33) \qquad\qquad J = \int\!\!\int R(x, y, z)\, dx\, dy,$$

where R is a rational function which becomes infinite only on $D_1, \ldots, D_k$. A *period* of J is the value of the integral extended over a cycle Γ_2 which does not meet any of the curves D_i. If $\Gamma_2 \backsim 0$, then the period is called a *residue*. If Γ_2 meets a curve D_i, the integral J over Γ_2 converges if D_i is a polar curve of first order of R. However, in this case the value of the integral in general varies continuously if the intersections of Γ_2 with D_i vary.

For double integrals of the second kind the condition that Γ_2 should not meet the curves D_i will be replaced later by the less restrictive condition: $(\Gamma_2 \cdot D_i) = 0$.

We consider the residues of J and we show how they can all be obtained from zéro-cycles of a special and particularly simple nature (Lefschetz, a; 4). Let Γ_2 be a zero-cycle not meeting the curves D_i

and let C_3 be the 3-chain bounded by Γ_2. Since the boundary of C_3 does not meet D_i, we may assume, deforming C_3 if necessary, that $C_3 \cdot D_i = \gamma_i$ is a 1-cycle, and that γ_i does not pass through the multiple points of the curve $D_1 + \cdots + D_k$. We isolate the neighborhood of γ_i in C_3 by a cycle Γ_2^i very close to γ_i. For the construction of Γ_2^i we may consider a pencil $\{D_t\}$ on F such that D_i is either a member D_0 of the pencil or a component of D_0. Then Γ_2^i is the locus of γ^t as t describes a small circle of center $t = 0$, where γ^t is the 1-cycle on D_t which is deformed into γ_i as $D_t \to D_0$. We shall refer to Γ_2^i as *a tubular cycle of axis* γ_i. If γ_i is deformed on D_i the residue of J relative to Γ_2^i remains unaltered, except when γ_i crosses certain points of D_i which give rise to *point residues* of J (see remark after theorem 5). These points are either the intersections of D_i with the other discontinuity curves or the multiple points of D_i, the reason being that as γ_i crosses such a point the deformed tubular cycle necessarily crosses a D_j ($j \neq i$) or D_i, and the corresponding residue may be altered.

It is clear that $\Gamma_2^i \sim 0$ on F. If we take away from C_3 the small tubes bounded by the cycles Γ_2^i and having two by two nothing in common, there remains a 3-chain not meeting the curves D_i and bounded by $\Gamma_2 - \sum \Gamma_2^i$. By the theorem of CAUCHY-POINCARÉ the residue of J relative to Γ_2 is then the sum of the residues relative to the cycles Γ_2^i. We therefore obtain all the residues of J if we consider only tubular cycles.

We have called *improper* (section 1, b) any integral of the type

$$(34) \qquad \iint \left(\frac{\partial U}{\partial x} + \frac{\partial V}{\partial y} \right) dx \, dy,$$

where U and V are rational functions of x, y and z and where the derivatives are evaluated by considering z as a function of x and y defined by the equation $f(x, y, z) = 0$. Thus, $\frac{\partial U}{\partial x} = \frac{f_z' U_x' - f_x' U_z'}{f_z'}$. The discontinuity curves of the integral are the polar curves of U and of V and the intersection K of the two surfaces $f = 0$ and $f_z' = 0$ outside the double curve of F.

Theorem 1. *The transform of an improper integral by a birational transformation of F is an improper integral.*

For the proof, which is immediate, see PICARD, a_2, p. 161; LEFSCHETZ, 4, p. 241.

Theorem 2. *If Γ_2 is any cycle not meeting the polar curve of U and the curve K, then*

$$\iint_{\Gamma_2} \frac{\partial U}{\partial x} dx \, dy = \iint_{\Gamma_2} \frac{\partial U}{\partial y} dx \, dy = 0.$$

In other words, the periods of the integral $\iint \frac{\partial U}{\partial x} dx \, dy$ *or of the integral* $\iint \frac{\partial U}{\partial y} dx \, dy$ *all vanish.* This theorem is an immediate consequence of the theorem of CAUCHY-POINCARÉ.

One should not conclude that theorem 2 holds also for the improper integral (34). If it is known that the integral (33) is improper, then there exist rational functions U and V such that $R = \dfrac{\partial U}{\partial x} + \dfrac{\partial V}{\partial y}$ (on F), but it does not necessarily follow that it is possible to determine U and V in such a manner that they should not possess polar curves other than those of R. See the example in PICARD (a_2, p. 229). In this circumstance resides the main difficulty of characterizing improper integrals.

However, if we consider only the *residues* of the integral (34), we have the following

Corollary. *An improper integral has no residues.* In fact, let Γ_2 be a tubular cycle whose axis γ lies on an irreducible discontinuity curve D of the integral (34). We may assume that γ does not pass through the intersections of D with the other discontinuity curves of the integrals $\displaystyle\iint \frac{\partial U}{\partial x}\, dx\, dy,\ \iint \frac{\partial V}{\partial y}\, dx\, dy.$ Then the cycle Γ_2 satisfies the conditions of theorem 2, q. e. d.

Theorem 3. *Double integrals of the second kind have no residues.*

Let the integral (33) be of the second kind. Keeping the notation of the proof of the preceding corollary, let A be a point of γ. There exists an improper integral (34) such that the integral

$$\left(R - \frac{\partial U}{\partial x} - \frac{\partial V}{\partial y}\right) dx\, dy$$

is finite in the neighborhood of A (section 1, b) and consequently *does not possess D as curve of discontinuity*[1]. The residue of this integral relative to Γ_2 therefore vanishes. The theorem now follows from the preceding corollary.

Corollary. *If $\Gamma_2 \sim \Gamma_2'$, the corresponding periods of any double integral of the second kind are equal.*

This corollary permits us to enlarge the class of 2-cycles for which it has a sense to speak of the corresponding periods of a given double integral of the second kind, in view of the following

Theorem 4. *If Γ_2 and Γ_2' are any two cycles on F, one of them, say Γ_2, can be replaced by an homologous cycle $\overline{\Gamma}_2$ whose arithmetic number of intersections with Γ_2' equals $(\Gamma_2 \cdot \Gamma_2')$* (LEFSCHETZ, a, p. 128).

Corollary. *If $(\Gamma_2 \cdot \Gamma_2') = 0$, there exists a cycle $\overline{\Gamma}_2 \sim \Gamma_2$ which does not meet Γ_2'.*

In virtue of the preceding corollaries it has a sense to speak of the period of a double integral J of the second kind relative to any cycle Γ_2, such that $(\Gamma_2 \cdot D_i) = 0$ for any discontinuity curve D_i of J, meaning

---

[1] We see that the integral $\displaystyle\iint \left(R - \frac{\partial U}{\partial x} - \frac{\partial V}{\partial y}\right) dx\, dy$ is finite not only in the neighborhood of the point A *but also in the neighborhood of a generic point of the curve D.* One may use this as the definition of double integrals of the second kind (LEFSCHETZ, 4, p. 243). The two definitions are equivalent.

by it the period relative to any cycle which is $\sim \Gamma_2$ and does not meet the curves D_i.

Put $R = \dfrac{A(x, y, z)}{B(x, y, z)}$ and let $S(x, y)$ be the resultant with respect to z of the two polynomials B and f. Then $S = PB + Qf$, where P and Q are polynomials, and hence $R = \dfrac{PA}{S(x, y)}$ on $f = 0$. Here $S(x, y) = 0$ is the equation of the projection on the (x, y)-plane of the curve of discontinuity of J. We may undertake a further reduction by breaking up R into partial fractions. The integral J is then expressed as the sum of integrals of the type:

$$(35) \qquad \iint \frac{A(x, y, z)}{\varphi(y)[g(x, y)]^\alpha} \, dx \, dy,$$

where A, g and φ are polynomials in their variables and g is irreducible. The axes being in generic position, we may assume that $g(x, y) = 0$ is the simple projection of the corresponding irreducible discontinuity curve D of the integral (35) and that the complete intersection of the surface F with the cylinder $g(x, y) = 0$ consists of the curve D:

$$(36) \qquad g(x, y) = 0, \qquad z = z(x, y) = \text{a rational function of } x, y,$$

and of another irreducible curve $\bar{D}$, on which the integral (35) is finite.

The following theorem is the converse of theorem 3:

Theorem 5. *If a double integral is without residues it is of the second kind* (PICARD, a$_2$, p. 203).

Proof. It is sufficient to prove the theorem for integrals of type (35). Dividing A, if necessary, by a convenient power of g, we may write the integral in the form

$$J = \iint \frac{B}{g^\beta} \, dx \, dy,$$

where B is a rational function which neither vanishes nor becomes infinite on D. If $\beta > 1$, we replace the integral J by the integral

$$J_1 = J - \iint \frac{1}{\beta - 1} \frac{\partial}{\partial x} \frac{B}{g'_x g^{\beta-1}} \, dx \, dy = \iint \frac{1}{(\beta - 1) g^{\beta-1}} \frac{\partial}{\partial x} \left(\frac{B}{g'_x} \right) dx \, dy.$$

We have subtracted from J an improper integral and this does not affect the residues (Theorem 2, corollary). J_1 is of the same type as J, except that β is replaced by $\beta - 1$. We may therefore assume that $\beta = 1$:

$$(37) \qquad J = \iint \frac{A(x, y, z)}{g(x, y)} \, dx \, dy, \quad A, \text{ a rational function.}$$

Let Γ_2 be a tubular cycle of axis γ on D. Integrating first over a cross section of the tube and then along the 1-cycle γ, we find that the period of J relative to Γ_2 equals $2\pi i \displaystyle\int_\gamma \frac{A}{g'_v} \, dx$, i. e. *equals the period relative to γ of the abelian integral*

$$(38) \qquad 2\pi i \int \frac{A}{g'_v} \, dx,$$

attached to the curve D, *whose equations are given by* (36). It follows that if J has no residues then the integral (38) is a rational function $2\pi i \, U(x, y)$, on D. Hence $A/g'_y = \dfrac{\partial U}{\partial x} = (g'_y U'_x - g'_x U'_y)/g'_y$, on $g = 0$, and consequently the integral

$$\iint \left[\frac{A}{g} - \frac{\partial}{\partial x}\left(\frac{U g'_y}{g}\right) + \frac{\partial}{\partial y}\left(\frac{U g'_x}{g}\right) \right] dx \, dy$$

is finite on D, q. e. d.

Corollary. *The double integrals of the second kind are invariant under birational transformations.*

Remark. The abelian integral (38) may be of the third kind and as such may possess besides cyclic periods also logarithmic periods. The corresponding residues of J are referred to as *cyclic residues* and *point residues* respectively. A point residue of J corresponds to a tubular cycle whose axis is a small circle on D drawn around a logarithmic singularity of the integral (38). These singularities are either at the intersections of D with the other discontinuity curves of J or at the multiple points of D.

Theorem 6. *Any double integral J of the second kind can be reduced by the subtraction of a suitable improper integral to an integral of the type*

$$(39) \qquad \int \frac{P(x, y, z)}{\varphi(y) f'_z} \, dx \, dy \,,$$

where P is an adjoint polynomial and $\varphi(y)$ is a polynomial in y (Picard, a_2, Chapter VII; Lefschetz, a, p. 141).

Proof. We express J as the sum of integrals of type (35). There exists an improper integral $\iint \left(\dfrac{\partial U}{\partial x} + \dfrac{\partial V}{\partial y}\right) dx \, dy$ such that

$$(40) \qquad \iint \left[\frac{A(x, y, z)}{\varphi(y) g^\alpha} - \frac{\partial U}{\partial x} - \frac{\partial V}{\partial y} \right] dx \, dy$$

is finite at an intersection of D with the curve $\overline{D}$, the residual intersection outside D of $f = 0$ with $g(x, y) = 0$, and hence is finite at a generic point of either curve D or $\overline{D}$. Let us decompose U and V into partial fractions and let U_1 and V_1 be the partial fractions which correspond to the irreducible factor g. It is clear that we may replace U and V by U_1 and V_1 respectively, without affecting the finiteness of the integral (40) along the curve $D + \overline{D}$. The integral (40) assumes then the form:

$$J_1 = \iint \frac{Q(x, y, z)}{\psi(y) g^\beta f'_z} \, dx \, dy \,,$$

where the polynomial Q necessarily vanishes on $D + \overline{D}$. Since $D + \overline{D}$ is the complete intersection of $g(x, y) = 0$ and $f = 0$, it follows by Noether's theorem that $Q = Af + Bg$, where A and B are polynomials.

Hence $J_1 = \iint \dfrac{B(x, y, z)}{\psi(y) g^{\beta-1} f'_z} \, dx \, dy$. Repeating this procedure we see that J_1 is of the form $\iint \dfrac{P_1(x, y, z)}{\psi(y) f'_z} \, dx \, dy$. Performing the same reduction for each discontinuity curve of the original integral J, we reduce J to an integral of type (39). Since the subtracted improper integrals are finite on the double curve of F, this must also be true for the reduced integral (39), and hence P is an adjoint polynomial, q. e. d.

The discontinuity curves of the integral (39) are curves $C_{\bar{y}}$ of the pencil $y = $ const., where $\bar{y}$ is a root of the equation $\varphi(y) = 0$, or $\bar{y} = \infty$. The intersection of $f = 0$ and $f'_z = 0$ outside the double curve of the surface F is not a discontinuity curve since on that curve the integrand is infinite only to the order $1/2$.

The study of the double integrals of the second kind is thus reduced to the study of the integrals of type (39). However, a further reduction is possible:

Theorem 7. *Any integral of the second kind of type (39) can be reduced by the subtraction of a suitable improper integral to an integral of the form*

(41) $$\iint \frac{P(x, y, z)}{f'_z} \, dx \, dy,$$

where P is an adjoint polynomial (PICARD, a_2, Chapter VII; LEFSCHETZ, a, p. 141).

This theorem will be proved in the next section (theorem 4, corollary). The immediate task before us is the study of the integrals of type (41). They will be referred to as the *normal integrals.*

10. Normal double integrals and the determination of the number of independent double integrals of the second kind. The normal integral

(41) $$J = \iint \frac{P(x, y, z)}{f'_z} \, dx \, dy$$

is finite everywhere except on C_∞. Its periods are defined for all cycles Γ_2 which do not meet C_∞. We shall call such cycles *finite cycles.*

We consider in particular the cycles $\sum \lambda_i \Delta_i + (C_a)$ of VI, 10. The condition for $\sum \lambda_i \Delta_i + (C_a)$ to be a cycle is $\sum \lambda_i \delta_i^a \sim 0$ on C_a, and in order that it be a finite cycle it is necessary and sufficient that the chain (C_a) should not contain the points at infinity $A_1, \ldots, A_n$, i. e. that (C_a) be a finite region $((C_a))$ (VI, 11). The condition $\sum \lambda_i \delta_i^a \sim 0$ on C_a implies

(42) $$\sum_{i=1}^{\mu} \lambda_i (\delta_i \cdot \gamma_h) = 0, \qquad h = 1, 2, \ldots, 2p,$$

where $\gamma_1, \ldots, \gamma_{2p}$ are $2p$ independent cycles on C_y. The condition that $\sum \lambda_i \delta_i^a$ should bound a finite region $((C_a))$ implies

(42′) $$\sum_{i=1}^{\mu} \lambda_i (\delta_i \cdot c_j) = 0, \qquad j = 2, \ldots, n,$$

where $c_2, \ldots, c_n$ are open arcs joining A_1 to $A_2, \ldots, A_n$ respectively. *Hence all the finite cycles $\sum \lambda_i \Delta_i + (C_a)$ are of the type*

$$(43) \qquad\qquad \sum_{i=1}^{\mu} \lambda_i \Delta_i + ((C_a)),$$

with the necessary and sufficient condition that the λ's satisfy (42) *and* (42′).

The matrix of the coefficients of the equations (42) and (42′) is of rank $2p - r + n - 1 = 2p - R_1 + n - 1$ (VI, 8). Consequently the number of independent solutions, or, what is the same thing, the number of arithmetically independent finite cycles of type (43) equals (VI, 10)

$$\mu - (2p - R_1 + n - 1) = R_2 - 1 + 2p - R_1.$$

If Γ_2 is an arbitrary cycle then $n \Gamma_2 - (\Gamma_2 \cdot C_y) C_y$ is a cycle $\overline{\Gamma}_2$ such that $(\overline{\Gamma}_2 \cdot C_\infty) = 0$, and hence $\overline{\Gamma}_2$ is homologous to a cycle which does not meet C_∞ (section 9, theorem 4, corollary). Hence *the number of finite cycles which are independent on F is $R_2 - 1$.* It follows that the number of arithmetically distinct finite cycles of type (43) which are homologous to zero is at least $2p - R_1$. It is not difficult to see *that this number is exactly $2p - R_1$ and that consequently any finite cycle is homologous on F to a finite cycle of type* (43). In fact, if $\sum \lambda_i \Delta_i + ((C_a))$ is a finite cycle ~ 0 on F, then there exists a 3-chain C_3 such that $C_3 \rightarrow \sum \lambda_i \Delta_i + ((C_a))$. The intersection $C_3 \cdot C_y$ is a cycle γ, since $F(C_3)$ does not meet a generic C_y, so that C_3 is the locus of γ. It follows that $\lambda_i = (\lambda \cdot \delta_i)$ (VI, 6, formula 6), and hence the number of independent homologies, $\sum \lambda_i \Delta_i + ((C_a)) \sim 0$ on F, is $2p - R_1$ (VI, 7), q. e. d.

Theorem 1. *Any finite cycle Γ_2 is reducible by a finite deformation into a cycle of type* (43) (PICARD, a_2, p. 335; POINCARÉ, 7; LEFSCHETZ, 4, and a, Note I).

Proof. By a finite deformation we understand a deformation which leaves invariant C_∞. We know that there exists a cycle $\overline{\Gamma}_2$ of type (43) such that $\Gamma_2 \sim \overline{\Gamma}_2$. The zero-cycle $\Gamma_2 - \overline{\Gamma}_2$ can be deformed by a finite deformation into a tubular cycle whose axis is a 1-cycle γ^∞ in C_∞ (section 9). Hence it is sufficient to prove the theorem for tubular cycles. Such a cycle, Γ_2, is the locus of a 1-cycle $\gamma = \gamma^y$ as y describes a large circle in the y-plane, where γ^y is the cycle in C_y whose position on C_∞ is the axis γ^∞ of Γ_2. This circle can be deformed into the boundary of the 2-cell E_2 into which the y-plane has been cut open by means of the cuts $l_i, i = 1, 2, \ldots, \mu$ (VI, 5). Hence Γ_2 can be deformed by a finite deformation into the locus of γ as y describes the boundary of E_2, i. e. both edges of the various cuts l_i. We have shown (VI, 6) that this locus can be deformed into the cycle $\sum (\gamma \cdot \delta_i) \Delta_i + (C_a)$. It remains to show that the deformation used in VI, 6 is finite. That deformation consisted essentially in deforming $\gamma' - \gamma$ into $(\gamma \cdot \delta_i) \delta_i$ over the 2-chain on C_y bounded by the zero-cycle $\gamma' - \gamma - (\gamma \cdot \delta_i)\delta_i$. It is, therefore, necessary to show that of the two co-residual chains on C_y, which are bounded by this zero-cycle, there is one which does not contain the points at infinity $A_1, A_2, \ldots, A_n$. The existence of such a chain follows immediately from the fact that as $y \to \alpha_i$, γ' tends to coincide with γ, δ_i shrinks to a point, while the points A_i may be supposed to he fixed throughout the deformation of C_y, q. e. d.

Corollary. *The normal integral J does not possess point residues and its residues relative to the invariant cycles on C_∞ all vanish.* In fact, if Γ_2 is a tubular cycle of axis γ^∞, then, as we have just shown, the cycle $\Gamma_2 - \sum (\gamma \cdot \delta_i)\varDelta_i - ((C_a))$ bounds a 3-chain which does not meet C_∞. If $\gamma^\infty \sim 0$ on C_∞ or if γ is invariant, then $(\gamma \cdot \delta_i) = 0$ and hence Γ_2 itself bounds such a 3-chain, q. e. d[1].

Remark. The significance of the $R_2 - 1 + 2p - R_1$ arithmetically independent finite cycles (43) is now clear: $R_2 - 1$ are independent on F and the remaining $2p - R_1$ are zero-cycles corresponding to the tubular cycles whose axes are $2p - R_1$ independent non-invariant 1-cycles on C_∞.

The integral

(44)
$$\int \frac{P}{f_z'} dx,$$

considered on C_y, is an abelian integral whose only possible logarithmic singularities are the n points at infinity $A_1, \ldots, A_n$. It possesses, therefore, at most $2p + n - 1$ distinct periods (functions of y), of which $n - 1$ are logarithmic periods. It can be shown by elementary considerations that these periods can be assigned arbitrarily on a generic C_y without contradicting the condition that P be a polynomial in y. We shall confine ourselves to stating this result under an equivalent form:

Theorem 2. *It is possible to assign $2p + n - 1$ integrals of type* (44) *whose period-determinant is $\neq 0$ (and which are therefore independent on a generic C_y).*

The period of the integral J given by (41), relative to the finite cycle (13), equals [section 5, (16)] $\sum \lambda_i \int_a^{\alpha_i} \Omega_i(y)\,dy$, where $\Omega_i(y)$ is the period of the integral (44) relative to δ_i.

Theorem 3. *There exists a normal integral J whose periods are arbitrarily preassigned numbers* (PICARD, a₂, p. 355).

Proof. We obtain an integral J with arbitrary periods by considering the integral
$$\iint \frac{\varphi(y)P(x,y,z)}{f_z'} dx\,dy,$$

---

[1] The non-existence of point residues follows also from the fact that C_∞ is the only discontinuity curve of J and that it is free from multiple points outside the double curve of F.

where P is arbitrary but fixed and where $\varphi(y)$ is an *arbitrary* polynomial in y. In fact, the period of this integral relative to the cycle (43) is given by

$$\sum_i \lambda_i \int_a^{\alpha_i} \varphi(y)\, \Omega_i(y)\, dy.$$

The impossibility of assigning arbitrarily these quantities for the $R_2 - 1 + 2p - R_1$ arithmetically independent finite cycles (43) would necessarily imply that a relation of the form

$$\sum_i \mu_i \int_a^{\alpha_i} y^k \Omega_i(y)\, dy = 0$$

holds for an *arbitrary* integer $k \geqq 0$ and for certain constants μ_i not all zero and independent of k. Then the following function of u

$$\sum_i \mu_i \int_a^{\alpha_i} \frac{\Omega_i(y)}{y - u}\, dy$$

vanishes identically. Let, for instance, $\mu_1 \neq 0$. As u turns about α_1 this function increases by $2\pi i\, \mu_1 \Omega_1(u)$. Hence $\Omega_1(u)$ is identically zero. This is impossible, since $\delta_1 \not\sim 0$ on C_y and therefore, by theorem 2, the period $\Omega_1(u)$ of the integral (44) can be assigned arbitrarily on a generic C_y, q. e. d.

Theorem 4. *A double integral of type* (39) *of the preceding section, without periods, is necessarily an improper integral* (Picard, a_2, p. 365).

Proof. Let

$$(45) \qquad J = \iint R(x, y, z)\, dx\, dy = \iint \frac{P(x, y, z)}{\varphi(y) f_z'}\, dx\, dy$$

be the given integral without periods, and let

$$(45') \qquad J_y = \int \frac{P(x, y, z)}{\varphi(y) f_z'}\, dx.$$

We fix $2p + n - 1$ independent abelian integrals of type (44):

$$J_y^h = \int \frac{P_h(x, y, z)}{f_z'}\, dx, \qquad h = 1, 2, \ldots, 2p + n - 1.$$

Let ω_k, ω_{hk} $(k = 1, 2, \ldots, R_1)$ be the periods of J_y, J_y^h relative to the R_1 invariant cycles of C_y, and let Ω_i, Ω_{hi} $(i = 1, 2, \ldots, \mu)$ be their periods relative to the vanishing cycles δ_i. We consider the following system of equations in the $2p + n - 1$ unknowns c_h:

$$(46) \qquad \sum_{h=1}^{2p+n-1} c_h \omega_k(y) = \int_a^y \omega_k(y)\, dy, \qquad k = 1, 2, \ldots, R_1;$$

$$(46') \qquad \sum_{h=1}^{2p+n-1} c_h \Omega_{hi}(y) = \int_{\alpha_i}^y \Omega_i(y)\, dy, \qquad i = 1, 2, \ldots, \mu.$$

We consider a finite cycle (43). From $((C_a)) \to \sum \lambda_i \delta_i^a$ follows $\sum \lambda_i \Omega_{hi}(a) = 0$, and more generally, since a is arbitrary,

(47) $$\sum_{i=1}^{\mu} \lambda_i \Omega_{hi}(y) = 0, \qquad h = 1, 2, \ldots, 2p + n - 1,$$

where $\Omega_{h1}(y), \ldots, \Omega_{h\mu}(y)$ are the periods of J_y^h relative to the cycles $\delta_1^y, \ldots, \delta_\mu^y$ into which $\delta_1^a, \ldots, \delta_\mu^a$ are deformed if y varies on a fixed path issued from $y = a$. Writing that the period of J relative to the considered finite cycle vanishes, we find $\sum_i \lambda_i \int_{\alpha_i}^{a} \Omega_i(y) \, dy = 0$. Also here it is permissible to consider a as arbitrary and to write

(47') $$\sum_{i=1}^{\mu} \lambda_i \int_{\alpha_i}^{y} \Omega_i(y) \, dy = 0.$$

The number of distinct relations (47) (or (47')) is $R_2 - 1 + 2p - R_1 = \mu - (2p - R_1 + n - 1)$. Hence of the μ equations (46') only $2p - R_1 + n - 1$ are independent, and these, together with the R_1 equations (46), determine the unknowns c_h, since the determinant of the coefficients is $\neq 0$. By a reasoning similar to the one employed in section 2 it follows that *the c's are rational functions of* y. The integral $J_y - \frac{\partial}{\partial y} \sum_h c_h J_y^h$ is without periods on a generic C_y and is therefore a rational function $U(x, y, z)$. If we put $V = \sum_h c_h P_h / f_z'$ we find that $R = \frac{\partial U}{\partial x} + \frac{\partial V}{\partial y}$, q. e. d.

Remark. A few special considerations are necessary if $\varphi(y)$ vanishes at a critical point $y = \alpha_i$ (LEFSCHETZ, a, p. 140).

Corollary. *An integral J of type (39) is reducible to a normal integral by the subtraction of a suitable improper integral* (this is theorem 7 of the preceding section). In fact, by theorem 3 there exists a normal integral J' of type (41) having the same periods as J. The integral $J - J'$ is of the same type as J and has no periods, hence is an improper integral, q. e. d.

We are now in position to derive rapidly the final result. Let Γ_2^j $(j = 1, 2, \ldots, \varrho_0)$ be ϱ_0 independent non-algebraic cycles whose algebraic number of intersections with any algebraic cycle equals zero (VI, 12). By theorem 3, Corollary, of the preceding section, it has a sense to speak of the periods of any double integral of the second kind relative to these cycles. We add $\varrho - 1$ other finite cycles $\Gamma_2^{\varrho_0+j}$ $(j = 1, 2, \ldots, \varrho - 1)$, so as to have $\varrho_0 + \varrho - 1 = R_2 - 1$ independent finite cycles. Let $D_1, \ldots, D_\varrho$ be a base for the algebraic cycles in the sense of SEVERI. The matrix $\|(\Gamma_2^{\varrho_0+h} \cdot D_k)\|$, $h = 1, 2, \ldots, \varrho - 1$, $k = 1, 2, \ldots, \varrho$, is of rank $\varrho - 1$, since no combination of the cycles $\Gamma_2^{\varrho_0+h}$ enjoys the property that its intersection number with *any* algebraic cycle equals zero. We may assume that

(48) $$|(\Gamma_2^{\varrho_0+h} \cdot D_k)| \neq 0, \qquad h, k = 1, 2, \ldots, \varrho - 1.$$

Theorem 5. *Given any algebraic curve D on F, it is possible to construct an improper integral of type (45) whose period relative to any finite cycle Γ_2 equals $-(\Gamma_2 \cdot D)$* (Lefschetz, a, p. 142).

Proof. Let ν be the order of D and let $\int V dx$ be an abelian integral of the 3rd kind attached to C_y, with logarithmic periods $+1$ at the points of $C_y \cdot D$ and $-\nu$ at one of the points at infinity of C_y. The construction of V involves conditions which in their totality are rational in y. Hence V is a rational function in x, y, z. The integral $\frac{\partial}{\partial y} \int V dx$ is of the second kind on a generic C_y and is therefore reducible to an integral J_y of type (45') by the subtraction of a suitable rational function of x, y and z. We have then $J_y = \frac{\partial}{\partial y} \int V dx + U = \int R(x, y, z) dx$. The desired double integral is the following: $\int\int R(x, y, z) dx dy = \int\int \left(\frac{\partial U}{\partial x} + \frac{\partial V}{\partial y} \right) dx \, dy$.

If we now replace D by D_k ($k = 1, 2, \ldots, \varrho - 1$) we deduce from (48) the following

Corollary. *There exists an improper integral of type (45) whose periods relative to the $\varrho - 1$ cycles $\Gamma_2^{\varrho_0+h}$ are arbitrarily preassigned numbers.*

The preceding theorem leads immediately to the following *characterization of the improper integrals of type* (45):

Theorem 6. *A necessary and sufficient condition that an integral J of type (45) be improper is that its periods relative to the ϱ_0 cycles Γ_2^i all vanish* (Lefschetz, a, p. 144).

In fact, by theorem 2 of section 9, this condition is necessary. To prove that it is sufficient we consider an improper integral J' of type (45) having the same periods as J relative to the cycles $\Gamma_2^{\varrho_0+h}$ (theorem 5, corollary). Then $J - J'$ is of the same type as J, is without periods and hence is improper (theorem 4), q. e. d.

Given any number of double integrals of the second kind, they are said to be *independent* if and only if no linear combination of these integrals is improper. With this definition we now may state the following

Final theorem. *The number of independent double integrals of the second kind equals ϱ_0.*

Proof. By theorem 3, we may form ϱ_0 normal integrals J_i without residues and possessing ϱ_0 distinct periods relative to the cycles Γ_2^i, giving rise to a period determinant $\neq 0$. These integrals are then of the second kind (section 9, theorem 3) and are independent by the preceding theorem 6. On the other hand, let J be any normal double integral. We may form a linear combination J' of the integrals J_i having the same periods as J relative to the ϱ_0 cycles Γ_2^i. The integral $J - J'$ is then necessarily improper, by theorem 6, q. e. d.

Appendix to Chapter VII

By David Mumford

The process of integrating a differential is not an algebraic one. However the study of the differential form itself can be carried out by purely algebraic methods and in all characteristics. In characteristic 0, this gives an algebraic theory of differentials essentially equivalent to the classical theory of integrals. To begin with, on any variety X, singular or not, one has the module of KÄHLER differentials Ω^1_X on X (GROTHEN-DIECK, 5, Ch. 0, § 20).

There are many interesting problems on the local structures of this module when X is *singular* (ZARISKI, 30; BERGER, 2).

However, when X is non-singular, Ω^1_X is a locally free $\mathcal{O}_X$-module, of rank n ($n = \dim X$), which can be called the sheaf *of simple differentials of first kind*. If Ω^r_X is the rth exterior power of Ω^1_X, then Ω^r_X is the sheaf of r-fold differentials (or r-forms) of first kind, and if we let $\Omega^0_X = \mathcal{O}_X$, then all these sheaves lie in a DE RHAM *complex:*

$$\Omega^{\cdot}_X : 0 \to \Omega^0_X \xrightarrow{d} \Omega^1_X \xrightarrow{d} \cdots \xrightarrow{d} \Omega^n_X \to 0 \ .$$

Unlike the topics treated heretofore, there is such a total dissimilarity between the results of the theory of differentials in $\operatorname{char} 0$ and in $\operatorname{char} p$ that I want to discuss these two cases quite separately. Therefore, let me first review developments in the $\operatorname{char} 0$ case, leaving $\operatorname{char} p$ to the second half of this appendix.

When the ground field $k = \mathbb{C}$, the whole problem of the relationship between differentials and topology has been clarified by the development of this connection a) for general differentiable manifolds and b) for general complex analytic manifolds. It has proved much more fruitful to understand these cases first, and then to relate these cases to the more rigid situation of algebraic differentials on an algebraic variety. For differentiable manifolds, the theory is due to DE RHAM (1); recent expositions can be found in HIRZEBRUCH (2), HÖRMANDER (1), § 7.5, HU (1), SINGER and THORPE (1). Let M be a differentiable manifold; let $\Omega^r_M(\text{diff})$ be the sheaf of complex differentiable r-forms on M. Then

(a) the DE RHAM complex

$$(*) \qquad 0 \to \mathbb{C} \to \Omega^0_M(\text{diff}) \xrightarrow{d} \Omega^1_M(\text{diff}) \xrightarrow{d} \cdots$$

is exact (POINCARÉ lemma),

(b) the sheaves Ω_M^r (diff) are flabby, hence have no higher cohomology groups.

Therefore, $H^r(M, \mathbb{C}) \cong \{$Space of closed r-forms$\}/\{$Space of exact r-forms$\}$. Supplementing this, if M is compact and is given a Riemannian metric, one can define differential operators $\delta: \Omega_M^{r+1}$ (diff) $\to \Omega_M^r$ (diff) adjoint to d, and applying techniques from the theory of elliptic differential operators, get

(c) an r-form ω is called *harmonic* if $d\omega = \delta\omega = 0$ (equivalently $(d\delta + \delta d)\,\omega = 0)$; then

$$\text{Space of closed } r\text{-forms} \cong \left\{\begin{matrix}\text{Space of}\\ \text{harmonic}\\ r\text{-forms}\end{matrix}\right\} \oplus \left\{\begin{matrix}\text{Space of}\\ \text{exact}\\ r\text{-forms}\end{matrix}\right\} \quad (\text{Hodge's theorem}).$$

This theory is due to Hodge (8); more recent expositions can be found in De Rham (1), Atiyah-Bott (1).

For complex analytic manifolds, the theory is principally due to Hodge (8), and Dolbeault (1). For a recent exposition, see Weil (6).

Let Z be a complex-analytic manifold; let Ω_Z^r (diff) be as above. Then Ω_Z^r (diff) splits canonically:

$$\Omega_Z^r (\text{diff}) \cong \Omega_Z^{r,0} \oplus \cdots \oplus \Omega_Z^{0,r}$$

(Weil, 6, p. 32), and $d = \partial + \bar{\partial}$, where $\partial: \Omega_Z^{i,j} \to \Omega_Z^{i+1,j}$, $\bar{\partial}: \Omega_Z^{i,j} \to \Omega_Z^{i,j+1}$. Let Ω_Z^r (hol) be the sheaf of holomorphic r-forms. Then

(d) the holomorphic De Rham complex

$(*)_h$ $\qquad\qquad 0 \to \mathbb{C} \to \Omega_Z^0 (\text{hol}) \xrightarrow{\partial} \Omega_Z^1 (\text{hol}) \xrightarrow{\partial} \cdots$

and the sequences:

$$0 \to \Omega_Z^r (\text{hol}) \to \Omega_Z^{r,0} \xrightarrow{\bar{\partial}} \Omega_Z^{r,1} \xrightarrow{\bar{\partial}}$$

are exact.

Now in $(*)_h$, the cohomology groups $H^i(\Omega_Z^i (\text{hol}))$ $(i > 0)$ are not zero. Therefore, $(*)_h$ gives us only the De Rham *spectral sequence*

$(\text{DR})_h$ $\qquad\qquad E_1^{p,q} = H^q(Z, \Omega_Z^p (\text{hol})) \Rightarrow H^*(Z, \mathbb{C})\,,$

and the second set of sequences give us:

$$H^q(Z, \Omega_Z^p (\text{hol})) \cong \left\{\begin{matrix}\text{Space of } \bar{\partial}\text{-closed}\\ (p, q)\text{-forms}\end{matrix}\right\} \bigg/ \left\{\begin{matrix}\text{Space of } \bar{\partial}\text{-exact}\\ (p, q)\text{-forms}\end{matrix}\right\}.$$

However if Z is compact and can be given a Kähler metric, one can define differentials operators $\delta: \Omega_Z^{i,j} \to \Omega_Z^{i-1,j}$, $\delta: \Omega_Z^{i,j} \to \Omega_Z^{i,j-1}$ adjoint to ∂ and $\bar{\partial}$ satsifying many significant identities. The theory asserts then:

(e) all the coboundaries in the spectral sequence $(\text{DR})_h$ are 0; if a differential form ω is called *harmonic* if equivalently $\partial\omega = \delta\omega = 0$ or $\bar{\partial}\omega = \delta\omega = 0$, then

$$\left\{\begin{matrix}\text{Space of } \bar{\partial}\text{-closed}\\ (p, q)\text{-forms}\end{matrix}\right\} \cong \left\{\begin{matrix}\text{Space of}\\ \text{harmonic}\\ (p, q)\text{-forms}\end{matrix}\right\} \oplus \left\{\begin{matrix}\text{Space of}\\ \bar{\partial}\text{-exact}\\ (p, q)\text{-forms}\end{matrix}\right\};$$

and finally putting everything together:

$$H^k(Z, \mathbb{C}) \cong \sum_{p+q=k} \begin{Bmatrix} \text{Space of harmonic} \\ (p, q)\text{-forms} \end{Bmatrix}.$$

HODGE's theorem in § 8 is just the special case of (e) asserting that the coboundaries in the spectral sequence:

$$d_1^{p,0} = d : H^0(\Omega_X^p) \to H^0(\Omega_X^{p+1})$$

(which in this case are just ordinary d of differential forms) are zero, i. e. all global holomorphic p-forms on a compact KÄHLER manifold are closed. This is false for some non-KÄHLER compact analytic manifolds.

Now let us return to the algebraic differentials. Let X be a non-singular algebraic variety over $\mathbb{C}$, and let Ω_X^r be the sheaf of algebraic differentials (of first kind). Now not only is Ω_X^r not flabby — something lost in the analytic case already — but the DE RHAM complex

$$\Omega_X^{\textbf{.}}: 0 \to \Omega_X^0 \xrightarrow{d} \Omega_X^1 \xrightarrow{d} \cdots \to \Omega_X^n \to 0$$

is not even exact at degrees ≥ 1 in the ZARISKI topology. However, there is still a spectral sequence associated to this complex:

$(\mathrm{DR})_a$ $\qquad\qquad E_1^{p,q} = H^q(X, \Omega_X^p) \Rightarrow H^*(X, \Omega_X^{\textbf{.}})$.

Its abutments are the so-called hypercohomology groups of the complex $\Omega_X^{\textbf{.}}$. The main theorem, which is due to ATIYAH-HODGE (1), and GROTHENDIECK (10) is that there is a natural isomorphism

(1) $\qquad\qquad\qquad H^k(X, \mathbb{C}) \cong H^k(X, \Omega_X^{\textbf{.}})$.

In particular, if X is *affine*, then

$H^k(X, \mathbb{C}) \cong \{\text{closed algebraic } k\text{-forms}\}/\{\text{exact algebraic } k\text{-forms}\}$.

On the other hand, suppose X is a *projective* variety. Then according to a theorem of SERRE (3),

$H^q(X/\text{classical top}, \Omega_X^p(\text{hol})) \cong H^q(X/\text{ZARISKI top}, \Omega_X^p)$.

So the 2 spectral sequences $(\mathrm{DR})_h$, $(\mathrm{DR})_a$ are isomorphic, and the coboundaries in $(\mathrm{DR})_a$ are all 0 This shows that $H^k(X, \Omega_X^{\textbf{.}})$ has a filtration, whose quotients are nothing but the groups $H^q(X, \Omega_X^p)$, with $p + q = k$. Using the theory of harmonic forms, we also found a natural splitting of this filtration (e above). This splitting does not in fact depend on the choice of KÄHLER metric on X, as we see in:

(f) If $H^k(X, \mathbb{C}) = F^0 \supset F^1 \supset \cdots \supset F^k \supset (0)$ is the filtration defined by the spectral sequence $(\mathrm{DR})_h$ and isomorphism (1), then let $H^{p,q}(X) = F^p \cap \overline{F}^q$, $(p + q = k)$. Then

$$H^k(X, \mathbb{C}) \cong \sum_{p+q=k} H^{p,q}(X) ,$$

$$H^{p,q}(X) \cong H^q(X, \Omega_X^p) .$$

Re: §§ 2, 3. Now look at the results in § 3 : the *Fundamental theorem* asserts first of all that the numbers

i) $\dim H^1(X, \mathcal{O}_X)$ = irregularity of X,

ii) $\dim H^0(X, \Omega^1_X)$ = number of simple differentials of 1st kind,

iii) $\frac{1}{2} \dim H^1(X, \mathbb{C}) = \frac{1}{2} R_1$

are equal. But this follows immediately, from (f) since $H^{0,1}$, $H^{1,0}$ are conjugate subspaces of $H^1(X, \mathbb{C})$, whose direct sum is the whole space. Without using KÄHLER theory, another proof that all 3 of these numbers are equal which is rather closer in spirit to the Italian ideas of § 2 can be based on showing that if

$$\Phi : X \to A$$

is the universal mapping of X into an abelian variety, i. e. A = Albanese variety of X, then all 3 of the above numbers equal $\dim A$.

i) $\dim H^1(X, \mathcal{O}_X) = \dim A$.

Proof. Use the fact that via Φ, the PICARD varieties P_1 of X and P_2 of A are isomorphic (LANG, 3). But P_2 and A are isogenous, hence of the same dimension (LANG, 3). And $\dim H^1(X, \mathcal{O}_X) = \dim P_1$ was discussed in Chapter 5, and can be proven algebraically.

ii) $\dim H^0(X, \Omega^1_X) = \dim A$.

Proof. In fact, the map Φ can be explicitly constructed by integrating the differentials of first kind and A can be constructed as the set of linear functionals on $H^0(X, \Omega^1_X)$ modulo those given by the periods. The only problem is to check that this A is an abelian variety and not merely a complex torus. One can either use the RIEMANN conditions (cf. MUMFORD, 9, § 3) for the algebraizability of tori, or the fact that the torus you get is a quotient of the Jacobian of a sufficiently ample curve C on X.

iii) $\dim H^1(X, \mathbb{C}) = 2 \dim A$.

Proof. Use LANG (3) to prove that almost all unramified abelian coverings $\pi : X' \to X$ are induced via unramified coverings of A. Then for almost all primes P,

$$\{x \in A \mid pX = 0\} \cong \begin{Bmatrix} \text{covering group of maximal unramified abelian} \\ \text{covering } \pi : A' \to A, \text{ with group killed by } p \end{Bmatrix}$$

$$\cong \begin{Bmatrix} \text{covering group of maximal unramified abelian} \\ \text{covering } \pi : X' \to X, \text{ with group killed by } p \end{Bmatrix}$$

$$\cong H_1(X, \mathbb{Z}/p\,\mathbb{Z}) .$$

Equating the ranks of these 2 groups, we get (iii).

Re: § 2. Differentials of second and third kind have been studied by WEIL (2) and HODGE-ATIYAH (1). Consider first rational, closed 1-forms ω (= simple differentials of 3rd kind). Then if $D = \sum_{i=1}^{n} D_i$ is its polar cycle, then integrating around each D_i, we get $2\pi i$ times a constant a_i, called the residue of ω at D_i. The a_i can be characterized purely algebraically as the unique set of numbers such that on every ZARISKI-open affine U in which each D_i has a local equation f_i,

$$\omega = \sum a_i \frac{df_i}{f_i} + dg + \eta \,,$$
$$g = \text{a rational function} \,,$$
$$\eta = \text{a 1-form, regular in } U \,.$$

A simple closed differential ω of 3rd kind is said to be *of 2nd kind* if all its residues are 0; or equivalently, if locally on X in ZARISKI topology $\omega = dg + \eta$, g a rational function, η a regular 1-form; or equivalently, if locally on X in the classical topology, $\omega = dg$, g a meromorphic function. The main theorem of § 2 can be strengthened to give:

Theorem[1]. *Let V be a non-singular, complete variety. Let $U \subset V$ be an affine open subset. Then in the isomorphism of $H^1(U, \mathbb{C})$ with {closed 1-forms regular on U}/$d[\Gamma(U, \mathcal{O}_V)]$, let the 1-form ω correspond to the cohomology class a. Then a extends to some $a' \in H^1(V, \mathbb{C}) \Leftrightarrow$ the residues of ω on $V - U$ are 0.*

As U gets smaller and smaller among affine open sets in X, we get a picture like this:

On the other hand, the degree of freedom in the choice of polar parts of differentials of 3rd kind is described in the result of WEIL (2):

Theorem. *Let V be a non-singular, projective n-dimensional variety. Let U_i be a classical covering of V, and let ω_i be a closed meromorphic 1-form in U_i such that $\omega_i - \omega_j$ is holomorphic in $U_i \cap U_j$. Then a differential ω of 3rd kind on V such that $\omega - \omega_i$ is holomorphic in U_i exists if and only if the $(2n - 2)$-cycle with complex coefficients[2]*

$$\sum_{\substack{\text{irred.} \\ \text{divisors } D}} \operatorname{res}_D(\omega_i) \cdot D$$

is homologous to 0.

---

[1] GRIFFITHS (3) shows also that in case $V - U = H$ is a sufficiently ample non-singular hyperplane section, then ∀ closed 1-forms ω on U, $\omega = \omega_1 + dg$, where ω_1 has only a *double* pole on H.

[2] Note that since $\omega_i - \omega_j$ is holomorphic, ω_i and ω_j have the same residues.

Re: § 9, 10. Defining residues of closed n-forms ω on their polar divisors becomes harder when n increases, and the polar divisor of ω becomes singular. The problem has been studied by LERAY (1), HODGE-ATIYAH (1), and GRIFFITHS (3). If, for instance, X is a non-singular n-dimensional variety, and $Y \subset X$ is a non-singular closed $(n-1)$-dimensional subvariety, then one constructs with algebraic differentials "residue" maps giving a long exact "Gysin" sequence:

$$\cdots \to H^i(X, \Omega_X^{\boldsymbol{\cdot}}) \to H^i(X-Y, \Omega_X^{\boldsymbol{\cdot}}) \xrightarrow{\text{residue}} H^{i-1}(Y, \Omega_Y^{\boldsymbol{\cdot}}) \to H^{i+1}(X, \Omega_X^{\boldsymbol{\cdot}}) \to \cdots.$$

In particular, if $X-Y$ is affine, and ω is a closed i-form, regular on $X-Y$, then this gives residue $(\omega) \in H^{i-1}(Y, \Omega_Y^{\boldsymbol{\cdot}}) \cong H^{i-1}(Y, \mathbb{C})$ whose vanishing is necessary and sufficient for ω to be the image of some $\eta \in H^i(X, \Omega_X^{\boldsymbol{\cdot}})$. If the poles of ω lie on singular divisors, HODGE-ATIYAH (1) explains how subtle the analog of residue is. For all divisors Y on a non-singular X, they introduce mysterious residue sheaves:

$$R^p(Y) = \left\{ \begin{array}{l} \text{sheaf of closed meromorphic} \\ p\text{-forms, poles on } Y \end{array} \right\} \Big/ d \left\{ \begin{array}{l} \text{sheaf of meromorphic} \\ (p-1)\text{-forms, poles on } Y \end{array} \right\}.$$

These play a central role, and they prove

i) $R^0(Y) \cong \mathbb{C}$ (constant sheaf).

ii) $R^1(Y) \cong \underset{i}{\oplus} \, \mathbb{C}_{Y_i}$, Y_i = components of Y and $\mathbb{C}_Z$ = constant sheaf $\mathbb{C}$ on Z.

iii) $\text{Supp}(R^p(Y)) \subset$ singular locus on Y, if $p \geqq 2$.

In the case $n = 2$, they give however a complete sheaf-theoretic development of the results of § 9 and 10, including a careful proof that 3 *a priori* distinct definitions of a double differential of 2nd kind are equivalent. In particular, with $U \subset X$ affine, we get the diagram:

$$H^2(X, \mathbb{C}) \Big/ \left\{ \begin{array}{l} \text{subspace spanned by} \\ \text{classes of the components} \\ D_i \text{ of } X - U \end{array} \right\} \; \lhook\joinrel\longrightarrow \; H^2(U, \mathbb{C})$$

$$\wr\| \qquad\qquad\qquad\qquad\qquad\qquad \wr\|$$

$$\dfrac{\left\{ \begin{array}{l} \text{2-forms, regular on } U \\ \text{of 2nd kind at } \infty \end{array} \right\}}{\{ d\eta \mid \eta \in \Gamma(U, \Omega_X^1) \}} \quad \subset \quad \dfrac{\left\{ \begin{array}{l} \text{2-forms regular} \\ \text{on } U \end{array} \right\}}{\{ d\eta \mid \eta \in \Gamma(U, \Omega_X^1) \}}.$$

As U shrinks, this leads to the corollary:

$$H^2(X, \mathbb{C}) \Big/ \left\{ \begin{array}{l} \text{subspace of algebraic} \\ \text{2-cycles} \end{array} \right\} \cong \dfrac{\{ \text{2-forms of 2nd kind} \}}{\{ d\eta \mid \eta \text{ any 1-form on } X \}}$$

which is ZARISKI's "final theorem". GRIFFITHS (3) also proves that if $X - U$ is a sufficiently ample non-singular hyperplane section, then the image of $H^{2,0} + H^{1,1} \subset H^2(X, \mathbb{C})$ (resp. *all* of $H^2(X, \mathbb{C})$) in $H^2(U, \mathbb{C})$ is represented by 2-forms ω of 2nd kind with only *double* poles on $X - U$ (resp. only *triple* poles on $X - U$). Of course, he proves the generalization of this to n-dimensions. This is very convenient in actually constructing a representation of the cohomology of X by algebraic differentials regular on U.

We do not have space to summarize all of the many ideas that GRIF-
FITHS (2, 4, 5) has introduced to extend the theory of integrals to
higher-dimensional varieties. We want to mention in passing only 2 of
his ideas which are most relevant to algebraic surfaces or to the methods of
this chapter. The first is his method of describing the "periods" of 2-forms
on a surface X. (Again, he generalizes this to k-forms on any variety, but
we will not enter into this.) Fix (i) a surface X *and* (ii) a polarization
of X, i. e., the subspace $\mathbb{C} \cdot \omega \subset H^2(X, \mathbb{C})$ of multiples of the class ω of a
hyperplane section. By definition, $H^2(X, \mathbb{C})_0$ is the orthogonal comple-
ment of $\mathbb{C} \cdot \omega$ with respect to cup product, and is called the primitive
2-cycles. Fix (iii) an isomorphism

$$\phi : H^2(X, \mathbb{Z}) \cap H^2(X, \mathbb{C})_0 \cong \mathbb{Z}^{B_2 - 1}$$

($B_2 = $ 2nd BETTI number of X). Let

   $G = $ the Grassmanian of $h^{2,0}$-dimensional subspaces of $\mathbb{C}^{B_2 - 1}$.

Then define

$$\Omega(X, \omega, \phi) \in G$$

to be the point represented by the subspace which is the image of the
map:

$$H^0(X, \Omega^2_X) \to H^2(X, \mathbb{C})_0 \cong [H^2(X, \mathbb{Z}) \cap H^2(X, \mathbb{C})_0] \otimes \mathbb{C} \overset{\phi_{\mathbb{C}}}{\cong} \mathbb{C}^{B_2 - 1}.$$

GRIFFITHS calls $\Omega(X, \omega, \phi)$ the "periods" of the 2-forms on X and he
proves that it varies holomorphically in G as X varies in a holomorphic
family. It has proved to be a very valuable and suggestive tool in studying
families of algebraic surfaces. We refer the reader to his papers for further
details.

   Re: § 5. The second idea of GRIFFITHS that we want to mention is his
generalization of the normal functions of POINCARÉ in an attempt to
prove HODGE's conjecture. HODGE's conjecture is a generalization of
what ZARISKI calls the existence theorem of LEFSCHETZ-POINCARÉ:

$$\text{HODGE } conjecture: \begin{bmatrix} \text{subspace of } H^{2k}(X, \mathbb{Q}) \\ \text{spanned by the classes} \\ \text{of algebraic } n - k\text{-} \\ \text{dimensional cycles} \end{bmatrix} = H^{2k}(X, \mathbb{Q}) \cap H^{k,k}$$

where X is any non-singular projective variety of dimension n.

   For $k = 1$, this has been proven. For higher k, GRIFFITHS has developed an attack quite analogous to the method of the text using a generic
pencil $\{H_t\}$ on X and a family of complex tori J_t associated to the
$(2k - 1)$st-cohomology groups of the H_t's (cf. WEIL, 6, p. 82; and
GRIFFITHS, 2). Although HODGE's conjecture is still open, a by-product
of his work is an example of a 1-dimensional algebraic cycle Z on a
3-dimensional X such that the fundamental class of Z in $H^4(X, \mathbb{Z})$ is 0,
but no multiple of Z is algebraically equivalent to 0.

The quickest sheaf-theoertic approach to the case $k = 1$ and in particular to LEFSCHETZ's existence theorem in the surface case is via the exact sequence:

$$0 \to \mathbb{Z} \to \mathcal{O}_X(\text{hol}) \xrightarrow{\text{exp}} \mathcal{O}_X^*(\text{hol}) \to 0$$

where $\exp(f) = e^{2\pi i f}$, and X is a non-singular projective variety. This gives us in particular:

$$H^1(X, \mathbb{Z}) \to H^1(\mathcal{O}_X(\text{hol})) \to H^1(\mathcal{O}_X^*(\text{hol})) \to H^2(X, \mathbb{Z}) \to H^2(\mathcal{O}_X(\text{hol})) .$$

Using SERRE's basic paper (3), we know not only that $H^i(\mathcal{O}_X(\text{hol})) \cong$ $\cong H^i(\mathcal{O}_X)$, but also that $H^1(\mathcal{O}_X^*(\text{hol}))$, which is the group of holomorphic line bundles over X, is isomorphic to $H^1(\mathcal{O}_X^*)$, the group of algebraic line bundles. This gives us:

$$\text{Pic}\,X = \begin{cases} \text{Group of divisors on } X, \\ \text{mod linear equivalence} \end{cases}$$

$$\parallel \qquad\qquad \alpha \downarrow$$

$$H^1(\mathcal{O}_X) \to H^1(\mathcal{O}_X^*(\text{hol})) \to H^2(X, \mathbb{Z}) \to H^2(\mathcal{O}_X) .$$

The composite map α is easily checked to take a divisor D to its cohomology class. Thus, just as with normal functions, we kill 2 birds with one stone:

a) We find that $\text{Pic}(X)$ contains the image of a q-dimensional vector space mod a countable subgroup $(q = \dim H^1(\mathcal{O}_X))$, hence X possesses q-dimensional non-linear algebraic families of divisors.

b) A cohomology class $\sigma \in H^2(X, \mathbb{Z})$ is the cohomology class of a divisor if and only if its image in $H^2(\mathcal{O}_X)$ is 0. (b) is a simple translation of LEFSCHETZ's theorem.

TATE (1, 2) has conjectured existence theorems to characterize the linear combinations of the cohomology classes of divisors in the *étale* cohomology groups $H^2(X_{\text{ét}}, \mathbb{Z}_l)$. In (3), he proved his conjecture for abelian varieties over finite fields, and, as a corollary, for products of 2 curves over finite fields.

Re: § 6. POINCARÉ's theorem asserts that if $Y \subset X$ are 2 abelian varieties, then an abelian subvariety $Z \subset X$ exists such that $X = Y + + Z$, $Y \cap Z$ finite. This was proven algebraically, in all characteristics, by WEIL (4) (cf. LANG, 3 and MUMFORD, 9, p. 173).

Re: § 7 (d). PICARD-SEVERI's theorem on the regularity of the adjoint system was greatly generalized in a very important paper of KODAIRA (4). In this paper, KODAIRA shows that if D is a divisor on a non-singular projective variety X such that nD is a hyperplane section for some $n \geq 1$, some projective embedding, then $H^i(\mathcal{O}_X(-D)) = (0)$, for $0 \leq i < \dim X$. By SERRE's duality, this is equivalent to $H^i(\Omega_X^n(D)) = (0)$, $0 < i \leq \dim X$. In fact, he even showed that is true if $\mathcal{O}_X(D)$ is replaced by a line bundle L, positive in a certain differential-geometric sense, on an arbitrary KÄHLER manifold X. This result, known as KODAIRA's Vanishing Theorem, is the basis of his proof (6) that KÄHLER manifolds whose fundamental 2-form Ω has integral periods are projective varieties. AKIZUKI-NAKANO (1) prove also that $H^i(\Omega^j_X(-D)) = (0)$ for $0 \leq i + j < \dim X$. The assertion $H^1(\mathcal{O}_X(-D)) = (0)$ can be proven by purely algebraic methods (GRAUERT, unpublished; MUMFORD, 7).

In conclusion, let me simply mention the very important work of MANIN (3), GRIFFITHS (5) and GROTHENDIECK (11) on the variation of the DE RHAM cohomology groups of a variety X, when X itself varies in an algebraic or analytic family in which case they construct the so-called GAUSS-MANIN connection. Their work generalizes the classical PICARD-LEFSCHETZ differential equations for the periods of a curve of variable moduli. See also KATZ-ODA (1), KATZ (1).

Now look at characteristic p. Almost without exception we find a dismal picture with all the beautiful results failing. There appear to be two reasons for this: (i) all the cohomology groups associated to differentials are groups killed by p, hence subtle p-torsion phenomena on X get mixed up with these groups. A more far-reaching theory requires cohomology groups over the WITT vectors $W(k)$ associated to the ground field k, which reduce to the DE RHAM groups after $\otimes_{W(k)} k$. (ii) Differentials are, of course, just linear approximations to the full non-linear integrals. But in char p, non-constant functions can have identically 0 derivatives. Therefore, what seems to be necessary are some kind of higher order infinitesimal objects which are closer to integrals. Suggestions in the line of (i) are due to SERRE (4), MONSKY-WASHNITZER (1), LUBKIN (1), ODA (1); in the line of (ii), we have GROTHENDIECK's theory of crystals (11). These new ideas are too inconclusive to report on at present, so I will merely try to show here that a problem exists by describing the known counter-examples.

Start with the DE RHAM complex of a non-singular projective surface:

$$\Omega^{\cdot}_X : 0 \to \mathcal{O}_X \xrightarrow{d} \Omega^1_X \xrightarrow{d} \Omega^2_X \to 0 \,.$$

We can consider the basic groups $H^q(X, \Omega^p_X)$, $0 \leq p, q \leq 2$; and the spectral sequence $H^q(X, \Omega^p_X) \Rightarrow H^*(X, \Omega^{\cdot}_X)$, just as in char 0. Let $h^{p,q} = \dim H^q(X, \Omega^p_X)$. Now from topological considerations, we know what is the "right" first BETTI number of X. In fact, if $A =$ the ALBANESE variety of X, then almost all abelian unramified coverings of X are induced from coverings of A (LANG, 1), hence if $p = \dim A$:

$$\dim_{\mathbf{Z}_l} (H^1(X_{et}, \mathbf{Z}_l)) = 2p \,, \quad \text{all } l \neq \text{char}(k) \,,$$

i. e. $2p =$ "right" first BETTI number. But we find
(a) the maps

$$\Phi^* : H^1(A, \mathcal{O}_A) \to H^1(X, \mathcal{O}_X) \,,$$

$$\Phi^* : H^0(A, \Omega^1_A) \to H^0(X, \Omega^1_X)$$

are always injective (IGUSA, 5) but need not be surjective. Since the left hand groups have dimension p, both of the inequalities $h^{0,1} > p$, $h^{1,0} > p$ are possible. $h^{0,1} \neq h^{1,0}$ is possible too (IGUSA, 6; SERRE, 4; MUMFORD, 1).

(b) There can exist non-closed regular 1-forms (MUMFORD, 1), hence the spectral sequence $H^q(X, \Omega^p_X) \Rightarrow H^*(X, \Omega^{\cdot}_X)$ may be non-trivial. The map

$$\Phi^* : H^1(A, \Omega^{\cdot}_A) \to H^1(X, \Omega^{\cdot}_X)$$

although injective, need not be surjective (ODA, 1).

(c) To make matters more subtle, there is a new map, CARTIER's *operator* (1)

$$C : \Omega^1_{X,\text{closed}}/d(\mathcal{O}_X) \xrightarrow{\sim} \Omega^1_X$$

which gives rise to even further obstructions to lifting co-cycles from X to A (cf. ODA, 1).

(d) The regularity of the adjoint $H^1(X, \Omega^2(H)) = (0)$ (cf. § 7 d) can be false, at least for some normal surfaces X, and divisors H such that $nH =$ a hyperplane section (MUMFORD, 7).

(e) If we define the 2nd BETTI number B_2 to be $\dim_{\mathbb{Z}_l} H^2(X_{et}, \mathbb{Z}_l)$ for all $l \neq \text{char}(k)$, then it can happen that

$$B_2 < h^{0,2} + \varrho$$

where $\varrho =$ base number (MUMFORD, 3).

Chapter VIII

Branch Curves of Multiple Planes and Continuous Systems of Plane Algebraic Curves

1. The problem of existence of algebraic functions of two variables (ENRIQUES, 18; ZARISKI, 3, VAN KAMPEN, 1). Let z be a k-valued algebraic function of two complex variables x and y, defined by an irreducible algebraic equation,

(1)
$$F(x, y, z) = 0.$$

The branch curve f,

(2)
$$f(x, y) = 0,$$

of the function z is found by eliminating z between $F = 0$ and $\partial F/\partial z = 0$ and by neglecting in the resultant certain factors which correspond to multiple curves of the surface $F = 0$ (*apparent branch curves*) The definition of f may be rendered exact by assuming that: (a) the polynomial f contains no multiple factors; (b) the curve f is the locus of the effective branch points $(x_1, y), (x_2, y), \ldots, (x_n, y)$ of the function $z = z(x, y)$, for y fixed, and of the lines $y = c = $ const. such that $y = c$ is an effective branch point of z if x is fixed and generic. It may be necessary to include the line at infinity of the projective plane (x, y) in the branch curve. However, we may always choose the coördinates x and y in such a manner that the line at infinity does not belong to the branch curve.

The determination of the *monodromy group* of the function z leads to the consideration of the *fundamental group G* of the residual space $P - f$, where P is the space (of four real dimensions) of the complex projective plane (x, y). The curve f is a two-dimensional orientable circuit immersed in P (f is a manifold if and only if each point of f is the origin of a single branch). We recall the definition of G (see, for instance, VEBLEN, a; LEFSCHETZ, e). Its elements are oriented 1-spheres in $P - f$ issued from a fixed point O. Two elements g_1 and g_2 are *equivalent*, in symbols $g_1 = g_2$, if g_1 can be deformed into g_2 in $P - f$, the point O remaining fixed; in other words, if $g_1 - g_2$ bounds a 2-cell in $P - f$. All equivalent 1-spheres correspond to one and the same element of G. In particular the 1-spheres g which bound a 2-cell in $P - f$, and which therefore can be deformed into the point O, all correspond to the element 1 of G, and we write $g = 1$. The meaning of the symbols

g_1g_2 and g^{-1} is obvious. In general $g_1g_2 \neq g_2g_1$. G is a discrete group, in general infinite.

Let $\{L_t\}$ be the pencil of lines in P of center O and let $A_1, A_2, \ldots, A_n$ denote the intersections of L_t with f, where n is the order of f. Let L_0 be a fixed line of the pencil whose intersections $A_1^0, \ldots, A_n^0$ with f are distinct. By a reasoning similar to the one employed in VI, 7 it can be shown that every element g of G can be deformed in $P - f$ into the "line" L_0, a two-dimensional manifold homeomorphic to a sphere. In $L_0 - A_1^0 - \cdots - A_n^0$ every g is equivalent to a product of loops $g_1, \ldots, g_n$ surrounding the points A_i^0 and having nothing in common but the point O. Hence *the elements $g_1, \ldots, g_n$ are generators of G* (LEF-SCHETZ; for a complete proof, see VAN KAMPEN, 1).

Let $L_{\alpha_1}, \ldots, L_{\alpha_\mu}$ be the lines of the pencil for which some of the intersections A_i coincide. Let $\gamma_1, \ldots, \gamma_\mu$ be a set of non-intersecting loops in the t-plane, issued from $t = 0$ and surrounding the points $\alpha_1, \ldots, \alpha_\mu$ respectively. As t describes the loop γ_j, the set of inter-sections A_i of the variable line L_t varies continuously, starting from and returning to its original position $\{A_1^0, \ldots, A_n^0\}$. Let A_i^0 be carried into the point $A_{s_{ij}}^0$, where $(s_{1j}, s_{2j}, \ldots, s_{nj})$ is a permutation of the indices $1, 2, \ldots, n$. At the same time each loop g_i is deformed continuously into a loop g_i', surrounding the point $A_{s_{ij}}^0$ and therefore expressible in the form $g_{ij}^{-1} g_{s_{ij}} g_{ij}$, where g_{ij} is some element of G. Since $g_i = g_i'$, we have the following *generating relations*:

$$(3) \qquad \varphi_{ij}(g_1, \ldots, g_n) = g_i^{-1} g_{ij}^{-1} g_{s_{ij}} g_{ij} = 1, \qquad \begin{array}{l} i = 1, 2, \ldots, n, \\ j = 1, 2, \ldots, \mu. \end{array}$$

Moreover, the following trivial relation holds:

$$(3') \qquad\qquad\qquad g_1 g_2 \cdots g_n = 1.$$

VAN KAMPEN (1) has proved *that the $\mu n + 1$ relations* (3) *and* (3') *constitute a complete set of generating relations for the group G*.

Let S_i be the substitution of the branches of the function z as the point (x, y) describes the loop g_i. By the monodromy theorem it follows that the monodromy group of z is generated by the substitutions $S_1, \ldots, S_n$ and that these substitutions satisfy the relations:

$$(4) \qquad\qquad\qquad \varphi_{ij}(S_1, \ldots, S_n) = 1,$$

$$(4') \qquad\qquad\qquad S_1 S_2 \ldots S_n = 1.$$

We now invert the order of the preceding considerations. We assign the curve f and we ask for *necessary and sufficient conditions of existence of algebraic functions z of x and y possessing f as branch curve.* Assuming the knowledge of the group G, ENRIQUES (18) solves this *existence problem for algebraic functions of two variables* by proving that:

Given arbitrarily n substitutions $S_1, \ldots, S_n$ on k letters, satisfying the relations (4) and (4'), there exists a corresponding class of algebraic k-valued functions z possessing f as branch curve.

For the proof we assume that the pencil $\{L_t\}$ is the pencil $x = $ const. The assigned substitutions $S_1, \ldots, S_n$ determine on a generic line $x = $ const. a class of algebraic k-valued functions $z_x(y)$ of y (existence theorem of RIEMANN). In view of the relations (4) and (4') it follows that this class of functions is *rationally determined* on a generic line of the pencil, since as x describes any closed path the branch points of $z_x(y)$ *and the substitutions* relative to the assigned set of loops g_i remain unaltered. We may also describe the situation in a slightly different manner, by observing that the assigned substitutions define a k-sheeted RIEMANN surface R_x over each line of the pencil and that in view of the relations (4) and (4') this RIEMANN surface R_x *remains invariant* as x describes any closed path. Our class of algebraic functions $z_x(y)$ is the class of rational functions on R_x. If we fix the *order* v of $z_k(x)$, we obtain, for a generic x, a class of functions $z_k(x)$ depending only on a finite number of continuous parameters $t_1, \ldots, t_\varrho$. If we impose ϱ arbitrary linear conditions $c_1 t_1 + \cdots + c_\varrho t_\varrho = 0$, where the coefficients c are generic and independent of x, we obtain, for a generic x, a finite number of functions $z_x^{(1)}(y), z_x^{(2)}(y), \ldots, z_x^{(\sigma)}(y)$, all k-valued. It is now sufficient to consider a convenient elementary symmetric function of these σ functions, in order to obtain a k-valued algebraic function z of x and y possessing f as branch curve. It may happen that the complete branch curve of z consists, in addition to f, also of a finite number of lines $x = $ const. However, this complication can always be avoided if the order v is taken sufficiently high.

The following comment on the theorem of ENRIQUES may be of interest. From a purely topological point of view the theorem of ENRIQUES says that the relations (4) and (4') give *a complete set of conditions for the existence of k-fold covering manifolds with f as branch curve* (4-dimensional Riemannian varieties consisting of k samples of the projective plane P connected in a proper manner along the curve f)[1] However, from this *does not follow immediately* the *completeness* of the set of generating relations (3) and (3') for the fundamental group G, proved by VAN KAMPEN. The group-theoretic consequence of the theorem of ENRIQUES is the following: Assuming that the set of relations (3) and (3') is not complete for G, let G_1 be the group defined by the generating relations (3) and (3'). The elements of G_1, which become equivalent to 1 in virtue of the generating relations of G which are not consequences of (3) and (3'), form an invariant subgroup Γ of G_1 admitting a finite

---

[1] The proof of this topological and consequently weaker form of the theorem of ENRIQUES is naturally independent of RIEMANN's existence theorem and follows simply from the *invariance* of the RIEMANN surface R_x.

set of generators, and G is the quotient group G_1/Γ. If $\psi(g_i)$ is any element of Γ, then it follows from the theorem of ENRIQUES, that if given substitutions $S_1, \ldots, S_n$ satisfy (4) and (4') they also satisfy the relation $\psi(S_i) = 1$. Now the elements of G_1 which correspond to the identical substitution in the isomorphism between G_1 and the finite group generated by $S_1, \ldots, S_n$, form an invariant subgroup H of G_1 of finite index. Conversely, any such invariant subgroup H defines a finite group of substitutions and the substitutions S_i. Hence we have the following property of the invariant subgroup Γ of G_1: *every invariant subgroup of G_1 of finite index contains Γ*. Since $G_1 = G$ if Γ contains only the element 1, the possibility of deriving from ENRIQUES' result the completeness of the relations (3) and (3') depends on the solution of the following question: *can an infinite discrete group G_1 contain elements different from 1 which are common to all the invariant subgroups of G_1 of finite index?*

2. Properties of the fundamental group G. The theorem of ENRI-QUES settles the algebraic part of the existence problem discussed in the preceding section. Its complete solution depends, however, on the determination of the fundamental group G. This topological problem is much more difficult and is not yet solved. In this problem one is naturally concerned with the topological invariants of the space $P - f$, or what is the same, with *the invariants of a plane algebraic curve under isotopic deformations in the complex projective plane*. These invariants are also projective invariants of the curve f and include, for instance, the order of the curve, the number and the nature of its singularities. Another projective invariant whose topological character is less obvious will be derived in section 3.

The dependence of the group G on the singularities of f is illustrated by the following theorems (ZARISKI, 3).

Theorem 1. *If the curve f, of order n, is irreducible and possesses only ordinary double points, then G is a finite cyclic group of order n* $(g_1 = g_2 = \cdots = g_n, g_1^n = 1)$.

The proof of this theorem is based upon the remark that *if a variable curve f tends to a limit curve $\bar{f}$ then some generating relations between the generators of the group G may be destroyed but no new relations are introduced.* Let f vary continuously and degenerate into n lines (this is always possible, see section 4)[1]. Then it can easily be shown that the fundamental group becomes an abelian group with $n - 1$ free generators. Hence also the original group G of $P - f$ must have been abelian. On the other hand the relations (3) of the preceding section show that if f is irreducible,

---

[1] Unfortunately, it is *not* known whether this is possible. See footnote to theorem 5, section 4; theorem 1 therefore has been proven only for those curves which do degenerate in this fashion.

then the generators $g_1, \ldots, g_n$ are conjugate elements of G. Consequently $g_1 = \cdots = g_n$ and by (3') $g_1^n = 1$.

In a similar manner one can prove the following:

Theorem 2. *With the same hypotheses on the singularities of f as in theorem 1, if f is reducible and is made of s irreducible components of orders $n_1, \ldots, n_s$, then G is abelian and is generated by s elements $g_1, \ldots, g_s$ connected by one generating relation $g_1^{n_1} \cdots g_s^{n_s} = 1$* (ZARISKI, 3).

The following theorem is useful in application:

Theorem 3. *Let $\{f\}$ be an irreducible continuous system of irreducible plane algebraic curves. If the curves of $\{f\}$ cut out on a generic line the totality, ∞^n, of sets of n points and if the curves of $\{f\}$ which pass through n generic points of the line form an irreducible subsystem of $\{f\}$, then the fundamental group of a generic curve of the system $\{f\}$ is cyclic of order n* (ZARISKI, 3).

It is possible to derive theorem 1 by using theorem 3, since it can be shown that the conditions of the theorem are satisfied if $\{f\}$ is the system of all curves of order n possessing a given number of ordinary double points (nodes).

The theorems 1 and 2 show that in order that the group G be non-cyclic it is necessary that f possess singularities other than nodes. The curves with nodes and cusps are of special theoretical interest, since the branch curve of a surface with ordinary singularities is in general a curve of this type (VI, 4)[1]. The curve of lowest order with a non-cyclic group G is the 3-cuspidal quartic: G is finite of order 12 and possesses two generators g_1, g_2 connected by the relations: $g_1^2 = g_2^2, g_1^4 = 1, (g_1 g_2)^3 = g_1^2$. An example of an irreducible curve with an infinite group G is the branch curve of the general cubic surface, a sextic with 6 cusps on a conic: G is generated by two elements u, v such that $u^2 = 1, v^3 = 1$ (ZARISKI, 3).

3. The irregularity of cyclic multiple planes (ZARISKI, 4, 5). Further topological properties of the curve

$$f(x, y) = 0$$

can be derived by studying the surface F,

(5) $$z^k = f(x, y),$$

where k is an arbitrary integer. This surface is a k-fold cyclic covering manifold whose branch curve is the given curve f and, if k is not divisible by n, the line at infinity of the projective plane P. Our method consists in evaluating the BETTI number R_1 of the surface F, first by topological considerations and then algebro-geometrically, as the double of its

---

[1] The branch curves of general surfaces of a given order have been characterized by B. SEGRE (4) in terms of the position of their nodes and cusps on algebraic curves of lower orders.

irregularity q. Confronting the two results we shall be able to derive some interesting conclusions.

Theorem 1. *If f is an irreducible curve and if k is the power of a prime number then* $R_1 = 0$ (ZARISKI 4).

We give a rough sketch of the proof. The fundamental group G of the branch curve ($f +$ the line at infinity) is generated by $n + 1$ elements $g_1 \ldots, g_n, \gamma$, where $g_1, \ldots, g_n$ are the generators of the fundamental group of $P - f$. In view of the relation $g_1 g_2 \ldots g_n \gamma = 1$ it follows that the elements $g_1, \ldots, g_n$ are also generators of G. The relations (3) show that if f is irreducible then $g_1, \ldots, g_n$ are *conjugate elements of* G. In other words, the quotient group of G with respect to its commutator subgroup is a cyclic infinite group. As a fundamental set of 1-cycles on F we may take the $(n-1)k$ cycles, γ_{ij}, where

$$\gamma_{ij} = g_1^{-i}(g_1 g_{j+1}^{-1}) g_1^i, \qquad \begin{matrix} i = 0, \ldots, k-1, \\ j = 1, 2, \ldots, n-1. \end{matrix}$$

If we express the fact that g_{j+1} is the conjugate of g_1 in G, $g_{j+1} = g^{(j)} g_1 (g^{(j)})^{-1}$, where $g^{(j)}$ is some element of G, we find a corresponding homology between the cycles γ_{ij}, of the form:

$$(6) \qquad \gamma_{0,j} + \Gamma_j - \varphi(\Gamma_j) \backsim 0, \qquad j = 1, 2, \ldots, n-1,$$

where Γ_j is a 1-cycle on F and $\varphi(\Gamma_j)$ is obtained from Γ_j by expressing Γ_j in terms of the cycles γ_{ij} and by replacing in this expression γ_{ij} by $\varphi(\gamma_{ij}) = \gamma_{i+1,j}$ ($\gamma_{kj} = \gamma_{0j}$). We have also the homologies obtained from (6) by the cyclic permutation $(\gamma_{0j}\gamma_{1j} \cdots \gamma_{k-1,j})$:

$$(7) \qquad \gamma_{ij} + \varphi^i(\Gamma_j) - \varphi^{i+1}(\Gamma_j) \backsim 0, \qquad \begin{matrix} i = 0, 1, \ldots, k-1, \\ j = 1, 2, \ldots, n-1. \end{matrix}$$

We have here a system of $k(n-1)$ homologies between the $k(n-1)$ cycles γ_{ij}. Let Δ be the matrix of the coefficients of this system. From the form of the homologies (7) is is immediately seen that Δ is made up of an array of square matrices A_α^β, $\Delta = \| A_\alpha^\beta \|$, $\alpha, \beta = 1, 2, \ldots, n-1$, where A_α^β is a circulant of k rows and k columns

$$A_\alpha^\beta = \| a_{\alpha,0}^\beta, a_{\alpha,1}^\beta, \ldots, a_{\alpha,k-1}^\beta \|.$$

Moreover it is easily seen that the following relations hold:

$$(8) \qquad \sum_{i=0}^{k-1} a_{\alpha,i}^\beta = \delta_\alpha^\beta, \qquad \delta_\alpha^\beta = 0 \text{ if } \alpha \neq \beta, \qquad \delta_\alpha^\alpha = 1.$$

If we put

$$f_\alpha^\beta(x) = a_{\alpha,0}^\beta + a_{\alpha,1}^\beta x \cdots + a_{\alpha,k-1}^\beta x^{k-1}, \qquad \alpha, \beta = 1, 2, \ldots, n-1,$$
$$f(x) = | f_\alpha^\beta(x) |,$$

it can be proved that

(9)   det. $\varDelta = f(1)f(\omega_1) \ldots f(\omega_{k-1})$,

where $1, \omega_1, \ldots, \omega_{k-1}$ are all the kth roots of unity. The formula (9) is a generalization of the well-known expression of a circulant. From (8) it follows that $f(1) = 1$ and from this it follows very easily that no primitive root of unity whose index is the power of a prime number can be a root of the polynomial $f(x)$. Hence if k is the power of a prime number then det. $\varDelta \neq 0$, q. e. d.[1]

In evaluating the irregularity q of the surface F we shall confine ourselves to the case in which the branch curve f possesses only nodes and cusps (ZARISKI, 5). We shall also assume temporarily that $k \geqq n$. In this case the singularities of the surface consist of a multiple line at infinity in the (x, y) plane and of isolated singularities at the singularities of the curve f. When these singularities are analyzed with reference to their postulation on the adjoint surfaces of F, it is found that the irregularity q of F can be expressed in the following form:

(10) $q = \alpha_0 s_{n-3} + \alpha_1 s_{n-4} + \cdots + \alpha_j s_{n-3-j} + \cdots,$

where α_j is an integer defined in terms of n, k and j, and where $s_{n-3-j}(\geqq 0)$ denotes the superabundance of the complete linear system $|C_{n-3-j}|$ of the plane curves of order $n - 3 - j$ in the plane (x, y), passing through the cusps of f. Having due regard to the manner in which the integers α_j are defined it is found that *if there exists an integer j such that $6j < n$ and $s_{n-3-j} > 0$, then q increases indefinitely as k increases.* Applying theorem 1 we derive immediately the following:

Theorem 2. *If f is an irreducible curve of order n possessing only nodes and cusps, then the superabundances*

$$s_{n-3}, s_{n-4}, \ldots, s_{n-3-\mu} \qquad \left(\mu = \left[\frac{n-1}{6}\right]\right)$$

all vanish (ZARISKI, 5).

Using this theorem it is found that most of the terms in the right-hand member of (10) vanish and the following final theorem is obtained:

Theorem 3. *If k and n are both divisible by 6 then $q = s_{n-3-n/6}$. In all other cases $q = 0$* (ZARISKI, 5).

This theorem holds also if $k < n$.

For the correct understanding of the preceding considerations it is necessary to bear in mind that we use the surface F only as a tool for

---

[1] We point out that in the proof of theorem 1 we did not make use of the algebraic character of the manifold F. As a matter of fact this theorem holds under the following more general hypotheses: P is any r-dimensional manifold, f is an $(r-2)$-dimensional circuit on P *such that the fundamental group of $P - f$ is generated by conjugate elements of the group.*

deriving properties of the curve f.[1] Thus the true significance of theorem 3 *lies in the fact that in virtue of this theorem the superabundance* $s_{n-3-n/6}$ *appears as a topological invariant of the curve* f (defined only for curves whose order n is divisible by 6). This is the only non-trivial invariant known at present.

The following example may serve as an illustration. The branch curve of a general cubic surface is a sextic with 6 cusps on a conic and for this curve the above invariant (s_2) equals 1. There exist, however, sextics with 6 cusps not on a conic (see section 5) and for these sextics $s_2 = 0$. These two types of sextics are therefore topologically distinct. A sextic with 6 cusps on a conic is therefore not a particular case of a sextic with 6 cusps in a general position (see section 5).

The theorem 2 has an interesting application in the problem of existence of plane curves with assigned PLÜCKER characters (see section 5).

4. Complete continuous systems of plane curves with d nodes (SEVERI, a, Anhang F; SEVERI, 19; ENRIQUES and CHISINI, a₃, Chapter III, pp. 355—384; ENRIQUES, 16). A continuous system of plane curves of a given order n with d nodes (ordinary double points) is *complete*, if it is not contained in a continuous system of higher dimension of curves of the same type. The importance of these systems is due to the fact that any algebraic curve of genus p is birationally equivalent to a plane curve of a sufficiently high order $n (n \geqq p + 2)$ with $d \left(= \dfrac{(n-1)(n-2)}{2} - p \right)$ nodes, and that consequently these systems reflect to a certain extent the properties of the variety V of classes of birationally distinct curves. Regardless of their applications to the study of this variety (ENRIQUES, SEVERI, l. c.), the continuous systems of plane curves with d nodes enjoy remarkable properties, which make the theory of these systems a topic of independent interest. The systematic treatment of these systems is due to SEVERI (a., Anhang F) and constitutes a very elegant application of the notion

---

[1] In accordance with this method of approach, we are not concerned with any particular value of k, but on the other hand we restricted the treatment (except in theorem 1) to branch curves possessing only nodes ans cusps. Proceeding in a different direction, DE-FRANCHIS (2) and COMESSATTI (1) classified the irregular cyclic multiple planes for $k = 2$ and $k = 3$ respectively, without putting *a priori* any condition on the type of branch curve. They obtain the following results:

(1) *A necessary and sufficient condition that the surface* $z^2 = f(x, y)$ *have irregularity q, is that the curve* $f = 0$ *be made up of* $2q + 2$ *or of* $2q + 1$ *curves, belonging to one and the same pencil. The surface possesses an hyperelliptic pencil of genus q, in* (1, 2) *correspondence with the above linear pencil* (DE-FRANCHIS, 2).

(2) *Outside of one exceptional case, a necessary and sufficient condition that the surface* $z^3 = f(x, y)$ *be irregular, is that the curve* $f = 0$ *(whose order may be assumed to be a multiple of* 3) *be made up of* $3h$ *curves belonging to one and the same pencil. If of these* $3h$ *curves only N are distinct, then the irregularity of the surface equals* $N - 2$, *and the surface possesses an irrational pencil of genus* $N - 2$, *in* (1, 3) *correspondence with the above linear pencil* (COMESSATTI, 1).

of the characteristic series and of the method of analytical branches ("falde analitiche"; see V, 3, p.101). Referring the reader for complete proofs and for detailed information to the excellent exposition of SEVERI and to the reports of COMESSATTI (a) and GODEAUX (a), we shall give here a very brief outline of the main results.

Theorem 1. *If C_0 is a plane algebraic curve (irreducible or not) of order n with d nodes, it belongs to one and only one complete continuous system Σ of curves of order n with d nodes, and the dimension of Σ is $3n + p - 1$* $\left(= \dfrac{n(n+3)}{2} - d\right)$, *where* $p = \dfrac{(n-1)(n-2)}{2} - d$ *is the genus of C_0 (the virtual genus if C_0 is reducible).*

The system $|E|$ of V, 3 (p.101) is now the system of all plane curves of order n, its representative linear space is an S_N, $N = \dfrac{n(n+3)}{2}$, and the hypersurface M in S_N is now the image of the (irreducible) system of all plane curves of order n with one node. The characteristic sets on C_0 are cut out by the curves of order n passing through the d nodes of C_0. The essential fact, from which the theorem follows immediately, is the *independence* of the d conditions imposed by the nodes of C_0 on the curves of order n constrained to pass through them. If C_0 is irreducible, the independence of these conditions follows from the fact that the characteristic series on C_0 is of order $3n + 2p - 2$ and is therefore non-special. If C_0 is reducible, the independence of the d conditions can be derived by means of NÖTHER's $Af + B\varphi$ theorem (SEVERI, a, Anhang F).

Corollary 1. *If C_0 is irreducible then the characteristic series of Σ on C_0 is complete.*

Let C_0 be reducible, $C_0 = C_0^{(1)} + \cdots + C_0^{(k)}$. If the dimension of Σ is compared with the dimensions of the systems $\Sigma^{(1)}, \ldots, \Sigma^{(k)}$ containing $C_0^{(1)}, \ldots, C_0^{(k)}$ respectively, the following corollary can be proved:

Corollary 2. *If C_0 is reducible then the generic curve of Σ degenerates into the same number of components as C_0.*

If C_0 is regarded as the limit of a variable curve C of order n with $\bar{d} < d$ nodes, then it is said briefly that the $\bar{d}$ nodes of C_0 which are very near to the $\bar{d}$ nodes of the variable curve C are *assigned nodes* of C_0, while the remaining $d - \bar{d}$ nodes of C_0 are considered as *virtually non-existent*. The curve C_0 is considered then as being of virtual genus $\bar{p} = \dfrac{(n-1)(n-2)}{2} - \bar{d}$.

Theorem 2. *A curve C_0 with d nodes, of which $d - \bar{d}$ are considered as virtually non-existent, determines uniquely a complete continuous system Σ, of dimension $3n + \bar{p} - 1$, of curves of order n with $\bar{d}$ and only $\bar{d}$ nodes.*

The proof is the same as that of theorem 1. We point out that if C_0 is reducible, then to different manners of choosing the $d - \bar{d}$ virtually non-existent nodes of C_0 there may correspond different systems Σ. To see this we say, as in II, 5, that C_0 is a *connected* curve, if for any decomposition of C_0 into two mutually exclusive parts, $C_0 = C_0' + C_0''$, *the intersection number* $(C_0' \cdot C_0'')$ *is positive, provided only those intersections of C_0' and C_0'' are considered which fall at the virtually non-existent nodes of C_0.* Then the following theorem holds:

Theorem 3. *A necessary and sufficient condition that the generic curve C of the system Σ of theorem 2 be irreducible, is that C_0 be connected* (Severi, a, Anhang F).

Let us assume, for simplicity, that C_0 consists of two irreducible components, $C_0 = C_0' + C_0''$, and that j of the $d - \bar{d}$ virtually non-existent nodes of C_0 are at the intersection of C_0' and C_0''. If n' and n'' are the orders of C_0' and C_0'' respectively, and if π' and π'' are their virtual genera, evaluated with reference to the set of the assigned nodes on either curve, then the dimensions of the systems determined by C_0, C_0' and C_0'' are respectively $r = 3n + \bar{p} - 1$, $r' = 3n' + \pi' - 1$, $r'' = 3n'' + \pi'' - 1$. In view of Noether's formula, $\bar{p} = \pi' + \pi'' + j - 1$, we find $r = r' + r'' + j$, and hence the generic curve of Σ does not degenerate, as C_0, into two components, if and only if $j > 0$, q. e. d.

As an illustration, we consider the case in which C_0 is of order 4 and degenerates into a cubic curve Γ with a node O and into a line L meeting Γ in 3 distinct points A_1, A_2, A_3. If O is considered as virtually non-existent, then C_0 is not connected and Σ is the system, ∞^{11}, of all reducible quartic curves which degenerate into an arbitrary cubic curve and into a line. If, however, we consider as virtually non-existent one of the double points A_i, then Σ is the system, also ∞^{11}, of all (rational) irreducible quartic curves with 3 nodes.

The theorem 3 includes Enriques' *principle of degeneration* (II, 5) for plane curves with nodes. In addition it gives also a sufficient condition that *a reducible curve possessing only nodes could be regarded as the limit of an irreducible curve.* These conditions for curves with higher singularities are not generally known. The problem is studied by Albanese (6) in the general case and also by B. Segre (2) for curves with nodes and cusps only, but the conclusions of these authors cannot be regarded as final in view of the extreme delicacy of the arguments involved.

Theorem 4. *There exist irreducible curves of order n with d nodes for any value of $d \leq \dfrac{(n-1)(n-2)}{2}$* (Severi, a, Anhang F; Snyder, 1).

We take for C_0 a curve which degenerates into n lines $a_1, \ldots, a_n$ and which therefore possesses $d_1 = \dfrac{n(n-1)}{2}$ nodes. If $d_1 - d \, (\geqq n - 1)$ of its nodes, *including, for instance, the $n - 1$ intersections of the line a_1*

with the remaining n — 1 lines, are considered as virtually non-existent, then C_0 remains connected and the theorem follows from the theorem 3.

Dealing with the example of quartic curves with 3 double points we have seen that these curves distribute themselves into two complete continuous systems, but that only one of these systems contains the irreducible quartics. More generally we have the following important

Theorem 5. *The irreducible plane algbraic curves of order n with d nodes form one irreducible complete continuous system* (ENRIQUES-CHISINI, a_3, p. 366; SEVERI, a, Anhang F).

Corollary. *The variety of the classes of birationally distinct algebraic curves of a given genus p is irreducible.*

SEVERI proves this theorem by showing that: (1) any complete continuous system Σ of irreducible curves of order n with d nodes contains curves which degenerate into n-lines; (2) as a consequence of (1) Σ contains a complete system of irreducible rational curves; (3) the rational curves of order n form *one* irreducible system Σ_1; (4) it is possible to vary a curve of Σ_1 in Σ_1 in such a manner that any $\dfrac{(n-1)(n-2)}{2} - d$ of its nodes (virtually non-existent) be carried into any other $\dfrac{(n-1)(n-2)}{2} - d$ nodes of the curve.

If it is only a question of establishing the preceding corollary, then it is sufficient to prove theorem 5 for n sufficiently high with respect to p. The proof is then very simple (ENRIQUES-CHISINI, l. c.) and we give it here in full because the considerations used in it lead incidentally to the evaluation of the number of moduli and to RIEMANN's existence theorem for n-valued algebraic functions with $2n + 2p - 2$ assigned branch points, when n is sufficiently high with respect to p.

Let $n > 2p - 2$ and let us consider a complete continuous system Σ of irreducible curves C of order n with $\dfrac{(n-1)(n-2)}{2} - p$ nodes. Such a system exists (theorem 4) and its dimension is $3n + p - 1$ (theorem 1). Let O be a fixed point in the (x, y) plane, say the point at infinity on the y-axis, and let C_0 be a generic curve of Σ not on O. The curves of Σ which touch the $2n + 2p - 2$ tangent lines to C_0, $x = \alpha_i$, passing through O, form one or more irreducible subsystems of Σ. Let Σ_1 be one of them containing C_0. The characteristic series determined on C_0 by Σ_1 is immediately seen to be cut out by the adjoint curves of C_0 passing through the $2n + 2p - 2$ points of contact with C_0 of the above tangent lines. This series is of order $(3n + 2p - 2) - (2n + 2p - 2) = n > 2p - 2$ and is therefore non-special. Hence

---

[1] SEVERI's proof of this is unfortunately false. At present, it is not known whether or not this theorem is true. The corollary however has been proven — see Appendix 2 to this chapter,

the characteristic series of Σ_1 on C_0 is *complete* and the dimension of Σ_1 equals $n - p + 1$. It follows that the sets of $2n + 2p - 2$ lines of the pencil of center O which are tangent lines of curves C of Σ, depend on $(3n + p - 1) - (n - p + 1) = 2n + 2p - 2$ arbitrary parameters, and hence *these $2n + 2p - 2$ tangents are arbitrary lines of the pencil.* Having proved that there exists at least one algebraic n-valued function $y(x)$ with $2n + 2p - 2$ arbitrary branch points $x = \alpha_i$, represented by a convenient curve C of Σ, we choose in the x-plane a set of $2n + 2p - 2$ non-intersecting loops g_i surrounding the points α_i, in such a manner that the corresponding transpositions on the branches of $y(x)$ are those of a LÜROTH-CLEBSCH arrangement: two transpositions $(1, 2)$; two transpositions $(2, 3)$; . . .; two transpositions $(n - 2, n - 1)$ and $2p + 2$ transpositions $(n - 1, n)$. We now observe that it is possible to carry, by deformation, the original loops g_i into any other set of $2n + 2p - 2$ loops g_i surrounding the branch points α_i, by letting these branch points describe convenient paths which carry the set $[\alpha_1, \ldots, \alpha_{2n+2p-2}]$ into itself, to within a permutation of the α_i's. It follows, by letting C vary continuously in Σ, that Σ contains curves C providing n-valued algebraic functions $y(x)$ with $2n + 2p - 2$ arbitrary branch points, *for any set of assigned transpositions* (RIEMANN's existence theorem).

The preceding considerations show that the classes of the functions $y(x)$ form an irreducible variety and from this follows the preceding corollary. To complete the proof of theorem 5 there remains only to observe that the totality of birationally equivalent curves of order n and of genus p, touching a set of given generic $2n + 2p - 2$ lines on O, form an irreducible continuous system, and that consequently Σ coincides with the totality of curves of order n with d nodes.

The dimension R of the variety V of the classes of birationally distinct curves of genus p equals the difference between the dimension $3n + p - 1$ of Σ and the dimension x of the subsystem $\bar{\Sigma}$ of Σ consisting of the totality of curves C which are birationally equivalent to a generic C_0. The dimension of this subsystem is found by observing that (ENRIQUES-CHISINI, p. 357): 1) any curve of $\bar{\Sigma}$ corresponds to a g_n^2 on C_0 whose sets are referred projectively to the lines of the plane; 2) given a g_n^2 on C_0, the corresponding C is determined to within an arbitrary collineation; 3) any g_n^2 $(n > 2p - 2)$ on C_0 is contained in a complete series g_n^{n-p} and a g_n^{n-p} contains $\infty^{3(n-p-2)}$ series g_n^2; 4) C_0 carries ∞^p complete series g_n^{n-p}; 5) if two distinct series g_n^2 or two distinct modes of referring a given g_n^2 to the lines of the plane lead to one and the same curve C, then C_0 admits a birational transformation into itself. From these remarks it follows that if we assume that a generic curve of genus p admits ∞^i birational transformations into itself, then $x = p + 3(n - p - 2) + 8 - i = 3n - 2p + 2 - i$. Consequently $R = (3n + p - 1) - x = 3p - 3 + i$.

It is well known that $i = 0$ if $p > 1$, $i = 1$ if $p = 1$ and $i = 3$ if $p = 0$. Hence $R = 3p - 3$ if $p > 1$, $R = 1$ if $p = 1$ and $R = 0$ if $p = 0$.

A curve of genus $p > 1$ possesses therefore $3p - 3$ independent birational invariants (moduli). One may ask if the coefficients of the equation of the curve contain rationally these invariants. In other words, is the variety V rational or at least birationally equivalent to an involution in a linear space S_{3p-3}? While it is not difficult to see that the answer is affirmative for low values of p ($p < 11$, SEVERI; for the proof see B. SEGRE (3); for $p = 3$ see ENRIQUES-CHISINI, a_3, p. 376), the general solution of this important problem is as yet not available.

Further important contributions to the theory of moduli have been made by SEVERI (21) and especially by B. SEGRE (1, 3). See also ZARISKI (1). For an account of the results of these authors, which is beyond the scope of our exposition, we refer the reader to the report of GODEAUX (a).

5. Continuous systems of plane algebraic curves with nodes and cusps (LEFSCHETZ, 1; B. SEGRE, 2; ZARISKI, 5). The importance attached to the study of plane curves with nodes and cusps clearly appears from the considerations developed in the first three sections of this chapter. We have seen that from the point of view of topological classification and of applications to the theory of multiple planes, it is these curves which constitute the significant case, while the case of curves with nodes is trivial. The problems of algebro-geometric character which one encounters in the theory of curves with nodes and cusps are the same as those discussed in the preceding section for curves with nodes only, namely problems relative to the *existence*, the *dimension* and the *unicity* of complete continuous systems. However, while for curves with nodes these problems have been completely solved, our information on curves with cusps and nodes consists only of a few partial results, which, far from providing the general solution of any of the above mentioned problems, give only an indication of the difficulties involved.

Let us first consider the problem of determining the dimension r' of a complete continuous system Σ of curves C of order n with d nodes and k cusps, or briefly, of curves (n, p, k), where p is the genus of C, assuming that such curves exist for the given values of n, p and k. The characteristic series of Σ on a generic curve C of Σ is cut out by the adjoint curves of C of order n *which touch at each cusp the cusp tangent of* C. This series is of order $3n + 2p - 2 - k$. If $k < 3n$, this series is non-special and hence complete, by a reasoning repeatedly used before. The dimension of Σ is then $3n + p - 1 - k = \dfrac{n(n+3)}{2}$ $- d - 2k$. If $k \geq 3n$, the characteristic series may be special. If i

is its index of speciality, then the dimension of Σ is $\leq 3n + p - 1 - k$ $+ i$. We put

(11) $$r = \frac{n(n+3)}{2} - d - 2k = 3n + p - 1 - k,$$

and we call r the *virtual dimension* of Σ. Then we may write (B. SEGRE, 2):

(12) $$r' = r + i - \omega,$$

where ω is *the deficiency of the characteristic series of* Σ. From the fact that in the representative space $S_{\frac{n(n+3)}{2}}$ of the curves of order n, the complete neighborhood of a C in Σ is represented by the intersection of d analytical branches of dimension $\frac{n(n+3)}{2} - 1$ and of k analytical branches of dimension $\frac{n(n+3)}{2} - 2$, it easily follows (B. SEGRE, 2) that

(13) $$r' \geq r.$$

That the inequality sign $>$ may hold is shown by the example of the following system of curves:

(14) $$\varphi(x, y) = [f_{2m}(x, y)]^3 + [f_{3m}(x, y)]^2 = 0,$$

where $f_{2m}(x, y)$ and $f_{3m}(x, y)$ are arbitrary polynomials of degrees $2m$ and $3m$ respectively (B. SEGRE, 2). In this example we have $n = 6m$, $d = 0$, $k = 6m^2$, since the intersections of the curves $f_{2m} = 0$ and $f_{3m} = 0$ are cusps of $\varphi = 0$. The effective dimension ϱ of the system (14) equals $\frac{2m(2m+3)}{2} + \frac{3m(3m+3)}{2} + 1$, or

(15) $$\varrho = \tfrac{1}{2}(13m + 2)(m + 1).$$

The index of speciality i of the characteristic series of the system (14) equals the superabundance of the linear system $|L|$ of curves of order $6m$, passing through the $6m^2$ cusps of a curve $\varphi = 0$ and touching there the cusp tangents. Since these cusp tangents coincide with the tangents of the curve $f_{3m} = 0$, $|L|$ cuts out on $f_{3m} = 0$ a complete series, and the index of speciality of this series equals i. This series is also cut out on $f_{3m} = 0$ by the residual curves of $|L|$ with respect to the curve $f_{2m} = 0$ counted twice, i. e. by the system of all curves of order $2m$. From this it follows that i equals the number of independent curves of order $m - 3$, i. e.

(16) $$i = \frac{(m-1)(m-2)}{2}.$$

In the present case we have $r = \dfrac{6m(6m+3)}{2} - 12m^2$, i. e.

(16') $$r = 6m^2 + 9m.$$

In view of (16) and (16') we find from (15)

$$\varrho = r + i > r, \quad \text{if} \quad m > 2.$$

Comparing with (12) we conclude that $\varrho = r'$, i. e. that Σ is a *complete system and that $\omega = 0$.*

It is not known if there exist complete continuous systems of curves with nodes and cusps whose characteristic series is incomplete ($\omega > 0$).

We now pass to the problem of existence of irreducible curves (n, p, k). The following upper limits for k in terms of n and p hold for $n > 6$ and can be obtained in a purely arithmetic manner (LEFSCHETZ, 1) from the condition that the 3 PLÜCKER relations between the 6 PLÜCKER characters of the curves (n, p, k) provide non-negative values for the dual characters m (class), τ (number of double tangents) and i (number of flexes):

$$(17) \begin{cases} (17') & k \leq \frac{3}{2}(n + 2p - 2), \quad \text{if} \quad 0 \leq p \leq \frac{n + 1 - \sqrt{4n+1}}{2}; \\ (17'') & \begin{cases} k \leq \frac{4(n+p) - 11 - \sqrt{24p - 8n + 25}}{2}, \\ \text{if} \quad \frac{n + 1 - \sqrt{4n+1}}{2} < p \leq \frac{(n-3)(n-2)}{2}; \end{cases} \\ (17''') & k \leq \frac{(n-1)(n-2)}{2} - p, \quad \text{if} \quad \frac{(n-3)(n-2)}{2} < p \leq \frac{(n-1)(n-2)}{2}. \end{cases}$$

The inequalities given by LEFSCHETZ are slightly different from our inequalities, because LEFSCHETZ *postulates the inequality $r \geq 8$*, where r is given by (11), in order to express the fact that if curves (n, p, k) exist they depend at least on the 8 parameters of an arbitrary collineation. However, since the effective dimension r' may be greater than the virtual dimension r, we have no proof that the inequality $r \geq 8$ is a necessary condition for the existence of curves (n, p, k).

The above upper limits for k provide *only necessary but not sufficient existence conditions* (ZARISKI, 5). To see this we make use of theorem 2 of section 3. By this theorem the complete linear system of curves of order $n - 3 - \mu$, where $\mu = \begin{bmatrix} n-1 \\ 6 \end{bmatrix}$, passing through the k cusps of an irreducible curve (n, p, k), is regular. Hence its virtual dimension $(n - 3 - \mu)(n - \mu)/2 - k$ must not be not less than -1, i. e.,

$$(18) \qquad\qquad k \leq \frac{(n - \mu)(n - 3 - \mu)}{2} + 2.$$

For large values of n this upper limit for k is greater than the upper limit provided by the inequalities (17). Hence for large values of n (18) is trivial. But for certain small values of n the inequality (18) leads to significant examples of *non-geometric* sets of PLÜCKER characters, i. e. to sets of non-negative integers n, d, k, m, τ, i, which satisfy the PLÜCKER relations and which do not correspond to effectively existing curves. For instance, for $n = 7$ we find from (18), $k < 11$, so that an irreducible curve of order 7 cannot possess 11 cusps. However, for $n = 7$ and $k = 11$ the PLÜCKER equations provide two sets of non-negative solutions:

$$(19) \begin{cases} n = 7, \quad d = 0, \quad k = 11; \quad\quad m = 9, \quad \tau = 7, \quad i = 17; \quad\quad p = 4; \\ n = m = 7; \quad\quad d = \tau = 1; \quad\quad k = i = 11; \quad\quad p = 3. \end{cases}$$

We point out that in these two examples the virtual dimension $r = \dfrac{n(n+3)}{2}$ $- d - 2k$ equals 13 and 12 respectively, so that *the inequality $r \geqq 8$ is not even a sufficient condition for the existence of irreducible curves (n, p, k).*

Another example of a non-geometric self-dual set of PLÜCKER characters is the following: $n = m = 8; k = i = 16; d = \tau = 0; p = 5$.

The existence of a great variety of irreducible curves (n, p, k) can be deduces from the following theorem (B. SEGRE, 2):

If there exists a complete continuous system Σ of irreducible curves (n, p, k) and if the characteristic series of Σ is non-special, then there exist also irreducible curves (n, p_1, k_1) for any values of p_1 and k_1 such that

$$0 \leqq k_1 \leqq k, \quad\quad p \leqq p_1 \leqq \frac{(n-1)(n-2)}{2} - k_1.$$

The system of curves (n, p_1, k_1) is uniquely determined by considering as *virtually non-existent* a certain number of nodes and cusps of a generic curve of Σ and as *virtual nodes* a certain number of cusps. The proof is the same as that of theorem 2 of section 4.

This theorem permits one to establish rigorously the existence of all curves (n, p, k) and (n, p_1, k_1), where p and k satisfy the inequalities (17′) and $p_1 \geqq p$. In fact, if we take for k its maximum value, which is $3/2(n + 2p - 2)$ if n is even and $\dfrac{3(n + 2p - 2) - 1}{2}$ if n is odd, we find $i = 0$ in the first case and $i = 1$ in the second case. The dual curves (m, τ, i) possess then either only nodes or a number of nodes and *one* cusp. The existence of these dual curves in the first case has been proved in the preceding section; in the second case their existence can be proved by a few additional elementary considerations (LEFSCHETZ, 1).

The preceding theorem can be applied since in the present case we are dealing with curves (n, p, k) for which $k < 3n$, and hence the characteristic series is now non-special[1]. We observe that in this manner it is possible to prove the existence of irreducible curves of order n for any set of non-negative PLÜCKER characters, provided $n \leqq 6$, or $n = 7$ and $k < 11$. Thus, in order of magnitude of n and k, the sets of characters (19) give indeed the first examples of non-geometric sets of PLÜCKER characters.

---

[1] The inequality $k < 3n$ follows immediately from the relation $k - 3n = i - 3m$, which is a consequence of the PLÜCKER relations.

Another essentially new fact encountered in the theory of curves (n, p, k) is that *the irreducible curves (n, p, k) do not necessarily form one irreducible complete system*. For instance, for $m = 1$ the curves (14) are sextics with 6 cusps on a conic, and form a complete continuous system (of dimension 15). *There exist however sextics with 6 cusps not on a conic* and these sextics therefore constitute another complete continuous system, also of dimension 15. To prove the existence of these sextics it is sufficient to consider the dual curve of a general cubic, which is a sextic with 9 cusps. These 9 cusps cannot lie on one and the same conic. Moreover the characteristic series is non-special $(k < 3n)$. We may therefore assign six of the cusps which are not on a conic and consider the remaining three cusps as virtually non-existent. We recall (section 3) that these two types of sextics are also topologically distinct.

Appendix 1 to Chapter VIII

By Shreeram Shankar Abhyankar

Let V be an irreducible normal projective algebraic variety over an algebraically closed ground field k of characteristic p (which may or may not be zero). Let W be a proper subvariety of V. Let K be the function field of V. Let Ω be an algebraic closure of K. We define:

$\Omega(V-W) =$ the family of all finite galois (i. e., separable normal) extensions L of K in Ω such that $\Delta(L/V) \subset W$. Here $\Delta(L/V)$ denotes the set of all (rational) points P of V such that, upon letting R be the local ring of P on V and S be the integral closure of R in L, we have that [the number of maximal ideals in S] $< [L:K]$.

$\Omega'(V-W) =$ the family of all finite galois extensions L of K in Ω such that $\Delta(L/V) \subset W$ and V is tamely ramified in L. V is said to be tamely ramified in L if for every point P of V and every maximal ideal N in S, where S is as above, upon letting H be the group of all K-automorphisms h of L such that $h(N) = N$, we have that the order of H is not divisible by p.

$\pi(V-W) = $ *(algebraic fundamental group of $V-W$)* $=$ the inverse system of groups $(g(L/K))_{L \in \Omega(V-W)}$ where $g(L/K)$ denotes the galois group of L over K.

$\pi'(V-W) =$ *(tame fundamental group of $V-W$)* $=$ the inverse system of groups $(g(L/K))_{L \in \Omega'(V-W)}$.

The classical Riemann existence theorem has been generalized by Grauert and Remmert (1) thus:

Existence theorem. *If k is the field of complex numbers and $q: X^* \to V - W$ is any finite unramified topological covering map, where X^* is connected, then there exists an irreducible normal projective algebraic variety X over k together with a rational covering map (in general ramified) $r: X \to V$ and a homeomorphism $s: X^* \to r^{-1}(V - W)$ such that $q(x) = r(s(x))$ for all $x \in X^*$.*

Contributions to this theorem, for the case when V is the projective plane, were made by Enriques in the work cited in Chapter VIII: § 1. In connections with this work of Enriques, in Chapter VIII: § 1. Zariski raised the question: is it true that for any group G we have $\cap T = 1$ where the intersection is taken over all normal subgroups T of G of finite index? Higman (1) has given an example of a finitely presented group G for which $\cap T \neq 1$.

By the Existence theorem we see that if k is the field of complex numbers then the inverse system of all finite homomorphic images of the topological fundamental groups $\pi_1(V - W)$ is isomorphic to $\pi(V - W)$. This motivates the definition of π. A reason why one would be particularly interested in the more restricted object π' is that in general, for $p \neq 0$, the structure of π turns out to be very complex and quite unanalogous to that of π_1.

We note the following example of the complexity of π given in ABHYANKAR (5). Take V to be the projective line and W to be a single point. We know that if $p = 0$ then $\pi(V - W) = \{1\}$. On the other hand, if $p \neq 0$ then: every finite group is isomorphic to a subgroup of some member of $\pi(V - W)$; moreover, given any finitely generated field extension L' of k of transcendence degree one, there exists a member L of $\Omega(V - W)$ and subfield L^* of L with $K \subset L^*$ such that L' is k-isomorphic to L^*.

For the higher dimensional case of $p \neq 0$, a very small amount of information concerning π (and its local analogue) is given in ABHYANKAR (1), parts I and II.

The tame fundamental group π' has been studied in ABHYANKAR (6, 8) by using the following tools: (1) galois theory of local rings (especially the concepts of splitting and inertia groups) which is due to HILBERT and KRULL; (2) results on the local analogue of π' given in ABHYANKAR (1); (3) theorem of purity of branch locus due to ZARISKI (23); (4) generalized theorem of BERTINI given by ZARISKI (5); and (5) degeneration principal of ZARISKI. We shall now state some of the results proved in ABHYANKAR (6).

Henceforth assume that V is nonsingular, let $n = \dim V$, let $W_1, \dots, W_t$ be the irreducible components of W, and assume that $t > 0$ and $\dim W_j = n - 1$ for $1 \leq j \leq t$.

Definition 1. Let P be a point of W, let R be the local ring of P on V, and let J be the principal ideal in R which defines W at P (note that J is its own radical). W is said to have a *normal crossing* at P if there exists a basis $(x_1, \dots, x_n)$ of the maximal ideal in the completion R^* of R such that $J R^* = x_1 \dots x_m R^*$ for some m. W is said to have a *strong normal crossing* at P if there exists a basis $(x_1, \dots, x_n)$ of the maximal ideal in R such that $J = x_1 \dots x_m R$ for some m.

Definition 2. Suppose $n = 2$. Then there exists a unique system $[(V_i, U_i, Q_i)_{0 \leq i \leq d}, (f_i)_{1 \leq i \leq d}]$ where: d is a nonnegative integer; for $0 \leq i \leq d$ we have that V_i is a nonsingular algebraic surface, U_i is curve (not necessarily irreducible) on V_i, and Q_i is the set of all points of U_i at which U_i does not have a strong normal crossing; $V_0 = V$, $U_0 = W$, $Q_d = \phi$; and for $1 \leq i \leq d$ we have that $Q_{i-1} \neq \phi$, $f_i : V_i \to V_{i-1}$ is the quadratic transform with center Q_{i-1}, and $U_i = $ the total transform

$f_i^{-1}(U_{i-1})$ of U_{i-1}. For any irreducible component X_0 of W we define

$$\nu(X_0, W; V) = (1/2) \sum_{i=0}^{d} \sum_{P \in X_i \cap Q_i} \mu(P) \,(\mu(P) + 1)$$

where X_i is the proper transform of X_i on V_i, and $\mu(P)$ is the multiplicity of X_i at P.

Theorem 1. *Assume that V is algebraically simply connected (i. e., $\pi(V) = \{1\}$). Also assume that one of the following three conditions is satisfied.*

(1) $\dim|W_j| > 1$ *for* $1 \leqq j \leqq t$; *and W has a strong normal crossing at each of its points*;

(2) $n = 2$; *and* $\dim|W_j| > 1 + \nu(W_j, W; V)$ *for* $1 \leqq j \leqq t$.

(3) $n = 2$; *for* $1 \leqq i < j \leqq t$ *we have that some positive integral multiple of W_i is linearly equivalent to some positive integral multiple of W_j; and the irreducible components $W_1, \ldots, W_t$ can be relabelled so that* $\dim|W_j| > > 1 + \nu(W_j, W_j \cup W_{j+1} \cup \cdots \cup W_t; V)$ *for* $1 \leqq j \leqq t$.

Then every member of $\pi'(V - W)$ is an abelian group which can be generated by t generators and whose order is not divisible by p. Moreover, there exists an abelian group G, which can be generated by t generators, and a family M of subgroups of G of finite index such that: $\bigcap_{N \in M} N = 1$; for any N_1 and N_2 in M we have $N_1 \cap N_2 \in M$; for any subgroups $N_1 \subset N_2$ of G with $N_1 \in M$ we have $N_2 \in M$; and the inverse system of groups $(G/N)_{N \in M}$ is isomorphic to $\pi'(V - W)$.

Remark 1. In Abhyankar (6), the above theorem was proved in the form in which condition (1) included the additional assumption that (1′) for $1 \leqq i < j \leqq t$ we have $W_i \cap W_j \neq \phi$, and condition (2) included the additional assumption that (2′) for $1 \leqq i < j \leqq t$ we have that W_i and W_j have a point in common at which W has a normal crossing. It was suggested to us by Serre that $(1) \Rightarrow (1′)$; in fact this clearly follows from the following lemma.

Lemma. *Let Y and Z be any irreducible $(n - 1)$-dimensional subvarieties of V such that $\dim|Y| > 1$ and $\dim|Z| \geqq 1$. Then $Y \cap Z \neq \phi$.*

From this Lemma we can also deduce that $(2) \Rightarrow (2′)$. To see this let the notation be as in definition 2. Let W_j' be the proper transform of W_j on V_r. Since $\dim|W_j| > 1 + \nu(W_j, W; V)$, as in Abhyankar (6), Part II, we see that $\dim|W_j'| > 1$. Therefore, given any i and j with $1 \leqq i < j \leqq t$, by the lemma there exists a point P_{ij}' which is common to W_i' and W_j'. Let P_{ij} be the point of V which corresponds to P_{ij}'. Then clearly $P_{ij} \in W_i \cap W_j$, and W has a strong normal crossing at P_{ij}.

The following proof of the lemma was indicated by Hironaka. Since $\dim|Y| \geqq 1$, there exists a positive divisor Y' on V such that $Y' \equiv Y$ (where $\equiv$ denotes linear equivalence) and $Y' \cap Z \neq \phi$; since $\dim|Z| \geqq 1$, there exists a positive divisor Z' on V such that $Z' \equiv Z$ and $Z' \cap Y \neq \phi$.

Suppose if possible that $Y \cap Z = \phi$. Then, in view of the relationship between linear equivalence and intersection cycles, we must have $Y' = Z + Y^*$ and $Z' = Y + Z^*$ where Y^* and Z^* are nonnegative divisors on V. Therefore $Y \equiv Z$. Let $e = \dim |Y|$, let V^* be the e-dimensional projective space, and let u be the rational map of V into V^* which corresponds to the linear system $|Y|$. Since $Y \equiv Z$ and $Y \cap Z = \phi$, we have that the linear system $|Y|$ is free from base points and hence u has no fundamental points on V. Now $Y = u^{-1}(Y_1)$ and $Z = u^{-1}(Z_1)$ where Y_1 and Z_1 are hyperplanes in V^*. Since Y is irreducible and $\dim |Y| > 1$, we have that $\dim u(V) > 1$; consequently $u(V^*) \cap Y_1 \cap Z_1 \neq \phi$ and hence $Y \cap Z \neq \phi$. This is a contradiction. Therefore $Y \cap Z \neq \phi$.

Theorem 2. *Assume that V is algebraically simply connected, $n = 2$, and the irreducible components $W_1, \ldots, W_t$ can be relabelled so that $\dim |W_j| > 1 + \nu(W_j, W_j \cup W_{j+1} \cup \cdots \cup W_t; V)$ for $1 \leq j \leq t$.*

Then every member of $\pi'(V - W)$ is a t-step solvable group which can be generated by t generators and whose order is not divisible by p (a group G is said to be t-step solvable if, upon letting $G_0 = G$ and $G_i = $ the commutator subgroup of G_{i-1} for all $i > 0$, we have $G_t = 1$). Moreover, there exists a t-step solvable group G, which can be generated by t generators, and a family M of normal subgroups of G of finite index such that: $\bigcap_{N \in M} N = 1$; for any N_1 and N_2 in M we have $N_1 \cap N_2 \in M$; for any normal subgroups $N_1 \subset N_2$ of G with $N_1 \in M$ we have $N_2 \in M$; and the inverse system of groups $(G/N)_{N \in M}$ is isomorphic to $\pi'(V - W)$.

Theorem 3 (corollary to theorem 1). *Assume that V is the n-dimensional projective space. Let $g_1^*, \ldots, g_t^*$ be the orders of the irreducible components of W (as hypersurfaces in V). Let $e = 1$ in case $p = 0$, and $e = $ the highest power of p which divides $g_1^*, \ldots, g_t^*$ in case $p \neq 0$. Let $g_i = g_i^*/e$ for $1 \leq j \leq t$. Let G be the abelian group on t generators $a_1, \ldots, a_t$ with the only relation $a_1^{g_1} \ldots a_t^{g_t} = 1$. Assume that either: $n \geq 2$ and W has a strong normal crossing at each of its points; or: $n = 2$ and the irreducible components $W_1, \ldots, W_t$ can be relabelled so that $\dim |W_j| > 1 + \nu(W_j, W_j \cup W_{j+1} \cup \cdots \cup W_t; V)$ for $1 \leq j \leq t$ (note that, since $n = 2$, we have $\dim |W_j| = (1/2) g_j^* (g_j^* + 3)$ where g_j^* is the order of W_j).*

Then every member of $\pi'(V - W)$ is an abelian group which can be generated by t generators and whose order is not divisible by p. Moreover, the inverse system of all finite homomorphic images of G is isomorphic to $\pi'(V - W)$.

Remark 2. Suppose that V is the projective plane. Then the above theorem 3 is an algebraic analogue of ZARISKI's result on π_1 stated as theorem 2 in § 2 of Chapter VIII. Since in ZARISKI's result no distinction is made between a normal crossing and a strong normal crossing, one may ask whether theorem 3 remains valid if we replace ν by ν^* where ν^* is defined thus: in definition 2 replace the phrase "strong normal crossing" by the phrase "normal crossing" and then define

$$v^*(X_0, W; V) = (1/2) \sum_{i=0}^{d} \sum_{P \in X_i \cap Q_i} \mu(P) \, (\mu(P) + 1) \, .$$

In Abhyankar (6), part III, this question is answered in the negative. Namely: assume that $p \neq 2$; let $t = 3$; let W_1 be the cubic curve with affine equation $y^3 + y^2 - x^2 = 0$; let W_2 be the line $y + 1 = 0$; and let W_3 be the line at infinity. Then $v^*(W_1, W_1 \cup W_2 \cup W_3; V) = 5$, $v^*(W_2, W_2 \cup W_3; V) = 0$, $v^*(W_3, W_3; V) = 0$, $\dim |W_1| = 9$, $\dim |W_2| = 2$, and $\dim |W_3| = 2$; hence $\dim |W_1| > 1 + v^*(W_1, W_1 \cup W_2 \cup W_3; V)$, $\dim |W_2| > 1 + v^*(W_2, W_2 \cup W_3; V)$, and $\dim |W_3| > 1 + v^*(W_3, W_3; V)$. However, there exists a member of $\pi'(V - W)$ which is nonabelian (hence in particular, if k is the field of complex numbers then $\pi_1(V - W)$ is nonabelian).

Zariski has shown in Chapter VIII, § 2, that if V is the complex projective plane and W is a three-cuspidal quartic then $\pi_1(V - W)$ is isomorphic to the group G' on two generators a and b connected by the relations: $a^2 = b^2$, $a^4 = 1$, $(a\,b)^3 = a^2$; note that the group G' is a nonabelian group of order 12. The following algebraic analogue of this result is proved in Abhyankar (6), part V, and (8).

Theorem 4. *If V is the projective plane, W is a three-cuspidal quartic, and $p \neq 2, 3$, then $\pi'(V - W)$ is isomorphic to the inverse system of all homomorphic images of the above mentioned group G'.*

Finally we note that, in Abhyankar (6), part VI, $\pi'(V - W]$ is computed when V is the projective plane and W is any curve (which may be reducible) whose order is at most four and which is a not a three-cuspidal quartic.

Appendix 2 to Chapter VIII

By David Mumford

Re: § 1. We follow the notation of the previous appendix. The most vexing side of the theory of π is the lack of any general methods for constructing non-solvable coverings $X' \to X$, hence for computing π in cases where it is not solvable. For example, almost all the information we have about $\pi(X)$ when X is a non-singular curve in char$p$ (not necessarily complete) has been obtained by lifting X to a curve $\tilde{X}$ of char0, and showing that all coverings $p : X' \to X$ which are only tamely ramified at points of the completion of X, can also be lifted to coverings $\tilde{p} : \tilde{X}' \to \tilde{X}$ in char0 (Grothendieck, 6, 1960–61; Murre, 5). Once in char0, we can rely on the old topological determination of π_1 and Riemann's Existence theorem. The result is this:

Theorem (Grothendieck). *If X is a non-singular curve of genus g and char$p$, and $X = \overline{X} - \{x_1, \ldots, x_n\}$ where $\overline{X}$ is complete, and if*
$\pi'(X)$ = *group of maximal covering of X, tamely ramified at $x_1, \ldots, x_n$,*
$\pi^*(X)$ = *group of maximal covering of X with order (covering group) prime to p,*
then $\pi'(X)$ is generated by elements $a_1, \ldots, a_g, b_1, \ldots, b_g, c_1, \ldots, c_n$ with at least the relation

$$(*) \qquad a_1 b_1 a_1^{-1} b_1^{-1} \ldots a_g b_g a_g^{-1} b_g^{-1} c_1 \ldots c_n = 1 \;,$$

and $\pi^*(X)$ is the pro-finite group which is the inverse limit of all finite groups generated by $a_1, \ldots, c_n$ such that (a) $*$ holds and (b) $p \nmid$ order.

Re: § 2. Edmunds (1) has proved:

Theorem. *Let X be a non-singular projective surface such that $\pi(X) = (1)$. Let $Y \subset X$ be an irreducible divisor with only ordinary double points such that*

$$\dim |Y| \geqq 2 \;.$$

Then the maximal solvable quotient of $\pi'(X - Y)$ is finite cyclic, of order prime to p.

Re: § 4. Severi's theory of plane curves, with d nodes can be reformulated using Kodaira-Spencer's theory of deformation (2). In fact, for every morphism $\Phi : C \to \mathbf{P}^2$, C a non-singular complete curve, possibly with several components, let

$$N = \Phi^*(T_{\mathbf{P}^2})/T_C$$

(T_X = tangent sheaf to X, for $X = \mathbb{P}^2$, C). Then (1) infinitesimal deformations of the pair (C, Φ) correspond to sections of N and (2) obstructions to extending deformations defined $\bmod t^n$ (t the parameter of the deformation) to deformations $\bmod t^{n+1}$ lie in $H^1(N)$. If the differential $d\Phi$ is nowhere 0, i. e., every singularity of $\Phi(C)$ is a union of non-singular branches, then it is easy to check that

$$N \cong \Omega^1_C(3H)$$

where H is the divisor class Φ^{-1}(line). Therefore $H^1(N) = (0)$, and there are no obstructions to deforming the pair (C, Φ). So by Kodaira-Spencer's theory, there is a non-singular space M of deformations of (C, Φ) whose dimension is

$$\dim M = \dim H^0(N)$$
$$= 3n + p - 1, \text{ if } \begin{cases} n = \text{degree of } \Phi(C) \\ p = 1 - \chi(\mathcal{O}_C) = \text{virtual genus of } C. \end{cases}$$

This generalizes theorem 1, § 4. This theory of Severi's is equally valid in char p (i. e., theorems 1–4, § 4) and gives a very elementary proof that every non-singular curve in char p is a specialization of a non-singular curve in char 0. In fact, it suffices to show that any equation $f(x, y, z)$ defining a plane curve with d nodes over a perfect field k of char p is the reduction $\bmod p$ of an equation $F(x, y, z)$ with coefficients over the Witt vectors $W(k)$, such that the char 0 curve $F(x, y, z) = 0$ still has d nodes. Let

$$f(x, y, z) = \sum a_{i,j} x^i y^j z^{n-i-j},$$

and consider the curve $f = 0$ as corresponding to the particular point $w_{ij} = a_{ij}$ in the space of all homogeneous equations f of degree n. Severi's theory tells us that there are d formal power series $P_l(\ldots, w_{ij}, \ldots)$ with coefficients in $W(k)$ such that for any curve $f' = 0$, with coefficients of the form $w_{ij} = a_{ij} + (\delta a)_{ij}$ lying in a complete valuation ring R, and with $(\delta a)_{ij}$ in the maximal ideal, then

$P_l(\ldots, (\delta a)_{ij}, \ldots) = 0 \Leftrightarrow f' = 0$ has a node specializing to the lth node of $f = 0$.

Moreover, his theory proves that the linear terms of $P_1, \ldots, P_d$ are independent over the field k. Then it is easy to deduce that the equations $P_l(\ldots, (\delta a)_{ij}, \ldots) = 0$, all l, have a solution with $(\delta a)_{ij} \in p \cdot W(k)$. Therefore there exists a curve $f' = 0$, defined over $W(k)$ with d nodes. A recent treatment of Severi's theory can be found in Popp (1).

Severi's proof of theorem 5, § 4, that there exists a *unique* family of irreducible curves of degree n with d nodes, is unfortunately incorrect, so the result remains open. The corollary however that the moduli space of curves of given genus is irreducible can be established in char 0, either by topological methods as sketched in the text, or as a simple consequence of the TEICHMÜLLER theory (cf. AHLFORS, 1; BERS, 1). It has been proven in char p by DELIGNE and MUMFORD (1), as a consequence of the result in char 0 *and* SEVERI's idea of using curves with nodes to enlarge the moduli space by adding certain "boundary components". As for the *local* structure of the moduli space of curves there are now many ways both in char 0 and in char p, for seeing that it is locally a $3g - 3$-dimensional space and that there are no obstructions to deforming curves (cf. Appendix to Chapter 5 for the precise definitions of local moduli spaces); in char 0, both the KODAIRA-SPENCER (2), and the TEICHMÜLLER theories apply, while in char p, GROTHENDIECK's (7) formal deformation theory applies. In all these theories, the number of moduli $3g - 3$ comes out either because it is the dimension of $H^1(X, T_X)$ (T_X = tangent sheaf), or because it is the dimension of $H^0(X, (\Omega_X^1)^{\otimes 2})$, which is dual to the first space via SERRE duality. The problem of whether the moduli space of curves is birationally isomorphic to $\mathbb{P}^{3g-3}$ is still unsolved, even in char 0; however MUMFORD (10) showed that in char 0 the moduli space has no rational mappings into abelian varieties.

A natural generalization of the classification of plane curves with only nodes is to the classification of surfaces in $\mathbb{P}^3$ with only nodes (= ordinary double points). There is an extensive literature on what is the *maximal* number of nodes that a surface of degree n can carry: cf. GALLARATI (1, 2), TOGLIATTI (1), KREISS (1). The exact answer is still not known.

The classification of plane curves C with d nodes and k cusps and the computation of $\pi_1(\mathbb{P}^2 - C)$ has unfortunately not been pursued. ZARISKI's techniques in § 3, theorem 1, are closely connected to certain techniques in knot theory (cf. CROWELL-FOX, 1; NEUWIRTH, 1). For instance the polynomial $f(X)$, which is the determinant appearing in the proof of theorem 1, § 3, is analogous to the ALEXANDER polynomial. If $\Gamma = \pi_1(\mathbb{P}^2 - C)$, $\Gamma' = [\Gamma, \Gamma]$, $\Gamma'' = [\Gamma', \Gamma']$, it would be interesting to investigate the structure of Γ'/Γ'' as a $\mathbb{Z}[\Gamma/\Gamma']$-module, since when Γ is instead a knot group, then Γ'/Γ'' is a finitely generated torsion module with a free resolution over $\mathbb{Z}[\Gamma/\Gamma']$:

$$0 \to F_1 \xrightarrow{\varphi} F_0 \to \Gamma'/\Gamma'' \to 0$$

and φ is the so-called ALEXANDER matrix. The main problem in § 5 as to whether the characteristic series of complete families of curves with nodes and cusps is complete would appear to have a *negative* answer. One would expect that the branch curves to suitably generic projections onto $\mathbb{P}^2$ of surfaces with obstructed moduli (cf. Appendix to Chapter 5) would provide counter-examples. However, this should be looked into.

Appendix A

Series of Equivalence

The theory of series of sets of points on an algebraic surface, intended as a generalization of the classical theory of linear systems of curves, has been created and developed recently by SEVERI in four papers (SEVERI, 23, 24, 25, 27). In this theory the analogue of a linear system is a *series of equivalence* of sets of points. The extension to algebraic varieties (SEVERI, 24, 25; TODD, 1; B. SEGRE, 8) leads to the introduction of *systems of equivalence* of V_k's on a V_r. While a satisfactory perspective of the theory will undoubtedly be provided for by its future developments and especially by its possible applications, at present a review of the main facts may be regarded as a timely and useful undertaking.

1. Equivalence between sets of points. We define a *series of intersection* σ_0 on a surface F as the series of sets of points of intersection of two curves C_1 and C_2, outside of certain fixed intersections, where C_1 and C_2 vary in two linear systems Σ_1 and Σ_2 respectively. We include in σ_0 also the limit — sets which arise when C_1 and C_2 have a common component. The number m of points in a set of σ_0 is the *degree* of σ_0. We do not exclude the case in which one of the two systems Σ_i, say Σ_1, consists of only one curve. In this case the sets of σ_0 do not cover the whole surface, and σ_0 is a linear series on the fixed curve C_1, if C_1 is irreducible. If C_1 is reducible, $C_1 = C_{11} + C_{12} + \cdots + C_{1k}$, then each set Γ of σ_0 is a collection of sets Γ_i, where Γ_i varies in a linear series of dimension $\geqq 0$ on C_{1i} as Γ varies in σ_0. Such series on reducible curves have been studied at length by SEVERI (24) and are called *curvilinear series of equivalence*. A set of such a series on a surface F is referred to as *a semifixed set*.

It follows clearly from the definition that series of intersection are invariant under birational transformations.

Any two sets of a series of intersection σ_0, which is not a curvilinear series of equivalence, are characteristic sets of some linear system (SEVERI, 25, p. 81). In fact, let Γ_a and Γ_b be two sets of σ_0, sets of intersection of C_1^a, C_2^a and of C_1^b, C_2^b respectively. The pencil determined in Σ_1 by C_1^a and C_1^b and the pencil determined in Σ_2 by C_2^a and C_2^b give rise to a series of intersection, ∞^2, which is a rational involution on F. This series can be regarded (in infinitely many ways) as the characteristic series of a net of curves, S, taking for S the homologue of any homaloidal net in the plane whose points are in $(1, 1)$ correspondence with the sets of the involution.

However, a series of intersection is not always totally contained in the characteristic series of some linear system, as can be shown by examples (SEVERI, 25, p. 74).

The definition of the relation of equivalence between sets of points (SEVERI, 27) can be based upon simple group-theoretic considerations (TODD, 1). In the same way as we have introduced virtual curves (II, 8), we may introduce besides effective sets of points also virtual sets $-A$, $A - B$. The totality of all effective and virtual finite sets of points on F forms an abelian group G with respect to addition (the operation consisting in uniting two sets into one). We define a subgroup H of G as follows: H contains: (1) all sets $A - B$ such that for some set C, $A + C$ and $B + C$ are effective sets belonging to a series of intersection; (2) all finite sums of such sets: $(A_1 - B_1) + (A_2 - B_2) + \cdots + (A_i - B_i)$.

Definition: *Two sets A, B of G are said to be equivalent, in symbols,* $A \equiv B$, *if $A - B$ is in H.*

In other words, A and B are equivalent if they belong to the same one of the cosets into which G is divided by its subgroup H. It follows from the definition that (1) if $A \equiv B$ and $B \equiv C$ then $A \equiv C$; (2) if $A \equiv B$ and $C \equiv D$ then $A \pm C \equiv B \pm D$.

It is a simple matter to show (TODD, 1) that the above definition of equivalence coincides with the following definition given by SEVERI (27, p. 420): *$A \equiv B$, if, after deleting a set K of common points, two sets $A - K$ and $B - K$ are obtained, which are either characteristic sets of some net of curves or become such after the addition of convenient semifixed sets of one and the same curvilinear series of equivalence.*

2. Series of equivalence. Let our surface F be in S_r and let σ_0 be a series of intersection on F, defined by two linear systems of curves, Σ_1 and Σ_2. These systems are cut out on F, outside of base curves, by two linear systems of hypersurfaces $|\Phi_1|$ and $|\Phi_2|$ respectively. The complete intersection of a Φ_1 and a Φ_2 is an $(r - 2)$-dimensional variety M, which varies in a continuous system as Φ_1 and Φ_2 vary in their respective linear systems. It is clear that *the series of intersection σ_0 is cut out on F by the continuous system of the varieties M, outside of fixed points and semifixed sets on the base curves of $|\Phi_1|$ and $|\Phi_2|$ on F.*

Let us now consider an arbitrary continuous system of $(r - 2)$-dimensional varieties M. It will in general not be possible to regard M as the complete intersection of two hypersurfaces which vary in two linear systems, so that in general the series cut out on F by the varieties M is not a series of intersection. However, it can be shown (SEVERI, 27,

p. 422) that in the most general case M *is the difference of two complete intersections*, i. e. that there exist three linear systems $|\Phi_1|, |\Phi_2|, |\Phi_3|$ such that $M = \Phi_1 \cdot \Phi_2 - \Phi_1 \cdot \Phi_3$. From this it follows immediately that the continuous series σ of sets of points cut out on F by the varieties M — and in defining σ we may, if we wish, disregard any number of fixed points and semifixed sets — consists of mutually equivalent sets. The series σ is called by SEVERI *a series of strict equivalence* (SEVERI, 27, p. 423). These series are invariant under CREMONA transformations of the S_r but not under birational transformations of F. We call attention to the fact that from the definition it follows that *any continuous series of sets of points totally contained in a series of strict equivalence is itself a series of strict equivalence.*

A series of equivalence (in the broad sense) is defined by SEVERI as follows (SEVERI, 27, p. 423): *a continuous series σ of sets of points on F is a series of equivalence if there exists a series of strict equivalence, σ_1, such that every set Γ of σ can be associated with a definite set Γ_1 of σ_1 in such a manner that as Γ varies in σ the set $\Gamma + \Gamma_1$ varies continuously in a series of strict equivalence σ_2.* In a more concise form, we may say that *a series of equivalence is defined as the difference of two series of strict equivalence, i. e.* as the series of residues of a series of strict equivalence with respect to another series of strict equivalence which is partially contained in the first series. From the definition it follows that the sum or the difference of two series of equivalence is again a series of equivalence. In particular, if we impose on a series of equivalence fixed points we obtain a series of equivalence, provided the sets passing through the fixed points form a continuous series.

Any continuous series of sets of points totally contained in a series of equivalence is itself a series of equivalence. We have pointed out above that this statement holds for series of strict equivalence. Let σ be a series of equivalence and let σ_1 and σ_2 be the two series of strict equivalence mentioned in the above definition. If σ' is a continuous series totally contained in σ, then, as the set Γ varies in σ', the associated set Γ_1 of σ_1 varies continuously. Hence, as Γ varies in σ', the associated set Γ_1 and also the set $\Gamma + \Gamma_1$ vary in series of strict equivalence totally contained in σ_1 and σ_2 respectively, q. e. d. It would be interesting to investigate the question as to whether *any continuous series of mutually equivalent sets is a series of equivalence.* In connection with this question we quote the following important theorem: *A rational series σ, ∞^ϱ, of sets of points on a surface is a series of equivalence* (SEVERI, 25, p. 105; 27, p. 495). This theorem is the analogue of ENRIQUES' theorem on rational systems of curves (V, 1) and holds also if σ is *unirational*, i. e. if the sets of σ are in (1, 1) correspondence with the sets of an involution in a linear space S_ϱ. As a corollary it follows that *if the surface F is rational then the totality of sets of n points of F is a series* *of equivalencee.* It is at present not known whether the converse is true (see section 4).

A difficult, as yet unsettled point of the theory, is the notion of a *complete series of equivalence.* If we do not wish to lose the essential property of a complete series containing a given set A of points, *its uniqueness*, we must define this series as *the totality of sets equivalent to* A. Since any two equivalent sets are contained in some continuous series of equivalence, this totality is itself a continuous series. However, it is conceivable that a complete series of equivalence may consist of an infinity of algebraically irreducible components, which may very well be of different dimensions. It is also not known whether the irreducible components of a complete series of equivalence are or are not themselves series of equivalence.

3. Invariant series of equivalence. The simplest example of a series of equivalence invariantively related to the given surface F is the *canonical series* S_c, defined as the series of sets of points of intersection of the canonical curves of F. This is a series of degree $p^{(1)} - 1$ and is invariant under birational transformations of F which do not possess fundamental curves.

Another invariant series of equivalence is *the series S_s of* SEVERI (SEVERI, 23). It is obtained as a generalization of the canonical series on an algebraic curve, by using the following characteristic property of this series: any set of the canonical series g_{2p-2}^{p-1} on a curve $f(x, y) = 0$ of genus p is the set of zeros of a differential of the first kind, $du = \frac{\varphi(x, y)}{f_y} dx$; or, what is the same, a canonical set is the Jacobian set of some abelian integral of the first kind. Let $u = \int \frac{A\,dx + B\,dy}{f_z}$ be a simple integral of the first kind attached to the surface F ($f = 0$). The equation $u = $ const. defines on F a *transcendental* pencil of curves, *free from base points* (SEVERI, 23, p. 279). The double points of the curves of the pencil, i. e. the points where $\frac{du}{dx} = \frac{du}{dy} = 0$, constitute the *Jacobian set* of the integral u. *The series S_s of* SEVERI *is the series, ∞^q, of the Jacobian sets of the ∞^q simple integrals of the first kind attached to F.* The series S_s is obviously a rational series and is therefore a series of equivalence (see preceding section). It can also be shown (SEVERI, 23, p. 303) that it is an *involutorial series.*

We proceed to express S_s as a combination of certain series of intersection. This will enable us to calculate the degree of S_s and will incidentally provide another proof that S_s is a series of equivalence. The Jacobian set Γ of u is the set of intersections of the two curves A_1 and B_1, cut out on the surface $f = 0$ by the surfaces $A = 0$ and $B = 0$ respectively, outside their common points at infinity and on the curve J

of intersection (outside the double curve of F) of $f = 0$ with $f'_z = 0$. Using the integrability conditions (31) and (31') of (VII, 8) it is found that the curves A_1 and B_1 meet outside of Γ in the following sets of points: (1) a canonical set K on the curve at infinity of F; (2) a set Q on J, such that the complete intersection of A_1 (or of B_1) with J is the set $Q + G_a(Q + G_b)$, where $G_a(G_b)$ is the set of double points (*the Jacobian set*) of the pencil of curves $x = $ const. ($y = $ const.). If $|C|$ denotes the system of plane sections of F, then A_1 and B_1 are total curves of the system $|C' + C|$, where $|C'|$ is the adjoint system of $|C|$, and J is a total curve of $|C' + 2C|$. We have therefore for Γ the following functional relation:

$$\Gamma \equiv (C' + C) \cdot (C' + C) - C' \cdot C - (C' + C) \cdot (C' + 2C) + G,$$

or (Severi, 23, p. 281),

(1) $$\Gamma \equiv G - C \cdot C - 2C' \cdot C,$$

where G is the Jacobian set of a pencil of curves C. If we denote by Δ and δ the number of points in the sets Γ and G respectively, we find from (1) the following numerical relation:

$$\Delta = \delta - n - 4\pi + 4,$$

or

(2) $$\Delta = I + 4,$$

where I is the invariant of Zeuthen-Segre (III, 8 and VI, 10).

The relation (1) can be used to define the series S_e also on regular surfaces, since it can be proved (Severi, 24, p. 49) that the complete series of equivalence defined by the (virtual or effective) set $G - C \cdot C - 2C \cdot C'$ is independent of the linear system $|C|$.

It is not difficult to see that a birational transformation in which exceptional curves are lost or are introduced, has opposite effects on *the series S_c and S_s, and that consequently the series $S_c + S_s$ is an absolutely invariant series of equivalence.* The double of this series, $S_e = 2(S_c + S_s)$, has been considered by Enriques (19), who gave it the following functional interpretation: *if Σ is an arbitrary net of curves C and if $(C)_\chi$ is the set of cusps of the cuspidal curves of the net, then the complete series of equivalence defined by the (virtual or effective) set $(C)_\chi - 12 C' \cdot C$ is independent of Σ and coincides with the series S_e.* The degree of S_e equals $\chi - 24\pi + 24$, where χ is the number of points in the set $(C)_\chi$ and where π is the genus of C. Since $\chi = 24(\pi + p_a)$ (Zeuthen, 1) it follows that the degree of S_e is $24(p_a + 1)$. Incidentally, from the relation $S_e = 2(S_c + S_s)$ follows the numerical relation $I = 12 p_a - p^{(1)} + 9$, due to Noether (2).

Numerous other examples of invariant and *covariant* series of equivalence have been given by Campedelli (2) and especially by B. Segre (6).

The invariant series considered by these authors are all expressible as linear combinations of the two series S_c and S_s and lead to a number of functional interpretations and numerative formulas.

4. Topological and transcendental properties of series of equivalence. Let σ be an algebraic series of sets of n points on a surface F and let V be the variety, birationally determined, whose points are in $(1, 1)$ correspondence with the sets of σ. As a variable point of V describes a cycle Γ_q on V $(0 \leqq q \leqq 4)$ the corresponding set of σ describes a cycle on F. We shall denote this cycle by $T(\Gamma_q)$, where T is the symbol of the correspondence between V and F.

If σ is a series of equivalence, the following properties hold (SEVERI, 27, p. 598).

(a) $T(\Gamma_1) \sim 0$ on F for any 1-cycle Γ_1 on V.
(b) $T(\Gamma_2)$ is an algebraic two-cycle on F for any 2-cycle Γ_2 on V.
(c) If $\Gamma_2 \approx 0$ on V then $T(\Gamma_2) \sim 0$ on F.

For the proof we first observe that zero-cycles of V go into zero-cycles on F. In fact, if $C_{q+1} \rightarrow \Gamma_q$, where C_{q+1} is a chain on V, then $T(C_{q+1}) \rightarrow T(\Gamma_q)$. Moreover, since the correspondence T is algebraic, algebraic 2-cycles on V go into algebraic 2-cycles on F. It follows that the properties (a), (b) and (c) hold if σ is a series of intersection, since in that case V, a rational variety, carries only zero 1-cycles, is without torsion and all its 2-cycles are algebraic. In the second place, if σ_1, σ_2 and σ are series on F such that $\sigma_1 + \sigma_2$ is totally contained in σ and such that the properties (a), (b), (c) hold for σ and σ_1, then it is obvious that these properties also hold for σ_2. The proof is immediate. Now, since any series of equivalence is totally contained, outside of fixed points, in a linear combination of series of intersection, it follows that the properties (a), (b) and (c) hold for any series of equivalence.

Let σ be a series enjoying the property (a) and let $P_1, P_2, \ldots, P_n$ be the n points of a variable set of σ. If u is a simple integral of PICARD of the first kind attached to F, the Picardian sum $v = u(P_1) + u(P_2) + \cdots + u(P_n)$ is a simple integral of the first kind attached to V. Since property (a) holds, it follows that v is without periods on V and hence v *is a constant*.

We shall say that a series σ is of *linear circulation zero* if it satisfies the property (a). *If a series σ is of linear circulation zero, then the sum of the values which any simple integral of the first kind attached to F takes at the n points of a set G_n of σ, remains constant as G_n varies in σ.* The most ample (complete) series enjoying this property have been called by ALBANESE *regular series* (ALBANESE, 7). We have then that *any series of equivalence is totally contained in a regular series* (SEVERI, 25, p. 91).

In a similar manner it is seen that if a series σ enjoys the property (b) and if $J = \int\int \varphi(x, y, z)\, dx\, dy$ is any double integral of the first kind attached to F, then $\int\int \sum_{i=1}^{n} \varphi(x_i y_i z_i)\, dx_i\, dy_i$ (where x_i, y_i, z_i are the coordinates of P_i) is a double integral of the first kind attached to V, without periods, since the periods of J on the algebraic cycles on F vanish (LEFSCHETZ; see VII, 5). Hence by the theorem of HODGE (VII, 8), *the sum* $\sum_{i=1}^{n} \varphi(x_i, y_i, z_i)\, dx_i\, dy_i$ *is identically zero*, and this property holds for any series of equivalence (SEVERI, 25, p. 125).

An immediate corollary of the preceding results is the following theorem: *if the totality of sets of n points of a surface F is a series of equivalence, then F is regular, of genus $(p_g = p_a)$ zero, and is without torsion* (SEVERI, 27, p. 599). In fact, the sets of a series of equivalence passing through assigned fixed points form a series of equivalence, provided these sets form a continuous series. Hence, under the hypothesis of the theorem, the totality of points of F is itself a series of equivalence (of degree $n = 1$). It is not known if the properties of F given in the concluding part of the above theorem are sufficient to characterize *rational* surfaces.

5. (Added in 2nd edition, by D. Mumford). In the period 1935 to 1950, SEVERI published many papers on series of equivalence and its generalizations to higher dimensions. It is hard to untangle everywhere what he conjectured and what he proved and, unfortunately, some of his conclusions are incorrect. To trace his work, see, forinstance SEVERI (16), (17), (23). The definition itself of series of equivalence was debated sharply at the International Congress at Amsterdam (vol. III, p. 545). The modern phase begins with the work of CHEVALLEY (3) and SAMUEL (3), (5): as in the case of divisors, it is much easier to first set up an equivalence relation (rational equivalence) on the group of *all* cycles and to investigate its properties, postponing the investigation of families of suitably *positive* cycles until later. Thus CHEVALLEY and SAMUEL defined the *Chow ring* $A(X)$ (of a non-singular, projective variety) to be the quotient of the group of cycles by rational equivalence, and proved that it has a natural ring structure defined by intersection. The invariant series of equivalence had been extended to higher dimensions by TODD (1) and EGER (1). In the modern phase, they were recognised to be the products of the Chern classes (HODGE (6), NAKANO (1)) and were defined in all char. by GROTHENDIECK (3). Returning to 0-cycles on a non-singular variety X, the concept of linear circulation 0 can be rephrased algebraically as follows: if

$$\phi: X \to \text{Alb } X$$

is the Albanese mapping, and $S^n(X)$ is the nth symmetric power of X, then ϕ induces maps

$$\phi^{(n)}: S^n(X) \to \text{Alb } X$$

and if $\alpha: T \to S^n(X)$ defines a continuous family of positive 0-cycles of degree n on X, parametrized by T, then we can call this family *regular* or of linear circulation 0, if $\phi^{(n)} \circ \alpha$ maps T to a single point. A natural conjecture is that if n is large enough, all the fibres of the map $\phi^{(n)}$ are irreducible varieties admitting no rational maps into abelian varieties themselves. This was proved by KOIZUMI (3), (4) (see also MATTUCK (1)). As for the influence of 2-forms, assuming char $= 0$, a regular 2-form ω on X induces a regular 2-form $\omega^{(n)}$ on $S^n(X)$ and for every continuous family $\alpha: T \to S^n(X)$ of rationally equivalent 0-cycles, it can be shown that $\alpha^*(\omega^{(n)}) = 0$ (MUMFORD (8)). From this it follows that if $p_g > 0$ (or more generally if $P_n > 0$, some n), then the quotient of the group of 0-cycles of degree 0 by rational equivalence is essentially infinite dimensional (MUMFORD (8), MATTUCK (2)).

Appendix B

Correspondences between Algebraic Varieties

The generalization to surfaces of the classical theory of correspondences between algebraic curves has been the object of recent investigations by ALBANESE (7) and SEVERI (25, 27). There is a very close connection between the results arrived at by these authors and the general theory of transformations of manifolds developed by LEFSCHETZ in several papers (LEFSCHETZ, 7, 8) and systematically expounded in LEFSCHETZ, e, Chapter VI. The principal theorems of the classical theory of correspondences between algebraic curves have been derived by LEFSCHETZ as special cases of his general topological theory of transformations of manifolds (LEFSCHETZ, 9; see VI, 13).

A similar application of this topological theory to correspondences between surfaces, and more generally between varieties V_d, has been made by TODD (2), who derives the principal results of SEVERI and ALBANESE in a very concise manner by studying the effect of the correspondence upon the associated RIEMANN matrices (period-matrices of the simple integrals of the first kind).

1. The fixed point formula of LEFSCHETZ (LEFSCHETZ, 7, 8; e, Chapter VI). Let T be an (α, β) correspondence between two non-singular algebraic varieties V_1, V_2, of dimension d, and let Γ be the algebraic V_d which maps the pairs of homologous points P_1, P_2 of V_1 and V_2 under T upon the product manifold $V_1 \times V_2$. T is said to be *irreducible*, if V_d is irreducible; *degenerate*, if either α or β or both α and β are zero. If, for instance, $\beta = 0$, then $T(P_1)$ is defined only for the points P_1 of a V_{d-h} on V_1 $(h > 0)$ and is a V_h on V_2.

Let Γ_s^i $(i = 1, 2, \ldots, R_s(V_1))$ form a minimal base, for homologies *modulo* the zero-divisors, of s-cycles on V_1, and let similarly Δ_s^j form a minimal base for the s-cycles on V_2. Then each p-cycle on $V_1 \times V_2$ depends on the cycles $\Gamma_s^i \times \Delta_{p-s}^j$ $(s = 0, 1, \ldots, p)$. In particular we have for the $2d$-cycle Γ an expression of the form:

$$(1) \qquad \Gamma \approx \sum_{i,j,s} f_{i,j}^{(s)} \Gamma_s^i \times \Delta_{2d-s}^j,$$

where the coefficients $f_{i,j}^{(s)}$ are integers.

T transforms every Γ_s^i into an s-cycle $T(\Gamma_s^i)$ on V_2. Let

$$(2) \qquad T(\Gamma_s^i) \approx \sum_j \varepsilon_{ij}^{(s)} \Delta_s^j,$$

and let similarly

$$(2') \qquad T^{-1}(\Delta_s^j) \approx \sum_i \zeta_{ji}^{(s)} \Gamma_s^i.$$

The matrices f, ε, ζ are connected by the following relations (LEF-SCHETZ, e, p. 266):

$$(3) \qquad \varepsilon^{(2d-s)} = \eta_1^{(s)\prime} f^{(s)},$$

$$(3') \qquad \zeta^{(2d-s)} = (-1)^s \eta_2^{(s)\prime} f^{(2d-s)},$$

where $\eta_1^{(s)}$ and $\eta_2^{(s)}$ are the matrices $\| (\Gamma_s^i \cdot \Gamma_{2d-s}^h) \|$, $\| (\Delta_s^j \cdot \Delta_{2d-s}^i) \|$, respectively.

If V_1 and V_2 coincide, we are dealing with a correspondence T between the points of one and the same variety V. Let Ω denote the identical transformation of V into itself and let Γ_0 be the corresponding $2d$-cycle on the product manifold of V by itself. The intersection number $(\Gamma \cdot \Gamma_0)$ gives the number of *signed fixed points* of the correspondence T. For algebraic varieties the algebraic number of intersections of two complementary algebraic cycles is equal to the arithmetic number of intersections (see VI, 3). Hence if T has a finite number of fixed points, this number is given by $(\Gamma \cdot \Gamma_0)$. If T possesses a curve D (necessarily algebraic) of fixed points, in addition to isolated fixed points, the intersection number $(\Gamma \cdot \Gamma_0)$ is referred to as the *virtual number of fixed points of* T. The effective number of isolated fixed points of T is then $(\Gamma \cdot \Gamma_0) - \varepsilon$, where ε is an integer representing the *numerical equivalence* of D in regard to the virtual number of isolated fixed points (see section 4). Using the intersection formulas (LEFSCHETZ, e, p. 241):

$$(\Gamma_s^i \times \Delta_{2d-s}^j \cdot \Gamma_{2d-\sigma}^h \times \Delta_\sigma^l) = 0, \quad \text{if} \quad \sigma \neq s,$$

$$(\Gamma_s^i \times \Delta_{2d-s}^j \cdot \Gamma_{2d-s}^h \times \Delta_s^l) = (-1)^s (\Gamma_s^i \cdot \Gamma_{2d-s}^h)(\Delta_{2d-s}^j \cdot \Delta_s^l),$$

and taking into account that for the identity Ω the matrices $\varepsilon^{(s)}$ are unit matrices, it is found immediately from (3):

$$(4) \qquad (\Gamma \cdot \Gamma_0) = \sum_{s=0}^{2d} (-1)^s \, trace \, \varepsilon^{(s)},$$

which is the fundamental *fixed point formula of* LEFSCHETZ.

2. The transcendental equations and the rank of a correspondence

(ALBANESE, 7; TODD, 2). Let q_i be the irregularity of V_i $(i = 1, 2)$ and let γ_i, δ_j $[i = 1, 2, \ldots, R_1 (= 2q_1); j = 1, 2, \ldots, R_2 = (2q_2)]$ form a base for the 1-cycles on V_1 and V_2 respectively. We write the equations (2) and (2') for $s = 1$, replacing for convenience the matrices $\varepsilon^{(1)}$ and $\zeta^{(1)}$ by A and B.

$$(5) \qquad\qquad T(\gamma_i) \approx \sum_j A_{ij}\,\delta_j,$$

$$(5') \qquad\qquad T(\delta_j) \approx \sum_i B_{ji}\,\gamma_i.$$

Let $u_1, \ldots, u_{q_1}$ be q_1 independent simple integrals of the first kind attached to V_1 and let similarly $v_1, \ldots, v_{q_2}$ be q_2 independent integrals on V_2. The sum

$$v_t(P_1) + v_t(P_2) + \cdots + v_t(P_\beta)$$

of the values which any integral v_t takes at the β points on V_2 corresponding to a point P of V_1, is evidently, considered as a function of the variable point P on V_1, a simple integral U_t of the first kind attached to V_1. Hence

$$(6) \qquad U_t = v_t(P_1) + \cdots + v_t(P_\beta) \equiv \sum_{r=1}^{q_1} a_{tr} u_r + c_t, \qquad t = 1, 2, \ldots, q_2,$$

where the a's and c's are constants. Similarly, considering the inverse correspondence, we have

$$(6') \qquad V_r = u_r(Q_1) + \cdots + u_r(Q_\alpha) \equiv \sum_{t=1}^{q_2} b_{rt} v_t + d_r, \qquad r = 1, 2, \ldots, q_1.$$

The period of U_t on γ_i equals the period of v_t on $T(\gamma_i)$. If, therefore, ω, τ denote the period matrices of the integrals u_r and v_t respectively, we have the following relations:

$$(7) \qquad\qquad \tau A' = a\omega,$$

$$(7') \qquad\qquad \omega B' = b\tau.$$

We define the *rank* ν of T as the rank of the matrix a. This character of T has been introduced by ALBANESE (7). His definition of the rank is different from the present one, but the two definitions are equivalent, as we shall see later.

Let α be the rank of the matrix A. Of the $2q_1$ cycles $T(\gamma_i)$ only α are independent on V_2. From the definition of ν it follows that T transforms the q_2 integrals v_t of V_2 into integrals U_t of V_1, of which only ν are independent. Hence there exist exactly $q_2 - \nu$ independent integrals v_t whose transforms are constants. These integrals necessarily have zero-periods on the cycles $T(\gamma_i)$, and hence by a property of reducible integrals derived on p. 174, $\alpha \leqq 2\nu$. On the other hand, there exist exactly $2q_1 - \alpha$ independent cycles on V_1 whose transforms on V_2 are zero-cycles. The periods of the integrals U_t, of which ν are independent, on these $2q_1 - \alpha$ cycles must therefore vanish, and hence by the same argument we have $\alpha \geqq 2\nu$. We conclude that $\alpha = 2\nu$, and that consequently V_1 possesses a regular system of ν reducible integrals with 2ν periods and that V_2 possesses a regular system of $q_2 - \nu$ reducible integrals with $2(q_2 - \nu)$ periods. By the theorem of POINCARÉ (VII, 6, p. 175), V_1 possesses a complementary regular system of $q_1 - \nu$ reducible integrals with $2(q_1 - \nu)$ periods and V_2 possesses a complementary regular system of ν integrals with 2ν periods. We may reduce simultaneously the RIEMANN matrices of V_1 and V_2 to a canonical form as follows. We may find non-singular rational square matrices P and Q such that $PAQ = \begin{pmatrix} E_{2\nu}, & 0 \\ 0 & 0 \end{pmatrix}$, where E_α denotes the unit matrix with α rows and columns. This reduction of A to a diagonal form amounts to choosing properly the bases γ_i, δ_j on V_1 and V_2. In a similar manner we may find non-singular complex square matrices p, q such that $paq = \begin{pmatrix} E_\nu, & 0 \\ 0 & 0 \end{pmatrix}$. This amounts to choosing properly the independent integrals u_r, v_t. The matrices ω, τ and the equations (5) and (6) assume then the form:

$$\omega = \begin{pmatrix} \omega_1 & 0 \\ \omega_3 & \omega_2 \end{pmatrix}, \qquad \tau = \begin{pmatrix} \omega_1 & \tau_3 \\ 0 & \tau_2 \end{pmatrix};$$

(5a) $T(\gamma_i) \approx \delta_i, \quad i = 1, 2, \ldots, 2\nu; \qquad T(\gamma_i) \approx 0, \quad i = 2\nu + 1, \ldots, 2q_1,$

(6a)
$$v_t(P_1) + \cdots + v_t(P_\beta) \equiv u_t + \text{const.}, \qquad t = 1, 2, \ldots, \nu,$$
$$v_t(P_1) + \cdots + v_t(P_\beta) \equiv \text{const.}, \qquad t = \nu + 1, \ldots, q_2.$$

By a further change of the bases and of the integrals, which does not affect the equations (5 a) and (6a) (see VII, 6), we may make ω_3 and τ_3 zero. Then

(8) $$\omega = \begin{pmatrix} \omega_1 & 0 \\ 0 & \omega_2 \end{pmatrix}, \qquad \tau = \begin{pmatrix} \omega_1 & 0 \\ 0 & \tau_2 \end{pmatrix},$$

showing the presence of common submatrix ω_1 (with ν rows and 2ν columns) of ω and τ.

The rank of the inverse correspondence T^{-1} is the rank of the matrix b or also half the rank of B. A fundamental theorem is the following: *if T is irreducible and non-degenerate, then T and T^{-1} have the same rank.* This theorem was proved by ALBANESE (7) for surfaces and was extended by TODD (2) to varieties of any dimension. ALBANESE's proof is based upon the consideration of the PICARD varieties of the two surfaces, while TODD derives the theorem by establishing the following matrix relation:

$$A'C = -D'B,$$

where C and D are suitably chosen principal matrices of ω and τ. The above matrix relation is in itself significant, but if it is only a question of proving that A and B (and that hence also T and T^{-1}) have the same rank, we may arrive very rapidly at the result by using the relations (3) and (3′). From these relations it follows that $\varepsilon^{(s)}$ and $\zeta^{(2d-s)}$ have the same rank, since the matrices $\eta_1^{(s)}$ and $\eta_2^{(s)}$ are non-singular. On the other hand it is not difficult to see that *in the case of algebraic varieties the matrices $\varepsilon^{(1)}$ and $\varepsilon^{(2d-1)}$ coincide for a suitable choice of the bases*, and that consequently $\varepsilon^{(1)}$ and $\zeta^{(1)}$, i. e. A and B, have the same rank. We prove this statement for the case of surfaces, F_1 and F_2. The proof can be easily extended to varieties.

We first show that if T is irreducible and non-degenerate and if C is a generic irreducible curve on F_1, then $T(C)$ is irreducible (Todd, 2). In fact, since T is irreducible there is at most a finite number of points on F_1 where T is degenerate, namely those with an infinity of transforms on F_2. Let D be the branch curve of T, i. e. the locus of points P for which two or more of the β homologous points on F_2 coincide. We consider on F_1 an irreducible curve C which does not pass through the fundamental points of T and which belongs to an infinite linear system of positive degree (for instance, a generic hyperplane section of F_1). The generators of the fundamental group of $F_1 - D$ are all on C (VI, 7; VIII, 1), and hence the group of monodromy of T on F_1 is the same as that of T on C, so that T must be irreducible on C, since it is irreducible on F_1, q. e. d.

We choose the 1-cycles γ_i, δ_j as follows; γ_i is the intersection of Γ_3^i with C and δ_j is the intersection of Δ_3^j with $T(C)$. The $2q_1$ cycles γ_i are independent, since the Γ_3^i are the loci of the γ_i as C varies in a pencil $\{C\}$ immersed in $|C|$ (VI, 9). Similarly the $2q_2$ cycles δ_j are independent, because $T(C)$ also belongs to an infinite linear system (from $C_1 \equiv C_2$ follows $T(C_1) \equiv T(C_2)$). Since the intersection of $T(\Gamma_3^i)$ with $T(C)$ is $T(\gamma_i)$, it follows that if $T(\gamma_i) \approx \sum_j A_{ij}\delta_j$, then $T(\Gamma_3^i) - \sum_j A_{ij}\Delta_3^j$ is a cycle Δ_3, such that $\Delta_3 \cdot T(C) \approx 0$ on F_2. From this it follows immediately (VI, 9) that $\Delta_3 \approx 0$ on F_2 and that consequently $T(\Gamma_3^i) \approx \sum_j A_{ij}\Delta_3^j$, i. e. the matrices $\varepsilon^{(1)}$ and $\varepsilon^{(3)}$ coincide, q. e. d.

The condition that T be irreducible is essential. If T is reducible, T and T^{-1} may very well have different ranks, as can be shown by examples (Todd, 2).

Let $\{D\}$ be a complete continuous system of curves on F_1 made up of ∞^{q_1} linear systems, and let Σ be the transform of $\{D\}$ under T. The equations (6a) show that of the q_2 Picardian sums relative to the intersections of a generic $T(D)$ with the irreducible curve $T(C)$, $q_2 - \nu$ remain constant as D varies continuously in $\{D\}$. From (VII, 4) it follows then immediately that *the curves of Σ belong to ∞^ν distinct linear* *systems* and that consequently *the subsystems of Σ made up of linearly equivalent curves arise from subsystems of $\{D\}$, each made up of $\infty^{a_1-\nu}$ distinct linear systems*. This property forms the basis of ALBANESE's definition of the rank of a correspondence (ALBANESE, 7).

From the form (8) of the matrices ω and τ it is seen that the existence of a correspondence T between V_1 and V_2 implies a specialization of these matrices, unless (1) $q_1 = q_2$ and the matrices ω, τ are isomorphic, or (2) $\nu = 0$ and the matrices a, A, b, B are zero. In this last case the correspondence is said to be *non-singular* or *of valence zero*. Since the RIEMANN matrix of a variety of dimension ≥ 2 is the most general of its type (VII, 7a), it follows that a correspondence between varieties of general moduli is necessarily non-singular.

3. The case of two coincident varieties. Correspondences with valence. If the two varieties V_1, V_2 coincide, $V_1 = V_2 = V$, then (7) becomes

$$(9) \qquad\qquad \omega A' = a\omega.$$

The equation (9) defines a *multiplication* of the RIEMANN matrix ω of V. It is well known that the only multiplications of a *general* RIEMANN matrix are of the scalar type, i. e. A and a are multiples of unit matrices. As in the case of curves, a correspondence T on V is said to have *a valence* if T defines a scalar multiplication of ω. If $A = -\nu E_{2q}$, $a = -\nu E_q$, then ν is *the valence* of T (ν, an integer).

A correspondence with valence zero is of rank zero and conversely. It follows that if T has valence zero then also T^{-1} has valence zero, even if T is reducible. The identical correspondence Ω has valence -1. The valence of the sum of two correspondences is the sum of their valences. It follows that if ν is the valence of T, then $T + \nu\Omega$ has valence zero. The inverse correspondence $T^{-1} + \nu\Omega$ having valence zero, it follows that T *and* T^{-1} *have the same valence.*

From the definition of valence it follows immediately that the valence of a product TS, where T and S have valences t and s respectively, is $-ts$. Correspondences whose valence is 1 are easily constructed on a surface F, for instance the symmetric $(n-1, n-1)$ correspondences defined by a net of curves of degree n. From this it follows that there exist on F correspondences having arbitrary, positive or negative, valence.

We point out explicitly the following characteristic property of a correspondence T with valence ν on a surface F:

If a curve A varies in a continuous system $\{A\}$ and if A' denotes the homologous curve $T(A)$, then $A' + \nu A$ varies in a linear system; in other words, if A_1 and A_2 are any two curves of $\{A\}$, then $A'_1 + \nu A_1 \equiv A'_2 + \nu A_2$ (ALBANESE, 7).

4. The principle of correspondence of ZEUTHEN-SEVERI. In the fixed point formula (4) of section 1, applied to a correspondence T of indices (α, β) on a surface F, we have $\varepsilon^{(0)} = \alpha$, $\varepsilon^{(4)} = \beta$, and *if T has valence zero*, then $\varepsilon^{(1)}$ and $\varepsilon^{(3)}$ are zero matrices. Therefore

(10) $N = \alpha + \beta + trace\ \varepsilon^{(2)}$,

where N is the virtual number of fixed points of T.

If T has valence v, then *trace $\varepsilon^{(1)}$* = *trace $\varepsilon^{(3)}$* = $-2qv$, where q is the irregularity of F, and we have

(10') $N = \alpha + \beta + 4vq + trace\ \varepsilon^{(2)}$.

Since correspondences with valence in the sense of ALBANESE are characterized only by their effect upon the 1-cycles of F, we know nothing about the matrix $\varepsilon^{(2)}$, and therefore, at least in the present state of the theory, formulas (10) and (10') cannot be improved upon. It is probable, however, that correspondences with valence operate on the 2-cycles of a surface in a very special manner. It would be important to investigate this question if it is desired to arrive at a formula for N admitting a simple algebro-geometric interpretation.

The situation is different in the case of *correspondences with valence in the sense of* SEVERI (SEVERI, 25, 27). SEVERI defines correspondences with valence zero as follows: *an (α, β) correspondence T between the points of an algebraic surface F has valence zero, if the set Y of β points which correspond to a point P of F, varies in a series of equivalence as P varies on F*. Correspondences with valence v are then defined by the usual condition that $T + v\Omega$ be of valence zero, where Ω is the identical correspondence. In other words *T has valence v if $Y + vP$ varies in a series of equivalence*. By the property (a) of section (4) of Appendix A it follows that if T has valence zero in the sense of SEVERI, then every 1-cycle on F goes under T into a zero-cycle, and hence T has valence zero also in the sense of ALBANESE. Consequently the correspondences with valence in the sense of SEVERI have also a valence (the same valence) in the sense of ALBANESE (but not conversely).

A very simple type of correspondence with valence in the sense of SEVERI is given by the *correspondences of* ZEUTHEN (ZEUTHEN, 2). In a correspondence of ZEUTHEN the homologous points of a generic point P of F, assumed to be in S_r, are cut out by an algebraic variety V_{r-2} of a certain order, having with F at P an intersection multiplicity $v \geqq 0$. This correspondence T has valence v. As P varies on F, V_{r-2} varies in a continuous system.

ZEUTHEN gave the following formula:

(11) $N = \alpha + \beta + \delta_1 - v(I + 2)$,

where N is the virtual number of fixed points of T, δ_1 is the order of the hypersurface generated by the variety V_{r-2} as P varies on a hyperplane section of F, and where I is the invariant of Zeuthen-Segre of F. Zeuthen's proof has been completed in some points by Severi, who has also extended relation (11) (with a suitable definition of δ_1) to arbitrary correspondences which have a valence in the sense of Severi. We shall now show how (11) can be derived from (10) and (10') (Todd, 2), first for correspondences of Zeuthen and then for arbitrary correspondences having a valence zero in the sense of Severi.

First let $v = 0$. As P varies on a 2-cycle Γ_2^i on F, V_{r-2} describes an $(r-1)$-cycle on the linear space S_r. Since on S_r every $(r-1)$-cycle is homologous to a multiple of a hyperplane, it follows that $T(\Gamma_2^i) \approx \mu_i E$, when E denotes a hyperplane section of F. In particular $T(E) = \delta_1 E$. Let $E \approx \sum \lambda_i \Gamma_2^i$. Then $T(\Gamma_2^i) \approx \mu_i E \approx \sum_j \mu_i \lambda_j \Gamma_2^j$, and hence *trace* $\varepsilon^{(2)} = \sum \mu_i \lambda_i = \delta_1$, whence (11) follows in view of (10).

Now let $v > 0$. Since $T + v\Omega$ has valence zero, we have *trace* $\varepsilon^{(2)} + vR_2 = \delta_1$, and substituting into (10') we find

$$N = \alpha + \beta + \delta_1 - v(R_2 - 4q) = \alpha + \beta + \delta_1 - v(I + 2), \quad \text{q. e. d.}$$

Severi calls $\delta_1 - v$ *the rank* of T (not to be confused with the rank of a correspondence in the sense of Albanese) and denotes it by δ. Introducing δ instead of δ_1, we find

(12) $$N = \alpha + \beta + \delta - v(I + 1).$$

As P varies on F the set of β points which correspond to P under a Zeuthen correspondence with valence zero, varies in a series of strict equivalence (Appendix A, section 2). However, this series is not the most general of its type, since to obtain the most general series of strict equivalence it is necessary to neglect a set of fixed points and a semi-fixed set on a fixed curve L. Hence the most general correspondence T, such that the set $T(P)$ varies in series of strict equivalence, is of the form $T = T' - T_0 - T_1$, where T' is a Zeuthen correspondence, and where T_0 and T_1 are degenerate correspondences. Here $T_0(P)$ is a fixed set and $T_1(P)$ is a semifixed set on L. For T_0 the matrix $\varepsilon^{(2)}$ is a zero-matrix. T_1 transforms any 2-cycle on F into a multiple of L. In particular, if $T_1(L) = \delta_1 L$, we call δ_1 *the rank* of T_1 (Severi, 27, p. 870), and we see immediately that for T_1, *trace* $\varepsilon^{(2)} = \delta_1$. If we then define as *the rank* of T the number $\delta = \delta' - \delta_1$, where δ' is the rank of T', we find immediately that (12) (where we put $v = 0$) holds also for T.

Any series of equivalence is the difference of two series of strict equivalence. It follows that any correspondence T with valence zero in the sense of Severi is the difference $T_1 - T_2$ of two correspondences of the type just considered. If we define the *rank* δ of T as the difference of the ranks of T_1 and T_2, we see that the formula $N = \alpha + \beta + \delta$ holds for any correspondence with valence zero in the sense of SEVERI. If now T is any correspondence with valence v in the sense of SEVERI, the formula (12) holds, provided the rank δ of T is defined by the relation: $\delta = \delta_1 - v$, where δ_1 is the rank of $T_1 + v\Omega$.

Formula (12) gives the effective number of fixed points of T only if T possesses a finite number of fixed points. If T possesses a curve D of fixed points, then it can be proved (SEVERI, 27, p. 689 and 874) that the number of isolated[1] fixed points of T is given by the formula:

$$(13) \quad u = \alpha + \beta + \delta - v(I + 1) - (\varrho + \nu - 2\pi - 2\omega + 2),$$

where ν and π are respectively the virtual degree and the virtual genus of D, ϱ denotes the number of infinitely near points P and $P' = T(P)$ such that the direction PP' (*principal direction*, see footnote) is tangent to the curve of a fixed generic pencil $\{E\}$ which passes through P, and where $\omega = (E \cdot D)$. The expression $\varrho - 2\omega$ is independent of the pencil $\{E\}$ (SEVERI, 27, p. 874). We see that the expression $\varrho + \nu - 2\pi - 2\omega + 2$ gives the numerical equivalence (see section 1) of the curve D in regard to the virtual number of fixed points of T.

---

[1] Let P_0 be a fixed point of a correspondence and let P be a variable point approaching P_0 along *a definite path*. If P' is the homologue of P which approaches P_0, then the line PP' has as limit a definite tangent line t of F at P_0, called a *principal line*. A fixed point P_0 is called a *perfect fixed point*, if all the tangent lines of F at P_0 are principal lines. An isolated fixed point, not on the curve D, is always a perfect fixed point; however, a finite number of perfect fixed points may also fall on the curve D. The number u in (13) is the number of perfect fixed points, whether isolated or on the curve D.

Bibliography

Treatises, monographs and reports

BAKER, H. F.: (a) Principles of geometry, Vol. 6. Introduction to the theory of algebraic surfaces and higher loci. Cambridge: University Press 1933.

BERTINI, E.: (a) Einführung in die projektive Geometrie mehrdimensionaler Räume. Wien 1924.

BERZOLARI, L.: (a) Algebraische Transformationen und Korrespondenzen. Enzyklop. d. math. Wiss. III$_2$ 11 (1932).

BLISS, G. A.: (a) The reduction of singularities of plane curves by birational transformations. Bull. Amer. Math. Soc. Vol. 29 (1923); (b) Algebraic Functions. Amer. Math. Soc., Colloquium Publ., Vol. 16. New York (1933).

CASTELNUOVO, G., and F. ENRIQUES: (a) Die algebraischen Flächen vom Gesichtspunkte der birationalen Transformationen aus. Enzyklop. d. math. Wiss. III$_2$ 6b (1914).

COBLE, A. B.: (a) Algebraic Geometry and Theta Functions. Amer. Math. Soc., Colloquium Publ., Vol. 10. New York 1929.

COMESSATTI, A.: (a) Sur la classification des courbes algébriques et sur le théorème d'existence de Riemann. Bull. Sci. math. II. s. Vol. 46 (1922); (b) Reelle Fragen in der algebraischen Geometrie. Jber. Deutsch. Math.-Vereinig. Vol. 41 (1932).

COOLIDGE, J. L.: (a) A treatise on algebraic plane curves. Oxford (1931).

ENRIQUES, F., and O. CHISINI: (a) Lezioni sulla teoria geometrica delle equazioni e delle funzioni algebriche, 3 volumes (referred to as a$_1$, a$_2$, a$_3$). Bologna 1915—1924.

ENRIQUES, F.: (a) Lezioni sulla teoria delle superficie algebriche, Part I (raccolte da L. CAMPEDELLI). Padova (1932).

GEPPERT, H.: (a) Die Klassifikation der algebraischen Flächen. Jber. Deutsch. Math.-Vereinig. Vol. 41 (1932).

GODEAUX, L.: (a) Résultats récents dans la théorie des modules des courbes algébriques. Bull. Sci. math. II. s. Vol. 55 (1931); (b) Les surfaces algébriques non rationelles des genres arithmétique e géométrique nuls. Actual. Sci. et Industr. 123 IV (1934).

JUNG, H. W. E.: (a) Algebraische Flächen. Hannover 1925.

LEFSCHETZ, S.: (a) L'Analysis situs et la géometrie algébrique. Paris 1924; (b) Report on curves traced on algebraic surfaces. Bull. Amer. Math. Soc. Vol. 29 (1923); (c) Topics in algebraic geometry. Report of the Committee on rational transformations. Bull. Nat. Res. Counc. No. 63. Washington 1928. Chapters XV, XVI, XVII; (d) Géometrie sur les surfaces et les variétés algébriques. Mem. Sci. math., Vol. 40 (1929); (e) Topology. Amer. Math. Soc., Colloquium Publ. 12. New York 1930.

McCAULEY, F. S.: (a) The algebraic theory of modular systems. Cambridge Tracts 1916 No. 19.

PICARD, E., and G. SIMART: (a) Théorie des fonctions algébriques de deux variables indépendantes, 2 volumes (referred to as a$_1$ and a$_2$). Paris 1897—1906.

Segre, C.: (a) Mehrdimensionale Räume. Enzyclop. d. math. Wiss. III₂ 7 (1912, 1921).

Severi, F.: (a) Vorlesungen über algebraische Geometrie. Leipzig 1921; (b) Die Geometrie auf einer algebraischen Fläche. Pascal Repertorium, II₂ (1922); (c) Trattato di geometria algebrica, Vol. 1. Bologna 1926; (d) Conferenze di geometria algebrica (raccolte da B. Segre). Roma 1928.

Veblen, O.: (a) Lectures on Analysis situs. Amer. Math. Soc., Colloquium Publ. 5. New York 1922.

Waerden, B. L., van der: (a) Moderne Algebra, 2 volumes (referred to as a₁ and a₂). Berlin 1930, 1931.

White, H. S.: (a) Linear systems of curves on algebraic surfaces. Amer. Math. Soc., Colloquium Publ. 1. New York 1905.

List of papers

Albanese, G.: (1) Intorno ad alcuni concetti e teoremi fondamentali sui sistemi algebrici di curve d'una superficie algebrica. Ann. Mat. pura appl. III. s. Vol. 24 (1915); (2) Transformazione birazionale di una curva algebrica qualunque in un'altra priva di punti multipli. Atti Accad. naz. Lincei, Rend. V. s. Vol. 33¹ (1924); (3) Transformazione birazionale di una superficie algebrica qualunque in un'altra priva di punti multipli. Rend. Circ. mat. Palermo Vol. 48 (1924); (4) Sul teorema fondamentale della base per la totalità delle curve d'una superficie algebrica. Atti Accad. naz. Lincei, Rend. VI. s. Vol. 5 (1927); (5) Formule fondamentali della geometria sopra una varietà algebrica. Ann. Mat. pura appl. IV. s. Vol. 4 (1927); (6) Sulle condizioni perchè una curva algebrica riducibile si possa considerare come limite di una curva irreducibile. Rend. Circ. mat. Palermo Vol. 52 (1928); (7) Corrispondenze algebriche fra i punti di due superficie algebriche, I; II. Ann. Scuola norm. super. Pisa II. s. Vol. 3 (1934).

Albert, A. A.: (1) On the construction of Riemann matrices. Ann. of Math. II. s. Vol. 35 (1934); (2) A solution of the principal problem in the theory of Riemann matrices. Ann. of Math. II. s. Vol. 35 (1934); (3) A note on the Poincaré theorem on impure Riemann matrices. Ann. of Math. II. s. Vol. 36 (1935).

Alexander, J. W.: (1) Sur les cycles des surfaces algébriques et sur une définition topologique de l'invariant de Zeuthen-Segre. Atti Accad. naz. Lincei, Rend. V. s. Vol. 23² (1914).

Bagnera, G., and M. de Franchis: (1) Le nombre ϱ de Picard pour les surfaces hyperelliptiques et pour les surfaces irrégulières de genre zéro. Rend. Circ. mat. Palermo Vol. 30 (1910).

Bertini, E.: (1) Sui sistemi lineari. Rend. Ist. Lomb. II. s. Vol. 15 (1882).

Black, C. W. M.: (1) The parametric representation of the neighborhood of a singular point of an analytic surface. Proc. Amer. Acad. Arts and Sci. Vol. 37 (1901—1902).

Brauner, K.: (1) Klassifikation der Singularitäten algebroider Kurven. Abh. math. Semin. Hamburg. Univ. Vol. 6 (1928).

Brown, A. B., and B. O. Koopman: (1) On the covering of analytical loci by complexes. Trans. Amer. Math. Soc. Vol. 34 (1932).

Burau, W.: (1) Kennzeichnung der Schlauchknoten. Abh. math. Semin. Hamburg. Univ. Vol. 9 (1932).

Campedelli, L.: (1) Sopra alcuni piani doppi notevoli con curva di diramazione del decimo ordine. Atti Accad. naz. Lincei., Rend. VI. s. Vol. 15 (1932); (2) Intorno ad alcune serie invarianti di gruppi di punti sopra una superficie. Atti Accad. naz. Lincei., Rend. VI. s. Vol. 17 (1933).

CASTELNUOVO, G.: (1) Ricerche generali sopra i sistemi lineari di curve piane. Mem. Accad. Sci. Torino II. s. Vol. 42 (1892); (2) Sui multipli di una serie lineare di gruppi di punti appartenenti ad una curva algebrica. Rend. Circ. mat. Palermo Vol. 7 (1893); (3) Sulla linearità delle involuzioni più volte infinite. Atti Accad. Sci. Torino Vol. 28 (1892—1893); (4) Sulla razionalità delle involuzioni piane. Math. Ann. Vol. 44 (1894); (5) Sulle superficie di genere zero. Mem. Soc. ital. Sci., detta dei 40 III. s. Vol. 10 (1896); (6) Alcuni risultati sui sistemi lineari di curve appartenenti ad una superficie algebrica. Mem. Soc. ital. Sci., detta dei 40 III. s. Vol. 10 (1896); (7) Alcune proprietà fondamentali dei sistemi lineari di curve tracciate sopra una superficie algebrica. Ann. Mat. pura ed appl. II. s. Vol. 25 (1897); (8) Sul genere lineare di una superficie e sulla classificazione a cui esso dà luogo (2 Notes). Atti Accad. naz. Lincei, Rend. V. s. Vol. 6 (1897); (9) Sugli integrali semplici appartenenti ad una superficie irregolare (3 Notes). Atti Accad. naz. Lincei, Rend. V. s. Vol. 14 (1905); (10) Sulle serie algebriche di gruppi di punti appartenenti ad una curva algebrica. Atti Accad. naz. Lincei, Rend. V. s. Vol. 15 (1906); (11) Sulle superficie aventi il genere aritmetico negative. Rend. Circ. mat. Palermo Vol. 20 (1905).

CASTELNUOVO, G., and F. ENRIQUES: (1) Sur quelques récents résultats dans la théorie des surfaces algébriques. Math. Ann. Vol. 48 (1897); (2) Sopra alcune questioni fondamentali nella teoria delle superficie algebriche. Ann. Mat. pura appl. III. s. Vol. 6 (1901); (3) Sulle condizioni di razionalità dei piani doppi. Rend. Circ. mat. Palermo Vol. 14 (1900); (4) Sur les intégrales simples de première espèce d'une surface on d'une variété algébrique a plusieurs dimensions. Ann. sci. Ecole norm. sup. III. s. Vol. 23 (1906).

CAYLEY, A.: (1) On the deficiency of certain surfaces. Math. Ann. Vol. 3 (1871).

CHISINI, O.: (1) La risoluzione delle singolarità di una superficie mediante trasformazioni birazionali dello spazio. Mem. Accad. Sci. Bologna VII. s. Vol. 8 (1921).

COMESSATTI, A.: (1) Sui piani tripli ciclici irregolari. Rend. Circ. mat. Palermo Vol. 31 (1911); (2) Fondamenti per la geometria sopra le superficie razionali dal punto di vista reale. Math. Ann. Vol. 73 (1912); (3) Sulla connessione delle superficie razionali reali. Ann. Mat. pura appl. III. s. Vol. 23 (1914); (4) Intorno alle superficie algebriche irregolari con $p_g \geq 2(p_a + 2)$ e ad un problema analitico ad esse collegato. Rend. Circ. mat. Palermo Vol. 46 (1922); (5) Sulla connessione delle superficie algebriche reali. Ann. Mat. pura appl. IV. s. Vol. 5 (1927—1928); (6) Sulla connessione e sui numeri base delle superficie algebriche reali. Rend. Semin. mat. Univ. Padova Vol. 3 (1932); (7) Sulla serie canonica d'una superficie algebrica. Atti Accad. naz. Lincei, Rend. VI. s. Vol. 16 (1932).

EHRESMANN, C.: (1) Sur la topologie de certains espaces homogènes. Ann. of Math. II. s. Vol. 35 (1934).

ENRIQUES, F.: (1) Una questione sulla linearità dei sistemi di curve appartenenti ad una superficie algebrica. Atti Accad. naz. Lincei, Rend. V. s. Vol. 2 (1893); (2) Ricerche di geometria sulle superficie algebriche. Mem. Accad. Sci. Torino II. s. Vol. 44 (1894); (3) Sulla massima dimensione dei sistemi lineari di curve di dato genere appartenenti ad una superficie algebrica. Atti Accad. Sci. Torino Vol. 29 (1894); (4) Un'osservazione relativa alla rappresentazione parametrica delle curve algebriche. Rend. Circ. mat. Palermo Vol. 10 (1896); (5) Introduzione alla geometria sopra le superficie algebriche. Mem. Soc. ital. Sci., detta dei 40 III. s. Vol. 10 (1896); (6) Sopra le superficie che posseggono un fascio di curve razionali. Atti Accad. naz. Lincei, Rend. V. s. Vol. 7 (1898); (7) Una proprietà delle serie continue di curve appartenenti ad una superficie algebrica regolare. Rend. Circ. mat. Palermo Vol. 13 (1899); (8) Sur les surfaces algébriques admettant des intégrales de différentielles totales de première espèce. Ann. Fac. Sci. Univ. Toulouse II. s. Vol. 3 (1901); (9) Intorno ai fondamenti della geometria sopra le superficie algebriche. Atti Accad. Sci. Torino Vol. 37 (1901); (10) Sulla proprietà caratteristica delle superficie algebriche irregolari. Rend. Accad. Sci. Bologna, nuova serie Vol. 9 (1904—1905); (11) Sulle superficie algebriche di genere geometrico zero. Rend. Circ. mat. Palermo Vol. 20 (1905); (12) Sulle superficie algebriche che ammettono un gruppo continuo di transformazioni birazionali in sè stesse. Rend. Circ. mat. Palermo Vol. 20 (1905); (13) Sopra le superficie algebriche di bigenere uno. Mem. Soc. ital. Sci., detta dei 40 III. s. Vol. 16 (1906); (14) Sui moduli delle superficie algebriche. Atti Accad. naz. Lincei, Rend. V. s. Vol. 17 (1908); (15) Sopra una involuzione non razionale dello spazio. Atti Accad. naz. Lincei, Rend. V. s. Vol. 21^1 (1912); (16) Sui moduli d'una classe di superficie e sul teorema d'esistenza per funzioni algebriche di due variabili. Atti Accad. Sci. Torino Vol. 47 (1912); (17) Sulla classificazione delle superficie algebriche e particolarmente sulle superficie di genere $p^1 = 1$ (2 Notes). Atti Accad. naz. Lincei, Rend. V. s. 23^1 (1914); (18) Sulla costruzione delle funzioni algebriche di due variabili possedenti una data curva di diramazione. Ann. Mat. pura appl. IV. s. Vol. 1 (1923); (19) Intorno ad alcune serie invarianti di gruppi di punti sopra una superficie algebrica. Atti Accad. naz. Lincei, Rend. VI. s. Vol. 16 (1932); (20) Intorno alle serie continue composte di involuzioni razionali di gruppi di punti sopra una superficie algebrica. Atti Accad. naz. Lincei, Rend. VI. s. Vol. 17 (1933).

ENRIQUES, F., and G. CASTELNUOVO: See G. CASTELNUOVO and F. ENRIQUES.

ENRIQUES, F., and F. SEVERI: (1) Mémoire sur les surfaces hyperelliptiques. Acta math. Vol. 32 (1909).

FRANCHIS, M. DE: (1) Sulla varietà ∞^2 delle coppie di punti di due curve o di una curva algebrica. Rend. Circ. mat. Palermo Vol. 17 (1903); (2) I piani doppi dotati di due o più differenziali totali di prima specie. Atti Accad. naz. Lincei, Rend. V. s. Vol. 13 (1904); (3) Sulle superficie algebriche le quali contengono un fascio irrazionale di curve. Rend. Circ. mat. Palermo Vol. 20 (1905).

FRANCHIS, M. DE, and G. BAGNERA: See G. BAGNERA and M. DE FRANCHIS.

GODEAUX, L.: (1) Sur les involutions douées d'un nombre fini des points unis, appartenant a une surface algébrique. Atti Accad. naz. Lincei, Rend. V. s. Vol. 23^1 (1914); (2) Sur certaines surfaces algébriques de diviseur supérieur à l'unité. Bull. Acad. Sci. Cracovie (1914); (3) Exemples de surfaces algébriques de diviseur supérieur à l'unité. Bull. Sci. math. II. s. Vol. 39 (1915); (4) Sur une surface algébrique de genres zéro et bigenre deux. Atti Accad. naz. Lincei, Rend. VI. s. Vol. 14 (1931).

HEGGARD, P.: (1) Sur l'Analysis Situs. Bull. Soc. Math. France Vol. 11 (1916).

HENSEL, K.: (1) Über eine neue Theorie der algebraischen Funktionen zweier Variablen. Acta math. Vol. 23 (1900).

HODGE, W. V. D.: (1) The isolated singularities of an algebraic surface. Proc. London Math. Soc. II. s. Vol. 30 (1930); (2) On multiple integrals attached to an algebraic variety. J. London Math. Soc. Vol. 5 (1930); (3) The geometric genus of a surface as a topological invariant. J. London Math. Soc. Vol. 8 (1933); (4) A Dirichlet problem for harmonic functionals, with applications to analytic varieties. Proc. London Math. Soc. II. s. Vol. 36 (1933).

HUMBERT, G.: (1) Théorie générale des surfaces hyperelliptiques (two memoirs). J. Math. pures appl. IV. s. Vol. 9 (1893); (2) Sur quelques points de la théorie des courbes et des surfaces algébriques. J. Math. pures appl. IV. s. Vol. 10 (1894).

HURWITZ, A.: (1) Über algebraische Korrespondenzen und das verallgemeinerte Korrespondenzprinzip. Math. Ann. Vol. 28 (1887); (2) Über Riemannsche Flächen mit gegebenen Verzweigungspunkten. Math. Ann. Vol. 39 (1891).

JUNG, H. W. E.: (1) Darstellung der Funktionen eines algebraischen Körpers zweier unabhängigen Veränderlichen x, y in der Umgebung einer Stelle $x = a$, $y = b$. J. reine angew. Math. Vol. 133 (1908); (2) Der Riemann-Rochsche Satz für algebraische Funktionen zweier Veränderlichen. Jber. Deutsch. Math.-Vereinig. Vol. 18 (1909); (3) Zur Theorie der Kurvenscharen auf einer algebraischen Fläche. J. reine angew. Math. Vol. 138 (1910); (4) Über die ausgezeichneten Kurven algebraischer Flächen. J. reine angew. Math. Vol. 142 (1913); (5) Algebraische Funktionen von zwei Veränderlichen. J. reine angew. Math. Vol. 165 (1931); (6) Sui gruppi di punti sopra una superficie e sulla serie di Severi, dal punto di vista della teoria dei corpi algebrici. Mem. Accad. Ital., Mat. Vol. 4 (1933).

KÄHLER, E.: (1) Über die Verzweigung einer algebraischen Funktion zweier Veränderlichen in der Umgebung einer singulären Stelle. Math. Z. Vol. 30 (1929); (2) Sui periodi degl'integrali multipli di una varietà algebrica. Rend. Circ. mat. Palermo Vol. 56 (1932); (3) Forme differenziali e funzioni algebriche. Mem. Accad. Ital., Mat. Vol. 3 (1932).

KAMPEN, E. R. VAN: (1) On the fundamental group of an algebraic curve. Amer. J. Math. Vol. 55 (1933).

KNESER, H.: (1) Die Integrale erster Gattung einer algebraischen Mannigfaltigkeit. Math. Ann. Vol. 107 (1932).

KOBB, G.: (1) Sur la théorie des fonctions algébriques de deux variables. J. Math. pures appl. IV. s. Vol. 8 (1892).

KOOPMAN, B. O., and A. B. BROWN: See A. B. BROWN and B. O. KOOPMAN.

KRONECKER, L.: (1) Grundzüge einer arithmetischen Theorie der algebraischen Größen. J. reine angew. Math. Vol. 92 (1882).

LEFSCHETZ, S.: (1) On the existence of loci with given singularities. Trans. Amer. Math. Soc. Vol. 14 (1913); (2) Sur les intégrales doubles des variétés algébriques. Ann. Mat. pura appl. III. s. Vol. 26 (1917); (3) Sur certains cycles à deux dimensions des surfaces algébriques. Atti Accad. naz. Lincei, Rend. V. s. Vol. 26[1] (1917); (4) Algebraic surfaces, their cycles and integrals. Ann of Math. II. s. Vol. 21 (1920); (5) On certain numerical invariants of algebraic varieties with applications to abelian varieties (Mémoire BORDIN). Trans. Amer. Math. Soc. Vol. 22 (1921); (6) Sur les intégrales multiples des variétés algébriques. J. Math. pures appl. IX. s. Vol. 3 (1924); (7) Intersections and transformations of complexes and manifolds. Trans. Amer. Math. Soc. Vol. 28 (1926); (8) Manifolds with a boundary and their transformations. Trans. Amer. Math. Soc. Vol. 29 (1927); (9) Correspondences between algebraic curves. Ann. of Math. II. s. Vol. 28 (1927); (10) Invariance absolue et invariance relative en géometrie algébrique. Rec. math. Soc. math. Moscou Vol. 39 No. 3 (1932).

LEFSCHETZ, S., and J. H. C. WHITEHEAD: (1) On analytical complexes. Trans. Amer. Math. Soc. Vol. 35 (1933).

LEVI, B.: (1) Sulla riduzione delle singolarità puntuali delle superficie algebriche dello spazio ordinario per transformazioni quadratiche. Ann. Mat. pura appl. II. s. Vol. 26 (1897); (2) Risoluzione delle singolarità puntuali delle superficie algebriche. Atti Accad. Sci. Torino Vol. 33 (1897); (3) Intorno alla composizione dei punti generici delle linee singolari delle superficiealgebriche. Ann. Mat. pura appl. III. s. Vol. 2 (1899); (4) Sulla transformazione dell'intorno di un punto per una corrispondenza birazionale fra due spazi. Atti Accad. Sci. Torino Vol. 35 (1899).

MARONI, A.: (1) Sulle superficie algebriche possedenti due fasci di curve algebriche unisecanti. Atti Accad. Sci. Torino Vol. 38 (1903).

NOETHER, M.: (1) Über die reduktiblen algebraischen Kurven. Acta math. Vol. 8 (1886); (2) Zur Theorie des eindeutigen Entsprechens algebraischer Gebilde. Math. Ann. Vol. 2 (1870); Vol. 8 (1875); (3) Über Flächen, welche Scharen rationaler Kurven besitzen. Math. Ann. Vol. 3 (1871); (4) Sulle curve multiple di superficie algebriche. Ann. Mat. pura appl. II. s. Vol. 5 (1871); (5) Extension du théorème de Riemann-Roch aux surfaces algébriques. C. R. Acad. Sci., Paris Vol. 103 (1886).

PAINLEVÉ, P.: (1) Sur les fonctions qui admettent un théorème d'addition. Acta Math. Vol. 27 (1903).

PEZZO, P. DEL: (1) Intorno ai punti singolari delle superficie algebriche. Rend. Circ. mat. Palermo Vol. 6 (1892).

PICARD, E.: (1) Sur la réduction du nombre des périodes des intégrales abéliennes, et, en particulieur, dans le cas des courbes du second genre. Bull. Soc. Math. France Vol. 11 (1883); (2) Sur les intégrales de différentielles totales algébriques de pre-mière espèce. J. Math. pures appl. IV. s. Vol. 1 (1885); (3) Sur les intégrales de différentielles totales de seconde espèce. J. Math. pures appl. IV. s. Vol. 2 (1886); (4) Mémoire sur la théorie des fonctions algébriques de deux variables. J. Math. pures appl. IV. s. Vol. 5 (1889); (5) Sur la théorie des groupes et des surfaces algébriques. Rend. Circ. mat. Palermo Vol. 9 (1895); (6) Sur les intégrales doubles de seconde espèce dans la théorie des surfaces algébriques. J. Math. pures appl. V. s. Vol. 5 (1899); (7) Sur les intégrales de différentielles totales de troisième espèce dans la théorie des surfaces algébriques. Ann. École norm. sup. III. s. Vol. 18 (1901); (8) Sur les périodes des intégrales doubles dans la théorie des fonctions algébriques de deux variables. Ann. École norm. sup. III. s. Vol. 19 (1902); (9) Sur quelques points fondamentaux dans la théorie des fonctions algébriques de deux variables. Acta math. Vol. 26 (1902); (10) Sur les relations entre la théorie des intégrales doubles de seconde espèce et les intégrales de différentielles totales. Ann. École norm. sup. III. s. Vol. 20 (1903); (11) Sur certaines surfaces algébriques dont les intégrales de différentielles totales sont algébro-logarithmiques. Ann. École norm. sup. III. s. Vol. 20 (1903); (12) Sur la formule générale donnant le nombre des intégrales doubles distinctes de seconde espèce relatives à une surface algébrique. Ann. École norm. sup. III. s. Vol. 22 (1905); (13) Sur quelques questions se rattachant à la connexion linéaire dans la théorie des fonctions algébriques de deux variables indépendantes. J. reine angew. Math. Vol. 129 (1905).

POINCARÉ, H.: (1) Sur la réduction des intégrales abéliennes. Bull. Soc. Math. France Vol. 12 (1884); (2) Sur les fonctions abéliennes. Amer. J. Math. Vol. 8 (1886); (3) Analysis Situs. J. École polytechn. II. s. Vol. 1 (1895); (4) Complément a l'Analysis Situs. Rend. Circ. mat. Palermo Vol. 13 (1899); (5) Second Complément a l'Analysis Situs. Proc. London Math. Soc. Vol. 32 (1900); (6) Sur les cycles des surfaces algébriques. J. Math. pures appl. V. s. Vol. 8 (1902); (7) Sur les périodes des intégrales doubles. J. Math. pures appl. VI. s. Vol. 2 (1906); (8) Sur les courbes tracées sur les surfaces algébriques. Ann. École norm. sup. III. s. Vol. 27 (1910); (9) Sur les courbes tracées sur une surface algébrique. Sitzungsber. Berlin. math. Ges. Vol. 10 (1911).

RHAM, G. DE: (1) Sur l'Analysis Situs des variétés à n dimensions. J. Math. pures appl. IX. s. Vol. 10 (1931); (2) Sur les périodes des intégrales de première espèce attachées à une variété algébrique. Comment. math. helv. Vol. 3 (1931).

ROSENBLATT, A.: (1) Sur les surfaces irrégulières dont les genres satisfont à l'inégalité $p_g > 2(p_a + 2)$. Rend. Circ. mat. Palermo Vol. 35 (1913).

SEGRE, B.: (1) Sui moduli delle curve poligonali e sopra un complemento al teorema di esistenza di Riemann. Math. Ann. Vol. 100 (1928); (2) Esistenza e dimensione di sistemi continui di curve piane algebriche con dati caratteri. Atti Accad.

naz. Lincei, Rend. VI. s. Vol. 10 (1929); (3) Sui moduli delle curve algebriche.
Ann. Mat. pura appl. IV. s. Vol. 7 (1930); (4) Sulla caratterizzazione delle curve
di diramazione nei piani multipli generali. Mem. Accad. Ital., Mat. Vol. 1
(1930); (5) Sulla completezza della serie caratteristica di un sistema continuo
di curve irreducibili tracciate su di una superficie algebrica. Rend. Circ. mat.
Palermo Vol. 55 (1931); (6) Determinazione geometrico-funzionale di gruppi
di punti, relativi a dati sistemi lineari di curve su di una superficie algebrica
(3 Notes). Atti Accad. naz. Lincei, Rend. VI. s. Vol. 18 (1933); (7) Sui moduli
delle superficie algebriche irregolari. Atti Accad. naz. Lincei, Rend. VI. s.
Vol. 29 (1934); (8) Nuovi contributi alla geometria sulle varietà algebriche.
Mem. Accad. Ital., Mat. Vol. 5 (1934).

SEGRE, C.: (1) Introduzione alla geometria sopra un ente algebrico semplicemente
infinito. Ann. Mat. pura appl. II. s. Vol. 22 (†894); (2) Intorno ad un carattere
delle superficie e delle varietà superiori algebriche. Atti Accad. Sci. Torino
Vol. 31 (1895); (3) Sulla scomposizione dei punti singolari delle superficie
algebriche. Ann. Mat. pura appl. II. s. Vol. 25 (1897).

SEVERI, F.: (1) Sulla deficienza della serie caratteristica di un sistema lineare di
curve appartenenti ad una superficie algebrica. Atti Accad. naz. Lincei, Rend.
V. s. Vol. 12 (1903); (2) Sulle corrispondenze fra i punti di una curva algebrica
e sopra certe classi di superficie. Mem. Accad. Sci. Torino II. s. Vol. 54 (1904);
(3) Sulle relazioni che legano i caratteri invarianti di due superficie in corrispondenza algebrica. Rend. Istit. Lombardo II. s. Vol. 36 (1903); (4) Osservazioni sui sistemi continui di curve appartenenti ad una superficie algebrica.
Atti Accad. Sci. Torino Vol. 39 (1903–1904); (5) Sul teorema di Riemann-
Roch e sulle serie continue appartenenti ad una superficie algebrica. Atti
Accad. Sci. Torino Vol. 40 (1904–1905); (6) Sulle curve algebriche virtuali
appartenenti ad una superficie algebrica. Rend. Istit. Lombardo II. s. Vol. 38
(1905); (7) Intorno alla costruzione dei sistemi completi non lineari che appar-
tengono ad una superficie algebrica. Rend. Circ. mat. Palermo Vol. 20 (1905);
(8) Sulle superficie algebriche che posseggono integrali di Picard della 2a specie.
Math. Ann. Vol. 61 (1905); (9) Sulla differenza tra i numeri degli integrali di
Picard della 1a e della 2a specie appartenenti ad una superficie algebrica. Atti
Accad. Sci. Torino Vol. 40 (1905); (10) Osservazioni varie di geometria sopra
una superficie e sopra una varietà. Atti Istit. Veneto Vol. 65 (1905–1906);
(11) Il teorema d'Abel sulle superficie algebriche. Ann. Mat. pura appl. III. s.
Vol. 12 (1906); (12) Intorno al teorema d'Abel sulle superficie algebriche ed
alla riduzione a forma normale degl'integrali di Picard. Rend. Circ. mat.
Palermo Vol. 21 (1906); (13) Sulla totalità delle curve algebriche tracciate
sopra una superficie algebrica. Math. Ann. Vol. 62 (1906); (14) Sulle superficie
algebriche che ammettono un gruppo continuo permutabile a due parametri.
Atti Instit. Veneto Vol. 67 (1907); (15) Sulla regolarità del sistema aggiunto
ad un sistema lineare di curve appartenente ad una superficie algebrica. Atti
Accad. naz. Lincei, Rend. V. s. Vol. 17^2 (1908); (16) La base minima pour la
totalité des courbes tracées sur une surface algébrique. Ann. École norm.
sup. III. s. Vol. 25 (1908); (17) Complementi alla teoria della base per la tofalità
delle curve di una superficie algebrica. Rend. Circ. mat. Palermo Vol. 30
(1910); (18) Transformazione birazionale di una superficie algebrica qualunque
in una priva di punti multipli. Atti Accad. naz. Lincei, Rend. V. s. Vol. 23^2
(1914); (19) Sulla classificazione delle curve algebriche e sul teorema d'esistenza
di Riemann. Atti Accad. naz. Lincei, Rend. V. s. Vol. 24^1 (1915); (20) Sulla
teoria degl'integrali semplici di 1a specie appartenenti ad una superficie
algebrica (7 Notes). Atti Accad. naz. Lincei, Rend. V. s. Vol. 30 (1921);
(21) Sul teorema di esistenza di Riemann. Rend. Circ. mat. Palermo Vol. 46

(1922); (22) Sugl'integrali algebrici semplici e doppi (4 Notes). Atti Accad. naz. Lincei, Rend. VI. s. Vol. 7 (1928); (23) La serie canonica e la teoria delle serie principali di gruppi di punti sopra una superficie algebrica. Comment. math. helv. Vol. 4 (1932); (24) Un nuovo campo di ricerche nella geometria sopra una superficie e sopra una varietà algebrica. Mem. Accad. Ital., Mat. Vol. 3, No. 5 (1932); (25) Nuovi contributi alla teoria delle serie di equivalenza sulle superficie e dei sistemi di equivalenza sulle varietà algebriche. Mem. Accad. Ital., Mat. Vol. 4 No. 6 (1933); (26) Über die Grundlagen der algebraischen Geometrie. Abh. math. Semin. Hamburg. Univ. Vol. 9 (1933); (27) La teoria delle serie di equivalenza e delle corrispondenze a valenza sopra una superficie algebrica (7 Notes). Atti Accad. naz. Lincei, Rend. VI. s. Vol. 17 (1933).

SEVERI, F., and F. ENRIQUES: See F. ENRIQUES and F. SEVERI.

SNYDER, V.: (1) Construction of plane curves of given order and genus, having distinct double points. Bull. Amer. Math. Soc. Vol. 15 (1908—1909).

TODD, J. A.: (1) Some group-theoretic considerations in algebraic geometry. Ann. of Math. II. s. Vol. 35 (1934); (2) Algebraic correspondences between algebraic varieties. Ann. of Math. II. s. Vol. 36 (1935).

WAERDEN, B. L. VAN DER: (1) Der Multiplizitätsbegriff der algebraischen Geometrie. Math. Ann. Vol. 97 (1927); (2) Topologische Begründung der Kalküls der abzählenden Geometrie. Math. Ann. Vol. 102 (1930); (3) Gradbestimmung von Schnittmannigfaltigkeiten einer beliebigen Mannigfaltigkeit mit Hyperflächen. Math. Ann. Vol. 108 (1933); (4) Über irreduzibile algebraische Mannigfaltigkeiten. Math. Ann. Vol. 108 (1933); (5) Algebraische Korrespondenzen und rationale Abbildungen. Math. Ann. Vol. 110 (1934).

WHITEHEAD, J. C. H., and S. LEFSCHETZ: See S. LEFSCHETZ and J. H. C. WHITEHEAD.

ZARISKI, O.: (1) Sull'impossibilità di risolvere parametricamente per radicali un'equazione algebrica $f(x, y) = 0$ di genere $p > 6$ a moduli generali. Atti Accad. naz. Lincei, Rend. VI. s. Vol. 3 (1926); (2) On a theorem of Severi. Amer. J. Math. Vol. 50 (1928); (3) On the problem of existence of algebraic functions of two variables possessing a given branch curve. Amer. J. Math. Vol. 51 (1929); (4) On the linear connection index of the algebraic surfaces $z^n = f(x, y)$. Proc. Nat. Acad. Sci., U. S. A. Vol. 15 (1929); (5) On the irregularity of cyclic multiple planes. Ann. of Math. II. s. Vol. 32 (1931); (6) On the topology of algebroid singularities. Amer. J. Math. Vol. 54 (1932).

ZEUTHEN, H. G.: (1) Études géometriques de quelques-unes des propriétés de deux surfaces dont les points se correspondent un-à-un. Math. Ann. Vol. 4 (1871); (2) Le principe de correspondance pour une surface algébrique. C. R. Acad. Sci. Paris Vol. 143 (1906).

Supplementary Bibliography for Second Edition

This bibliography was originally intended to be a complete bibliography on all work done on algebraic surfaces since the 1st edition of this book. This turned out to be more easily said than done, but we did try to assemble a fairly complete bibliography including of course all references needed for the new appendices. We hope the result also has some value as a place to browse.

ABHYANKAR, S. S.: (1) On the ramification of algebraic functions, I. Amer. J. Math. 77 (1955); II. Trans. Amer. Math. Soc. 89 (1958); (2) On the valuations centered in a local domain. Amer. J. Math. 78 (1956); (3) Local uniformization on algebraic surfaces over ground fields of characteristic $p \neq 0$. Ann. of Math. 63 (1956), 491—526; (4) Simultaneous resolution for algebraic surfaces. Amer. J. Math. 78 (1956); (5) Coverings of algebraic curves. Amer. J. Math. 79 (1957); (6) Tame coverings and fundamental groups of algebraic varieties. Amer. J. Math. I, 81 (1959); II, 82 (1960); III, 82 (1960); IV, 82 (1960); V, 82 (1960); VI, 82 (1960); (7) Ramification theoretic methods in algebraic geometry. Ann. Math. Studies. Princeton: Princeton Univ. Press, 1959; (8) Cubic surfaces with a double line. Mem. Coll. Sci. Kyoto 32 (1960); (9) Reduction to multiplicity less than p in a p-cyclic extension. Math. Ann. 154 (1964); (10) Uniformization in a p-cyclic extension of a 2-dimensional regular local domain. Wiss. Abh. Nordrhein-Westf. 33 (1966); (11) Nonsplitting of valuations in extensions of 2-dimensional regular local domains. Math. Ann. (1966); (12) An algorithm on polynomials in one indeterminate. Ann. Mat. Pura Appl. 171 (1966); (13) Resolution of singularities of embedded surfaces. New York: Academic Press 1966.

AHLFORS, L.: Lectures on quasiconformal mappings. New York-London: Van Nostrand 1966.

AKIZUKI, Y., and H. MATSUMURA: On the dimension of algebraic system of curves with nodes on a non-singular surface, Mem. Coll. Sci. Univ. Kyoto, Ser. A. Math. 30 (1957).

AKIZUKI, Y., and S. NAKANO: Note on Kodaira-Spencer's proof of Lefschetz's theorem. Proc. Japan Acad. 30 (1954).

ALBANESE, G.: (1) Corrispondenze algebriche fra i punti di due superficie algebriche. Ann. Scuola Norm. Sup. Pisa II, 3 (1934); (2) II of above, ibid. (1934).

ANDREOTTI, A.: (1) Questioni di equivalenza relative alle curve riducibili e ai punti base di un fascio di curve sopra una superficie algebrica. I. Atti Accad. Naz. Lincei. Rend. 4 (1948); (2) Sulle superficie di Kummer e die Weddle. Atti Accad. Naz. Lincei. Rend. 6 (1949); (3) Sopra le superficie algebriche che posseggono trasformazioni birazionali in se. Univ. Roma Ist Naz. Alta Mat. Rend. Mat. e Appl. 9 (1950); (4) Sopra le varieta die Picard d'una superficie algebrica e sulla classificazione delle superficie irregolari. Atti. Accad. Naz. Lincei. Rend. 10 (1951); (5) Sopra il problema dell'uniformizzazione per alcune classi di superficie algebriche. Rend. Accad. Naz. dei Lincei 2 (1951); (6) Les problemes de classification dans la theorie des surfaces algebriques irregulieres. Deux. Coll. de Geom. Alg. Liege, 1952; (7) Recherches sur les surfaces irregulieres. Acad. Roy. Belg. Cl. Sci. Mém. Coll. 8° 4 (1952); (8) Recherches sur les surfaces algebriques irregulieres. Acad. Roy. Belg. Cl. Sci. Mém. Coll. 8° 7 (1952).

ANDREOTTI, A., and T. FRANKEL: The Lefschetz theorem on hyperplane sections. Ann. of Math. 69 (1959).

ANDREOTTI, A., and P. SALMON: Anelli con unica decomponibilita in fattori primi ed un problema di intersezioni complete. Monatsh. Math. 61 (1957).

ARTIN, M.: (1) Grothendieck topologies. Mimeo. notes publ. by Harvard Univ. 1962; (2) On isolated rational singularities of surfaces. Amer. J. Math. 88 (1966); (3) The implicit function theorem in algebraic geometry. In: Algebraic Geometry. London: Oxford University Press, 1969; (4) Algebraization of formal moduli. I. In: Global Analysis. Princeton: Princeton University Press 1969. (5) Some numerical criteria for contractibility. Am. J. Math. 84 (1962).

ATIYAH, M. F.: Vector bundles over an elliptic curve. Proc. London Math. Soc. 7 (1957).

ATIYAH, M., and R. BOTT: A Lefschetz fixed point formula for elliptic complexes: I. Ann. of Math. 86 (1967).

ATIYAH, M., and F. HIRZEBRUCH: Analytic cycles on complex manifolds. Topology 1 (1962).

ATIYAH, M., and W. V. D. HODGE: Integrals of the 2nd kind on an algebraic variety. Ann. of Math. 62 (1955).

AUSLANDER, M., and O. GOLDMAN: The Brauer group of a commutative ring. Trans. Amer. Math. Soc. 97 (1960).

AZUMAYA, G.: On maximally central algebras. Nagoya Math. J. 2 (1951).

BAILY, W.: (1) On the moduli of jacobian varieties. Ann. of Math. 71 (1960); (2) On the theory of θ-functions, the moduli of abelian varieties, and the moduli of curves. Ann. of Math. 75 (1962).

BARSOTTI, I.: (1) Algebraic correspondences between algebraic varieties. Ann. of Math. 52 (1950); (2) Intersection theory for cycles of an algebraic variety. Pacific J. Math. 2 (1952).

BERGER, R.: (1) Über eine Klasse unvergabelter lokaler Ringe. Math. Ann. (1962); (2) Differentialmoduln eindimensionaler lokaler Ringe. Math. Z. 81 (1963).

BERS, L.: On moduli of Riemann surfaces. Mimeo. notes from Eidg. Techn. Hochsch., Zürich 1964.

BOMBIERI, E.: The pluricanonical map of a complex surface. Springer Lecture Notes, vol. 155 (1970).

BOREL, A., and A. HAEFLIGER: La classe d'homologie fondamentale d'un espace analytique. Bull. Soc. Math. France 89 (1961).

BOREL, A., and J. P. SERRE: Le theoreme de Riemann-Roch (d'après Grothendieck). Bull. Soc. Math. France 86 (1958).

BOTT, R.: On a theorem of Lefschetz. Michigan Math. J. 6 (1959).

BRIESKORN, E.: (1) Examples of singular normal complex spaces which are topological manifolds. Proc. Nat. Acad. Sci. U.S.A. 55 (1966); (2) Über die Auflösung gewisser Singularitäten. Math. Annalen 166 (1966); (3) Rationale Singularitäten komplexer Flächen. Inv. Math. 4 (1968); (4) Die Monodromie der isolierten Singularitäten von Hyperflächen. Manuscr. Math. 2 (1970).

CAMPEDELLI, L.: (1) Intorno alle superficie algebriche su cui esistono curve di genere π e grado $n \geq 2\pi - 2$. Atti Accad. Naz. Lincei Rend. VIs 18 (1933); (2) Sul computo dell'invariante di Zeuthen Segre. Atti Accad. Naz. Lincei Rend. VIs 19 (1934); (3) Ancora sul computo. Atti Acad. Naz. Lincei Rend. VIs 19 (1934); (4) Intorno alle superfici ellittiche con un fascio di curve di genere due. Rend. Sem. Mat. Univ. Padova 6 (1935).

CARTIER, P.: (1) Une nouvelle operation sur les formes differentielles. C. R. Acad. Sci. Paris Sér. A—B 244 (1957); (2) Calcul differentiel sur les varietes algebriques en caracteristique non nulle. C. R. Acad. Sci. Paris Sér. A—B 245 (1957); (3) Questions de rationalité des diviseurs. Bull. Soc. Math. France, 86 (1958).

CHATALET, F.: Variations sur un theme de H. Poincaré. Ann. Sci. École Norm. Sup. 61 (1944).

CHERN, S.: On the characteristic classes of complex sphere bundles and algebraic varieties. Amer. J. Math. **75** (1953).

CHEVALLEY, C.: (1) Varietes de Picard. Seminaire Chevalley, Secr. Math. Paris 1958/59; (2) Fondements de la geometrie algebrique. Secr. Math. Paris 1958; (3) Anneaux de Chow et applications. Seminaire Chevalley. Secr. Math. Paris 1958; (4) Sur la theorie de la variete de Picard. Amer. J. Math. **82** (1960).

CHISINI, O.: (1) Altre curve di diramazione dei piani n-pli. Atti Accad. Naz. Lincei Rend. VIs, **29** (1939); (2) Sulla funzione algebrica nell'intorno di un punto cuspidale della curva di diramazione. Ist. Lombardo Accad. Sci. Lett. Rend. A. IIIs **73** (1940).

CHOW, W.-L.: (1) Algebraic systems of positive cycles in an algebraic variety. Amer. J. Math. **72** (1950); (2) On the fundamental group of an algebraic variety. Amer. J. Math. **74** (1952); (3) On Picard varieties. Amer. J. Math. **74** (1952); (4) The Jacobian variety of an algebraic curve. Amer. J. Math. **76** (1954); (5) On equivalence classes of cycles in an algebraic variety. Ann. Math. **64** (1956); (6) On the principle of degeneration in algebraic geometry. Ann. of Math. **66** (1957); (7) On the connectedness theorem in algebraic geometry. Amer. J. Math. **81** (1959); (8) On meromorphic maps of algebraic varieties. Annals of Math. **89** (1969).

CHOW, W., and S. LANG: On the birational equivalence of curves under specialization. Amer. J. Math. **79** (1957).

CHOW, W., and B. L. VAN DER WAERDEN: Zur Algebraischen Geometrie, IX. Math. Ann. **113** (1937).

CLEMENS, C.: Picard-Lefschetz theorem for families of non-singular algebraic varieties acquiring ordinary singularities. Trans. Amer. Math. Soc. **136** (1969).

CONFORTO, F.: Le superficie razionali. Bologna: Nicola Zanichelli 1939.

CROWELL, R., and R. FOX: An Introduction to Knot Theory. Ginn 1963.

DEHN, M.: Die Gruppe der Abbildungsklassen. Acta Math. **69** (1938).

DELIGNE, P., and D. MUMFORD: The irreducibility of the space of curves of given genus. Publ. I.H.E.S. **36** (1970).

DOLBEAULT, P.: Sur la cohomologie des varietes analytique complexe. C. R. Acad. Sci. Paris Sér. A—B **226** (1948).

DOUADY, A.: Le probleme des modules pour les varietes analytiques complexes. Sem. Bourbaki **277**, Secr. Math. Paris, 1964/65.

EDMUNDS, G.: Coverings of node curves. J. London Math. Soc. **1** (1969).

EGER, M.: Les systèmes canoniques d'une varieté algebrique à plusieurs dimensions. Ann. Sci. École Norm. Sup. **60** (1943).

EICHLER, M.: Introduction to the theory of algebraic numbers and functions. New York: Academic Press 1966.

ENRIQUES, F.: (1) Sulle superficie ellittiche di genere zero. Atti Accad. Naz. Lincei Rend. VIs **19** (1934); (2) Sulla classificazione delle superficie algebriche particolarmente di genere zero. Rend. Semin. Mat. Fac. Sci. Univ. Roma IIIs, **1** (1934); (3) La proprieta caratteristica delle superficie algebriche irregolari e le curve infinitamente vicine. Atti Accad. Naz. Lincei Rend. VIs, **23** (1936); (4) Curve infinitamente vicine sopra una superficie algebrica. Rend. Semin. Mat. Roma IVs, **1** (1936); (5) Sulle singolarita che nascono per proiezione di una superficie o varieta algebrica. Scritti Mat. Luigi Berzolari (1936); (6) Sulla proprieta caratteristica delle superficie algebriche irregolari. Atti Accad. Naz. Lincei VIs, **27** (1938); (7) Sur la proprieté caracteristique des surfaces algebriques irregulieres. C. R. Acad. Sci. Paris Sér. A—B **208** (1939); (8) Des courbes paracanoniques appartenant a une surface algébrique irregulière. Bull. Soc. Roy. Liège **8** (1939); (9) Sur l'extension du théorème de Riemann-Roch aux surfaces algébriques. Bull. Sci. Math. IIs, **64** (1940); (10) Sur le théorème de Riemann-Roch concernant les surfaces algébriques et sur les systèmes des courbes canoniques et pluricanoniques. Rev. Acad. Ci. Madrid **40** (1946); (11) Le Superficie Algébriche. Bologna: Nicola Zanichelli 1949.

FRANCHETTA, A.: (1) Sulle curve eccezionali riducibili di prima specie. Boll. Un. Mat. Ital. II s. **2** (1940); (2) Sulle curve eccezionali. Boll. Un. Mat. Ital. II s. **3** (1940); (3) Sulle curve appartenenti a una superficie generale d'ordine $n \geqq 4$ dell' S_3. Atti Accad. Naz. Lincei Rend. **3** (1947); (4) Sui sistemi pluricanonici di una superficie algebrica. Univ. Roma Ist. Naz. Alta Rend. Mat. e Appl. **8** (1949); (5) Sulle curve riducibili appartenenti ad una superficie algebrica. Univ. Roma Ist. Naz. Alta Mat. Rend. Mat. e Appl. **8** (1949); (6) Sui sistema aggiunto ad una curva riducibile. Atti Accad. Naz. Lincei. **6** (1949); (7) Sui modelli pluricanonici delle superficie algebriche. Univ. Roma Ist. Naz. Alta Mat. Rend. Mat. e Appl. **9** (1950).

DE FRANCHIS, M.: (1) Dimostrazione del teorema fondamentale sulle superficie iperellittiche. Atti Accad. Naz. Lincei Rend. VI s. **24** (1936); (2) Sulla classificazione delle superficie iperellittiche. Scritti. mat. Luigi Berzolari (1936); (3) I sistemi canonici e pluricanonici e le forme algebricodifferenziali di prima specie. Ann. Mat. Pura Appl. IV s. **19** (1940).

FULTON, W.: Hurwitz schemes and irreducibility of moduli of alg. curves. Annals of Math. **90** (1969).

GALLARATI, D.: (1) Intorno a certe superficie algebriche aventi un elevato numero di punti singolari isolati. Atti Accad. Naz. Lincei Rend **11** (1951); (2) Una superficie dell' attavo ordine con 160 nodi. Acc. Ligure, 1957.

GODEAUX, L.: (1) Les surfaces algébriques non rationnelles de genres arithmétique et géometrique nuls. Paris: Hermann 1934; (2) Les involutions cycliques appartenant à une surface algébrique. Paris: Hermann 1935; (3) Sur les surfaces algébriques possedant un système lineaire simple. Bull. Acad. Roy. Belg. V s. **22** (1936); (4) Sur les points de diramation des surfaces algebriques multiples C. R. Acad. Sci. Paris Sér. A—B **205** (1937); (5) Sur les surfaces multiples ayant un nombre fini de points de diramation. Ann. Sci. École Norm. Sup. III s. **55** (1938).

GORENSTEIN, D.: An arithmetic theory of adjoint plane curves. Trans. Amer. Math. Soc. **72** (1952).

GRAUERT, H., and R. REMMERT: Komplexe Räume. Math. Ann. **136** (1958).

GRIFFITHS, P.: (1) On the periods of integrals on algerbaic manifolds. Rice Univ. Studies **54** (1968); (2) Periods of integrals on algebraic manifolds, I, II. Amer. J. Math. **90** (1968); (3) Some results on algebraic cycles on algebraic manifolds. In: Algebraic Geometry. London: Oxford Univ. Press 1969; (4) On the periods of certain rational integrals. Ann. of Math. **90** (1969); (5) Report on variations of Hodge structure. Bull. Amer. Math. Soc. **76** (1970).

GROTHENDIECK, A.: (1) Théorèmes de dualite pour les faisceaux algébriques cohérents. Sem. Bourbaki, exp. 149. Secr. Math. Paris (1957); (2) Sur une note de Mattuck-Tate. J. Reine Angew. Math. **200** (1958); (3) La théorie des classes de Chern. Bull. Soc. Math. France **86** (1958); (4) The cohomology theory of abstract algebraic varieties. Proc. Internat. Congress Math. New York: Cambridge Univ. Press 1960; (5) Éléments de la géometrie algébrique. Publ. I.H.E.S. **4, 8, 11,** etc. 1960; (6) Séminaire de Geometrie Algébrique. Mimeo. notes publ. by I.H.E.S.: 1960—1961: Revêtements étale, 1962: Cohomologie locale des faisceaux, 1963—1964: Cohomologie étale des schemas (with ARTIN and VERDIER), 1963: Schémas en groupes (with DEMAZURE), 1965—1966: Fonctions L et cohomologie l-adique, 1966—1967: Théorème de Riemann-Roch, 1968: (on monodromy theorems and vanishing cycles) (with DELIGNE); (7) Techniques de construction II: Le théorème d'existence en théorie formelles des modules. Sem. Bourbaki, exp. 195. Secr. Math. Paris (1960); (8) Techniques de construction IV: Les schémas de Hilbert. Sém. Bourbaki, exp. 221. Secr. Math. Paris

(1961); (9) Techniques de construction V: Les schémas de Picard. Sém. Bourbaki, exp. 232. Secr. Math. Paris (1961); (10) On the De Rham cohomology of algebraic varieties. Publ. I.H.E.S. **29** (1966); (11) Crystals and De Rham cohomology. In: Dix Exposes. Amsterdam-London: North-Holland, 1968; (12) Le groupe de Brauer, I, II, III, in Dix exposes sur la cohomologie des schemas. Amsterdam-London: North-Holland 1968; (13) Standard conjectures on algebraic cycles. In: Algebraic Geometry. London: Oxford Univ. Press 1969; (14) Sur la classification des fibrés holomorphes sur la sphère de Riemann. Ann. J. Math. **79** (1957).

HARTSHORNE, R.: (1) Residues and duality. Springer Lecture Notes **20** (1966); (2) Cohomological dimension of algebraic varieties. Annals of Math. **88** (1968); (3) Curves with high self-intersection on algebraic surfaces. Publ. I.H.E.S. **36** (1969).

HIGMAN, G.: A finitely generated infinite simple group. J. London Math. Soc. **26** (1951).

HIRONAKA, H.: (1) On the arithmetic genera and the effective genera of algebraic curves. Mem. Coll. Sci. Univ. Kyoto **30** (1957); (2) A generalized theorem of Krull-Seidenberg on parametrized algebras of finite type. Amer. J. Math. **82** (1960); (3) An example of a non-Kahlerian deformation. Ann. of Math. **75** (1962); (4) On resolution of singularities, Proc. Int. Congress, Stockholm, 1962; (5) Resolution of singularities of an algebraic variety over a field of characteristic O. Ann. of Math. **79** (1964); (6) Characteristic polyhedra of singularities. J. Math. Kyoto Univ. **7** (1968); (7) Smoothing of algebraic cycles of small simensions. Am. J. Math. **90** (1968); (8) Normal cones in analytic Whitney stratifications. Publ. I.H.E.S. **36** (1969). (9) Some numerical characters of singularities. J. Math. Kyoto Univ. **9** (1970).

HIRZEBRUCH, F.: (1) Arithmetic genera and the theorem of Riemann-Roch for algebraic varieties. Proc. Nat. Acad. Sci. U.S.A. **40** (1954); (2) Topological methods in algebraic geometry, 3rd Ed. Berlin-Heidelberg-New York: Springer 1966.

HODGE, W. V. D.: (1) The geometric genus of a surface as a topological invariant. J. London Math. Soc. **8** (1933); (2) Note on the theory of the base for curves on an algebraic surface. J. London Math. Soc. **12** (1937); (3) Algebraic correspondences between surfaces. Proc. London Math. Soc. II s. **44** (1938); (4) Note on the conditions for a p-cycle of an algebraic manifold to be of rank k. Proc. Cambridge Philos. Soc. **43** (1947); (5) Harmonic integrals on algebraic varieties. Proc. Cambridge Philos. Soc. **44** (1948); (6) The characteristic classes on algebraic varieties. Proc. London Math. Soc. **1** (1951); (7) The topological invariants of algebraic varieties. Proc. Int. Congress Math., Cambridge, Mass., 1950, vol. 1. Amer. Math. Soc. 1952; (8) Harmonic Integrals. London: Cambridge University Press 1952.

HODGE, W. V. D., and D. PEDOE: Methods of algebraic geometry. Vol. I, II, III. London: Cambridge University Press 1947, 1952, 1954.

HORMANDER, L.: Introduction to complex analysis in several variables. New York-London-Melbourne-New Dehli-Toronto: Van Nostrand 1966.

HU, S.: Differentiable manifolds. New York: Holt, Rinehart and Winston, 1969.

IGUSA, JUN-ICHI: (1) Algebraic correspondences between algebraic varieties. J. Math. Soc. Japan **3** (1951); (2) Some remarks on the theory of Picard varieties. J. Math. Soc. Japan **3** (1951); (3) On the Picard varieties attached to algebraic varieties. Amer. J. Math. **74** (1952); (4) On the arithmetic normality of the Grassmann variety. Proc. Nat. Acad. Sci. U.S.A. **40** (1954); (5) A fundamental inequality in the theory of Picard varieties. Proc. Nat. Acad. Sci. U.S.A. **41** (1955); (6) On some problems in abstract algebraic geometry. Proc. Nat. Acad U.S.A. **41** (1955); (7) Fibre systems of jacobian varieties, I and II. Amer. J. Math. **78** (1956); (8) Abstract vanishing cycle theory. Proc. Japan Acad. **34** (1958);

(9) Betti and Picard numbers of abstract algebraic surfaces. Proc. Nat. Acad. U.S.A. **46** (1960).

JONGMANS, F.: (1) Les limitations du nombre des modules des surfaces algebriques. Acad. Roy. Belg. Bull. Cl. Sci. **31** (1945), (1946); (2) Contributions à la théorie des variétés algébriques. Mém. Soc. Roy. Sci. Liège Coll. 8° **7** (1947); (3) Memoire sur les surfaces et les variétés algébriques à courbes-sections de genre quatre. Acad. Roy. Belg. Cl. Sci. Mém. Coll. 8° **23** (1949); (4) Extension d'une borne inferieure pour le genre lineaire des surfaces algébriques. Bull. Soc. Roy. Sci. Liège **19** (1950); (5) Sur l'étude des surfaces algébriques caracterisees par la condition $p_g \geqq 2(p_a + 2)$. Acad. Roy. Belg. Bull. Cl. Sci. **36** (1950).

JUNG, H. W. E.: (1) Das arithmetische Geschlecht. Math. Z. **37** (1933); (2) Sui gruppi di punti sopra una superficie e sulla serie di'Severi. Mem. Accad. Ital. Mat. **4** (1933); (3) Zur Theorie der algebraischen Funktionen zweier Veränderlicher I. J. Reine Angew. Math. **180** (1939); (4) II of above (Lineare Integrale). J. Reine Angew. Math. **181** (1939); (5) III of above (Über die Zahl δ der Zeuthen-Segreschen Invariante). J. Reine Angew. Math. **181** (1939); (6) Einführung in die Theorie der algebraischen Funktionen zweier Veränderlicher. Berlin: Akademie-Verlag 1951.

KAS, A.: (1) On obstructions to deformations of complex analytic surfaces. Proc. Nat. Acad. Sci. U.S.A. **58** (1967); (2) On deformations of a certain type of irregular algebraic surface. Amer. J. Math. **90** (1968).

KATZ, N.: (1) On the differential equations satisfied by period matrices. Publ. I.H.E.S. **35** (1968); (2) Seminar on degeneration of algebraic varieties. Mimeo. notes, Inst. Adv. Study 1969.

KATZ, N., and T. ODA: On the differentiation of De Rham cohomology classes with respect to parameters. J. Math. Kyoto Univ. **8** (1968).

KLEIMAN, S.: (1) Towards a numerical theory of ampleness. Ann. of Math. (1966); (2) Algebraic cycles and the Weil conjectures. In: Dix Exposes. Amsterdam-London: North-Holland 1968; (3) Geometry on Grassmannians. Publ. I.H.E.S. **36** (1969).

KODAIRA, K.: (1) The theorem of Riemann-Roch on compact analytic surfaces. Amer. J. Math. **73** (1951); (2) Arithmetic genera of algebraic varieties. Proc. Nat. Acad. Sci. U.S.A. **38** (1952); (3) On the theorem of Riemann-Roch for adjoint systems on Kählerian varieties. Proc. Nat. Acad. Sci. U.S.A. **38** (1952); (4) On a differential-geometric method in the theory of analytic stacks. Proc. Nat. Acad. Sci. U.S.A. **39** (1953); (5) Some results in the transcendental theory of algebraic varieties. Ann. of Math. **59** (1954); (6) On Kahler manifolds of a restricted type. Ann. of Math. **60** (1954); (7) On compact analytic surfaces, in Analytic Functions. Princeton. Princeton Univ. Press, N. J. 1960; (8) On compact complex analytic surfaces, I. Ann. of Math. **71** (1960); II, Ann. of Math. **77** (1963); III, Ann. of Math. **78** (1963); (9) A theorem of completeness for analytic systems of surfaces with ordinary singularities. Ann. of Math. **74** (1961); (10) On the structure of compact complex analytic surfaces. Amer. J. Math., I, **86** (1964); II, **88** (1966); III, **89** (1967); IV, **90** (1968); (11) On characteristic systems of families of surfaces with ordinary singularities in a projective space. Amer. J. Math. **87** (1965); (12) On a certain type of irregular algebraic surface. J. d'Analyse math. **19** (1967); (13) Pluricanonical systems on algebraic surfaces of generadet type. J. Math. Soc. Japan **20** (1968).

KODAIRA, K., and D. SPENCER: (1) On arithmetic genera of algebraic varieties. Proc. Nat. Acad. Sci. U.S.A. **39** (1953); (2) On deformations of complex analytic structures. Ann. of Math. **67** (1958); (3) A theorem of completeness of characteristic systems of complete continuous systems. Amer. J. Math. **81** (1959).

KOIZUMU, S.: (1) On the differential forms of the first kind on algebraic varieties. J. Math. Soc. Japan 1 (1949); (2) On the differential forms of the first kind on algebraic varieties. II. J. Math. Soc. Japan 2 (1951); (3) On Albanese varieties. Illinois J. Math. 4 (1960); (4) On Albanese varieties. II. Mem. Coll. Sci. Univ. Kyoto Ser. A. Math. 32 (1960).

KREISS, O.: (1) Über syzygetische Flächen. Annali di Mat., 1955.

KURANISHI, M.: New proof for existence of locally complete families of complex structures. Proc. Conf. Complex Anal. Berlin-Heidelberg-New York: Springer 1965.

LANG, S.: (1) Unramified class field theory over function fields in several variables. Ann. of Math. 64 (1956); (2) Introduction to algebraic geometry. New York: Wiley-Interscience 1958; (3) Abelian varieties. New York: Wiley-Interscience 1959.

LANG, S., and A. NERON: Rational points of abelian varieties over function fields. Amer. J. Math. 81 (1959).

LANG, S., and J.-P. SERRE: Sur les revêtements non-ramifies des variétés algébriques. Amer. J. Math. 79 (1957).

LANG, S., and J. TATE: Galois cohomology and principal homogeneous spaces. Amer. J. Math. 80 (1958).

LERAY, J.: Le calcul differentiel et integral sur une variété analytique complexe. Bull. Soc. Math. France 87 (1959).

LICHTENBAUM, S.: Curves over discrete valuation rings. Amer. J. Math. 90 (1968).

LIPMAN, J.: Rational singularities. Publ. I.H.E.S. 36 (1969).

LOJASIEWICZ, S.: Triangulation of semi-analytic sets. Ann. Scuola Norm. Sup. Pisa 18 (1964).

LUBKIN, S.: A p-adic proof of Weil's conjectures. Ann. of Math. 87 (1968).

MANIN, J. I.: (1) Algebraic curves over fields with differentiation. Izv. Akad. Nauk S.S.S.R. Ser. Mat. 22 (1958); (2) The Hasse-Witt matrix of an algebraic curve. Izv. Akad. Nauk S.S.S.R. Ser. Mat. 25 (1961); (3) Rational points of algebraic curves over function fields. Izv. Akad. Nauk S.S.S.R. 27 (1963); (4) On arithmetic of rational surfaces. Doklady Acad. Nauk, 152 (1963); (5) Rational surfaces over perfect fields. Publ. I.H.E.S. 30 (1966).

MATSUMURA, H.: Geometric structure of the cohomology rings in abstract algebraic geometry. Mem. Coll. Sci. Kyoto 32 (1959).

MATSUMURA, H., and M. NAGATA: On the algebraic theory of sheets of an algebraic variety. Mem. Coll. Sci. Univ. Kyoto. Ser. A. Math. 30 (1957).

MATSUSAKA, T.: (1) The theorem of Bertini on linear systems in modular fields. Mem. Coll. Sci. Univ. Kyoto Ser. A. Math. 26 (1950); (2) Specialization of cycles on a projective model. Mem. Coll. Sci. Univ. Kyoto Ser. Math. A. 26 (1951); (3) On the algebraic construction of the Picard variety. Japan J. Math. 21 (1951) and 22 (1952); (4) On the algebraic construction of the Picard variety. Proc. Japan Acad. 28 (1952); (5) On algebraic families of positive divisors and theri associated varieties on a projective variety. J. Math. Soc. Japan 5 (1953); (6) Some theorems on Abelian varieties. Nat. Sci. Rep. Ochanomizu Univ. 4 (1953); (7) On the theorem of Castelnuovo-Enriques. Nat. Sci. Rep. Ochanomizu Univ. 4 (1954); (8) The criteria for algebraic equivalence and the torsion group. Amer. J. Math. 79 (1957); (9) Polarized varieties and fields of moduli. Amer. J. Math. 80 (1958).

MATSUSAKA, T., and D. MUMFORD: Two fundamental theorems on deformations of polarized varieties. Amer. J. Math. 86 (1964).

MATTUCK, A.: (1) The irreducibility of the regular series on an algebraic variety. Illinois J. Math. 3 (1959); (2) Ruled surfaces and the Albanese mapping. Bull. Am. Math. Soc. 75 (1969).

MATTUCK, A., and J. TATE: On the inequality of Castelnuovo-Severi. Abh. Math. Sem. Univ. Hamburg **22** (1958).

MOISEZON, B. G.: Projective imbeddings of algebraic manifolds. Dokl. Akad. Nauk S.S.S.R. **141** (1961).

MONSKY, P., and G. WASHNITZER: Formal cohomology: I and II. Ann. of Math. **88** (1968).

MUHLY, H. T.: The irregularity of an algebraic surface and a theorem on regular surfaces. Bull. Amer. Math. Soc. **55** (1949).

MUHLY, H. T., and O. ZARISKI: Hilbert's characteristic function and the arithmetic genus of an algebraic variety. Trans. Amer. Math. Soc. **69** (1950).

MUMFORD, D.: (1) Pathologies of modular geometry. Amer. J. Math. **83** (1961); (2) Topology of normal singularities and a criterion for simplicity. Publ. I.H.E.S. (1961); (3) Further pathologies in algebraic geometry. Amer. J. Math. **84** (1962); (4) Geometric Invariant Theory. Berlin-Heidelberg-New York: Springer 1965; (5) Lectures on curves on an algebraic surface. Princeton, N. J.: Princeton University Press 1966; (6) Enriques' classification of surfaces, I. In: Global Analysis. Princeton, N. Y.: Princeton University Press 1969; (7) Pathologies III. Amer. J. Math. **89** (1967); (8) Rational equivalence of 0-cycles on surfaces. J. Math. Kyoto Univ. **9** (1969); (9) Abelian varieties. London: Oxford University Press 1970; (10) Abelian quotients of the Teichmuller modular group. J. Analyse Math. **28** (1967).

MURRE, J. P.: (1) On a uniqueness theorem for certain kinds of birational transformations. Nederl. Akad. Wetensch. Proc. Ser. A. **21** (1959); (2) On divisors on products of three factors. Nieuw Arch. Wisk. **8** (1960); (3) On Chow varieties of maximal, total, regular families of positive divisors. Amer. J. Math. **83** (1961); (4) Contravariant functors from preschemes to abelian groups. Publ. I.H.E.S. **23** (1964); (5) An introduction to Grothendieck's theory of the fundamental group. Mimeo. notes publ. by the Tata Inst., Bombay 1967.

NAGATA, M.: (1) A general theory of algebraic geometry over Dedekind domains. I. The notion of models. Amer. J. Math. **78** (1956) ;(2) A treatise on the 14th problem of Hilbert. Mem. Coll. Sci. Univ. Kyoto. **30** (1956); (3) On the imbedding problem of abstract varieties in projective varieties. Mem. Coll. Sci. Univ. Kyoto **30** (1956); (4) On the imbeddings of abstract surfaces in projective varieties. Mem. Coll. Sci. Univ. Kyoto **30** (1957); (5) Existence theorems for nonprojective complete algebraic varieties. Illinois J. Math. **2** (1958); (6) A general theory of algebraic geometry over Dedekind domains. II. Separably generated extensions and regular local rings. Amer. J. Math. **80** (1958); (7) On the closedness of singular loci. I.H.E.S. Publ. Math. 1959; (8) A general theory of algebraic geometry over Dedekind domains. III. Absolutely irreducible models, simple spots. Amer. J. Math. **81** (1959); (9) On rational surfaces. I and II. Mem. Coll. Sci. Univ. Kyoto **32** (1960) and **33** (1961).

NAKAI, Y.: (1) On the characteristic linear systems of algebraic families. Illinois J. Math. **1** (1957); (2) A property of an ample linear system on a non-singular variety. Mem. Coll. Sci. Univ. Kyoto **30** (1957); (3) Ramifications, differentials and differente on algebraic varieties of higher dimensions. Mem. Coll Sci. Univ. Kyoto **32** (1960); (4) Non-degenerate divisors on an algebraic surface. J. Sci. Hiroshima Univ. Ser. A—I **24** (1960).

NAKANO, S.: Tangent vector bundles and Todd canonical systems of an algebraic variety. Mem. Coll. Sci. Kyoto **29** (1955).

NARASIMHAN, M. S., and R. SIMHA, Manifolds with ample canonical class. Inv. Math. **5** (1968).

NERON, A.: (1) La théorie de la base pour les diviseurs sur les variétés algébrique. Coll. Geom. Alg. Liège, 1952; (2) Problèmes arithmétiques et géometriques rattaches à la notion de rang d'une courbe algébrique dans un corps. Bull. Soc.

Math. France **80** (1952); (3) Modèles minimaux des variétés abéliennes. Publ. I.H.E.S., **21**.

NERON, A., and P. SAMUEL: La variété de Picard d'une variété normale. Ann. Inst. Fourier (Grenoble) **4** (1954).

NEUWIRTH, L.: Knot groups. Ann. Math. Studies **56**. Princeton, N. J.: Princeton Univ. Press 1965.

NORTHCOTT, D. G.: (1) The number of analytic branches of a variety. J. London Math. Soc. **25** (1950); (2) An application of local uniformization to the theory of divisors. Proc. Cambridge Philos. Soc. **47** (1951); (3) On the genus formula for plane curves. J. London Math. Soc. **30** (1955); (4) Abstract dilatations and infinitely near points. Proc. Cambridge Phil. Soc. **52** (1956).

ODA, T.: The first De Rham cohomology group and Dieudonné modules. Ann. Sci. École Norm. Sup. **2** (1969).

OGG, A.: Cohomology of abelian varieties over function fields. Ann. of Math. **76** (1962).

PICARD, E.: Sur les periodes des integrales doubles et sur une classe d'équations differentielles lineaires. Ann. Sci. École Norm. Sup. III s. **50** (1933).

POPP, H.: Zur reduktionstheorie algebraischer Funktionskörper vom Transcendenz-grad 1. Arch. Math. (Basel) **17** (1966).

PORTEOUS, I. R.: Todd's canonical classes, to appear in Liverpool Symposium on Singularities.

RAYNAUD, M.: Caractéristique d'Euler-Poincaré d'un faisceaux et cohomologie des variétés abeliennes. Sem. Bourbaki, exp. 286, Secr. Math. Paris 1964—65.

REES, D.: On a Problem of Zariski. Illinois J. Math. **2** (1958).

DE RHAM, G.: Variétés différentiables. Paris: Hermann 1954.

ROQUETTE, P.: Arithmetischer Beweis der Riemannschen Vermutung in Kongruenz-zetafunktionen. J. Reine Angew. Math. **191** (1953).

ROSENLICHT, M.: (1) Equivalence relations on algebraic curves. Ann. of Math. **56** (1952). (2) Differentials of the second kind for algebraic function fields of one variable. Ann. of Math. **57** (1953); (3) Generalized Jacobian varieties. Ann. of Math. **59** (1954); (4) A universal mapping property of generalized jacobian varieties. Ann. of Math. **66** (1957).

ŠAFAREVICH, I.: (1) Principal homogeneous spaces defined over a function field. Trudy Mat. Inst. Steklov. **64** (1961); (2) Algebraic surfaces (with others). Proc. Steklov. Inst. Math. **75** (1965); (3) Lextures on minimal models. Tata Lecture notes, Bombay (1966).

SAMUEL, P.: (1) La notion de multiplicite en algèbre et en géometrie algébrique. I and II. J. Math. Pures Appl. **30** (1951); (2) Singularites des varietes algébriques. Bull. Soc. Math. France **79** (1951); (3) Méthodes d'algèbre abstraite en géometrie algébrique. Berlin-Göttingen-Heidelberg: Springer 1955; (4) Algèbricité de certains points singuliers algebroides. J. Math. Pures Appl. **35** (1956); (5) Rational equivalence of arbitrary cycles. Amer. J. Math. **78** (1956); (6) Relations d'équivalence en géometrie algébrique. Proc. Int. Congress Math. 1958. New York: Cambridge Univ. Press 1960.

SCHLESSINGER, M.: (1) Infinitesimal deformations of singularities. Ph. D. Thesis, Harvard, 1964; (2) Functors of Artin rings. Trans. Amer. Math. Soc. **130** (1968).

SCHWARZENBERGER, R. L. E.: (1) Vector bundles on algebraic surfaces. Proc. London Math. Soc. **11** (1961); (2) Vector bundles on the projective plane. Proc. London Math. Soc. **11** (1961).

SCOTT, D. B.: (1) Point-curve correspondences. II. Proc. Cambridge Philos. Soc. **42** (1946); (2) Point-curve correspondences. III. Proc. Cambridge Philos. Soc. **45** (1949); (3) The united curve of a point-curve correspondence on an algebraic surface, and some related topological characters of the surface. Proc. London

Math. Soc. **51** (1950); (4) On the fundamental theorem for point-point correspondences with valency on an algebraic surface. Pont. Acad. Sci. Acta **14** (1950); (5) Correspondences of dimensions two and three between algebraic surfaces. Proc. London Math. Soc. **2** (1952).

SEGRE, B.: (1) Sulla serie caratteristica d'una superficie sopra una varieta algebrica a quattro dimensioni. Atti. Accad. Naz. Lincei Rend. VI s. **17** (1932); (2) Determinazione geometrico-funzionale di gruppi di punti covarianti. Atti. Accad. Naz. Lincei, Rend. VI s. **18** (1933); (3) Sui moduli delle superficie algebriche irregolari. Atti. Accad. Naz. Lincei. Rend. VI s. **19** (1934); (4) Intorno alle parti fisse del sistema canonica sopra una superficie algebrica, Restratto di Rend. Accad. Sci. Ist. Bologna (1936); (5) Intorno ad un teorema di Hodge sulla teoria della base. Ann. Mat. Pura Appl. IV s. **16** (1937); (6) Un teorema fondamentale della geometria sulle superficie algebriche ed il principio di spezzamento. Ann. Mat. Pura Appl. IV s. **17** (1938); (7) Sur un théorème fondamental de géometrie sur les surfaces algébrique. C. R. Acad. Sci. Paris Sér. A—B **208** (1939); (8) Sul massimo numero di nodi delle superficie di dato ordine. Boll. Un. Mat. Ital. **2** (1947); (9) Un nuovo metodo per lo scioglimento delle singilarita. Atti Accad. Naz. Lincei. **3** (1947); (10) Sullo scioglimento delle singolarita delle varieta algebriche. Ann. Mat. Pura Appl. **33** (1952); (11) Nuovi metodi e risultati nella geometria sulla varieta algebriche. Ann. Mat. Pura Appl. **35** (1953); (12) Sul massimo numero di nodi delle superficie algebriche. Atti Accad. Ligure Sci. Lett. (1952 and 1953); (13) Dilatazioni e varieta canoniche. Ann. Mat. Pura Appl. **37** (1954).

SEIDENBERG, A.: The hyperplane sections of normal varieties. Trans. Amer. Math. Soc. **69** (1950).

SEMPLE, J. G., and L. ROTH: Introduction to Algebraic Geometry. Clarendon Press: Oxford 1949.

SERRE, J.-P.: (1) Un theoreme de dualite. Comm. Helv. **29** (1955); (2) Faisceaux algébriques cohérents. Ann. of Math. **61** (1955); (3) Géometrie analytique et géometrie algebrique. Ann. Inst. Fourier (Grenoble) **6** (1955—56); (4) Sur la topologie des variétés algébrique en caracteristique p. Symp. of Alg. Top., Mexico, 1956; (5) Groupes algébriques et corps de classes. Paris: Hermann 1959; (6) On the fundamental group of a unirational variety. J. London Math. Soc. **34** (1959); (7) Analogues Kahleriennes des certaines conjectures de Weil. Ann. of Math. **71** (1960); (8) Examples de variétés projectives en caractéristique p non relevables en caracteristique zero. Proc. Nat. Acad. Sci. U.S.A. **47** (1961).

SESHADRI, C. S.: (1) Variété de Picard d'une variété complete. Ann. Math. Pura Appl. **67** (1962); (2) Space of unitary vector bundles on a compact Riemann surface. Ann. of Math. **82** (1965).

SEVERI, F.: (1) Serie di equivalenza sopra una superficie algebrica. Atti Accad. Naz. Lincei Rend. **17** I, II and III (1933); (2) Corrispondenze a valenza sopra una superficie algebrica. Atti Accad. Naz. Lincei Rend. **17** I, II, III (1933); (3) Serie di equivalenza sulle superficie e sistemi di equivalenza sulle varieta algebriche. Mem. Accad. Ital. Mat. **4** (1933); (4) La théorie générale des correspondences entre deux surfaces algébriques. C. R. Acad. Sci. Paris Sér. A—B **198** (1934); (5) La base per le varieta in una data e correspondenze fra i punti di due superficie. Mem. Accad. Ital. **5** (1934); (6) Le involuzioni razionali sopra una superficie come serie die equivalenza: I and II. Atti Accad. Naz. Lincei Rend. VI s. **19** (1934); (7) Caratterizzazione geometrica, topologica e transcendente delle serie di equivalenza sopra una superficie. Atti Accad. Naz. Lincei Rend. VI s. **20** (1935); (8) Ancora sulla caratterizzazione topologica e transcendente delle serie di equivalenza. Atti Accad. Naz. Lincei Rend. **20** (1935); (9) Un altra proprieta fondamentale delle serie di equivalenza sopra una superficie. Atti Accad. Naz. Lincei Rend. **21** (1935); (10) The series of sets of points on an algebraic surface. Proc. Imp. Acad. Jap. **12** (1936); (11) Il rango di una corrispondenza a valenza sopra una superficie. Boll. Un. Mat. Ital. **15** (1936); (12) Equivalenza d'una curva come gruppo virtuale parziale d'una serie d'equivalenza. Scritti mat. off. a Luigi Berzolari (1936); (13) I fondamenti della geometria numerativa. Ann. Mat. Pura Appl. **19** (1940); (14) Caretterizzazione topologica delle superficie razionali e delle rigate. Vierteljschr. Naturforsch. Ges. Zürich **85**, **32** (1940); (15) Sulla classificazione delle rigate algebriche. Rend. Mat. e Appl. V s. **2** (1941); (16) Serie, Sistemi d'Equivalenza e Corrispondenze Algebriche sulle Varieta Algebriche. Rome: Edizioni Cremonese **1** (1942); (17) Ulteriori sviluppi della teoria delle serie di equivalenza sulle superficie algebriche. Pont. Acad. Sci. Comment. **6** (1942); (18) Über die Darstellung algebraischer Mannigfaltigkeiten als Durchschnitte von Formen. Abh. Math. Sem. Hansischen Univ. **15** (1943); (19) Le varieta multiple diramate e il loro teorema di esistenza. Mem. Mat. Inst. "Jorge Juan" **4** (1946); (20) Teoremi di regolarita sopra una superficie algebrica. Univ. Roma Ist. Naz. Alta Mat. Rend. Mat. e Appl. **6** (1947); (21) Fondamenti di Geometria Algebrica. Padova: CEDAM 1948; (22) Sulle molteplicita d'intersezione delle varieta algebriche ed analitiche e sopra una teoria geometrica dell'eliminazione. Math. Z. **52** (1950); (23) Problemes resolus et problemes nouveaux dans la theorie des systemes d'equivalence. Proc. Int. Cong. Math., 1954, Amsterdam, Vol. III; (24) Fondamenti per la geometria sulle varieta algebriche, III. Ann. Mat. Pura Appl. **41** (1956); (25) Il teorema di Riemann-Roch per curve, superficie e varieta; questioni collegate. Erg. der Math. und ihr. Grenz. Berlin-Göttingen-Heidelberg: Springer 1958; (26) Geometria dei sistemi algebrici sopra una superficie e sopra una varieta algebrica. Rome: Ediz. Cremonese **2** (1958).

SILOV, G. E.: Singular points of algebraic curves in the plane. Uspehi Mat. Nauk **5** (1950).

SINGER, I., and J. THORPE: Lecture notes on elementary topology and geometry. Glenview, Il.: Scott, Foresman 1967.

SNAPPER, E.: Cohomology groups and genera of higher-dimensional fields. Mem. Amer. Math. Soc. **28** (1957).

SPENCER, D. C.: Cohomology and the Riemann-Roch theorem. Proc. Nat. Acad. Sci. U.S.A. **39** (1953).

TATE, J.: (1) Algebraic cycles and poles of zeta functions in Arithmetical algebraic geometry. New York-London: Harper and Row, 1963; (2) On the conjectures of Birch and Swinnerton-Dyer and a geometric analog. Sem. Bourbaki, exp. 306, Secr. Math., Paris, 1965—66; (3) Endomorphisms of abelian varieties over finite fields. Invent. Math. **2** (1966); (4) Residues of differentials on curves. Ann. Sci. École Norm. Sup. **1** (1968).

TODD, J. A.: The geometrical invariants of algebraic loci. Proc. London Math. Soc. **43** (1937) and **45** (1939).

TOGLIATTI, E.: Sulle superficie algebriche col massimo numero di punti doppi. Rend. Sem. Mat. Torino **9** (1950).

VACCARO, G.: Le superficie razionali prive di curve eccezionali di prima specie. Atti Accad. Naz. Lincei. Rend. **4** (1948).

DU VAL, P.: (1) On isolated singularities of surfaces which do not affect the conditions of adjunction, I, II, III. Proc. Cambridge Philos. Soc. **30** (1934); (2) The fixed part of the canonical system on an algebraic surface. Proc. Cambridge Philos. Soc. **34** (1938); (3) On absolute and non absolute singularities of algebraic surfaces. Rev. Fac. Sci. Univ. Istanbul **11** (1944), (1946); (4) On regular surfaces of genus three. Canad. J. Math. **3** (1951); (5) On surfaces whose canonical system is hyperelliptic. Canad. J. Math. **4** (1952); (6) Regular surfaces of genus two. I. Canad. J. Math. **5** (1953).

VAN DER WAERDEN, B. L.: (1) Über einfache Punkte von algebraischen Mannigfaltigkeiten. Math. Z. **51** (1948); (2) Birationale Transformation von linearen Scharen auf algebraischen Mannigfaltigkeiten. Math. Z. **51** (1948); (3) Infinitely near points. Nederl. Akad. Wetensch., Proc. Indagat. Math. **12** (1950).

VESENTINI, E.: Sui comportamente effettive delle curve polari nei punti multipli. Ann. Mat. Pura Appl. **34** (1953).

WALKER, ROBERT J.: Reduction of the singularities of an algebraic surface. Ann. of Math. **36** (1935).

WALLACE, A.: Homology theory on algebraic varieties. Oxford-London-New York-Braunschweig: Pergamon 1958.

WASHNITZER, G.: (1) The characteristic classes of an algebraic fiber bundle, I. Proc. Nat. Acad. Sci. U.S.A. **42** (1956); (2) Geometric syzygies. Amer. J. Math. **81** (1959).

WEIL, A.: (1) Foundations of Algebraic Geometry. AMS Coll. Publ. **29** (1946); (2) Sur la théorie des formes differentielles attachées á une varéité analytique complexe. Comment. Math. Helv. **20** (1947); (3) Courbes algébriques et les variétés qui s'en deduisent. Paris: Hermann 1948; (4) Variétés abeliennes et courbes algébriques. Paris: Hermann 1948; (5) Sur les critères d'équivalence en géometrie algébrique. Math. Ann. **128** (1954); (6) Varietes Kahleriennes. Paris: Hermann 1958.

ZAPPA, G.: (1) Sulla degenerazione delle superficie algebriche in sistemi di piani distinti, con applicazioni allo studio de llerigate. Atti Accad. Naz. Mem. **13** (1942); (2) Sull'esistenza, sopra le superficie algebriche, di sistemi continui completi infiniti, la cui curve generica e a serie caratteristica incompleta. Pont. Acad. Sci. Acta **9** (1945); (3) Invarianti numerici d'una superficie algebrica e deduzione della formula di Picard-Alexander col metodo dello spezzamento in piani. Univ. Roma Ist. Naz. Alta Mat. Rend. Mat. e Appl. **5** (1946); (4) Sui sistemi continui di curve sopra una rigata algebrica. Giorn. Mat. Battaglini **77** (1947); (5) Alla ricerca di nuovi significati topologici dei generi geometrico e aritmetico di una superficie algebrica. Ann. Mat. Pura Appl. **30** (1949); (6) Sopra una probabile diseguaglianza tra i caratteri invariantivi di una superficie algebrica. Rend. Mat. e Appl. **14** (1955).

ZARISKI, O.: (1) Some results in the arithmetic theory of algebraic varieties. Amer. J. Math. **60** (1938); (2) Polynomial ideals defined by infinitely near points. Amer. J. Math. **60** (1938); (3) The reduction of the singularities of an algebraic surface. Ann. of Math. **40** (1939); (4) Local uniformization on algebraic varieties. Ann. of Math. **41** (1940); (5) Pencils on an algebraic variety and a new proof of a theorem of Bertini. Trans. Amer. Math. Soc. **50** (1941); (6) Simplified proof for the resolution of singularities of a surface. Ann. of Math. **43** (1942); (7) Foundations of a general theory of birational correspondences. Trans. Amer. Math. Soc. **53** (1943); (8) Reduction of the singularities of an algebraic 3-dimensional variety. Ann. of Math. **45** (1944); (9) The theorem of Bertini on the variable singular points of a linear system of varieties. Trans. Amer. Math. Soc. **56** (1944); (10) The concept of a simple point of an abstract algebraic variety. Trans. Amer. Math. Soc. **62** (1947); (11) Analytical irreducibility of normal varieties. Ann. of Math. **49** (1948); (12) A simple analytical proof of a fundamental property of birational transformations. Proc. Nat. Acad. Sci. U.S.A. **35** (1949); (13) A fundamental lemma from the theory of holomorphic functions on an algebraic variety. Ann. Mat. Pura Appl. **29** (1949); (14) Quelques questions concernant la theorie des fonctions holomorphes sur une variete algebrique. Collect. Int. du C.N.R.S. No. 24, 1950; (15) Sur la normalite analytique des varietes normales. Ann. Inst. Fourier Grenoble **2** (1950); (16) The fundamental ideas of abstract algebraic geometry. Proc. Int. Congr., Cambridge, 1950; (17) Theory and applications of holomorplic functions on algebraic varieties over arbitrary ground fields. Mem. Amer. Math. Soc. **5** (1951); (18) Complete linear systems on normal varieties and a generalization of a lemma of Enriques-Severi. Ann. of Math. **55** (1952); (19) Le probleme de la reduction des singularites d'une variéte algebrique. Bull. Sci. Math. **73** (1954); (20) Applicazioni geometriche della teoria delle valuatazioni. Univ. Roma Ist. Naz. Alta Mat. Rend. Mat. e Appl. **13** (1954); (21) Interpretations algebrico-geometriques du quatorzieme probleme de Hilbert. Bull. Sci. Math. **78** (1954); (22) Scientific report of the 2nd Summer Inst., Part III: Algebraic sheaf theory. Bull. Amer. Math. Soc. **62** (1956); (23) On the purity of the branch locus of algebraic functions. Proc. Nat. Acad. Sci. U.S.A. **44** (1958); (24) Introduction to the problem of minimal models. Publ. Math. Soc. Japan **4** (1958); (25) The problem of minimal models in the theory of algebraic surfaces. Amer. J. Math. **80** (1958); (26) On Castelnuovo's criterion of rationality $p_a = P_2 = 0$. Illinois J. Math. **2** (1958); (27) Proof that any birational class on non-singular surfaces satisfies the descending chain condition. Mem. Coll. Sci. Univ. Kyoto. **32** (1959); (28) La risoluzione delle singolarita delle superficie immerse, I and II. Rend. Accad. Lincei **31** (1961); (29) The theorem of Riemann-Roch for high multiples of an effective divisor on an algebraic surface. Ann. of Math. **76** (1962); (30) Characterization of plane algebroid curves whose module of differentials has maximum torsion. Proc. Nat. Acad Sci. U.S.A. **56** (1966); (31) The connected ness theorem for birational transformations. A Symposium in honor of S. Lefschetz. Princeton, N. J.: Princeton University Press, 1955; (32) An Introduction to the Theory of Algebraic Surfaces. Springer Lecture Notes **83** (1969).

ZARISKI, O., and S. F. BARBER: Reducible exceptional curves of the first kind. Amer. J. Math. **57** (1935).

ZARISKI, O., and H. T. MUHLY: Hilbert's characteristic function and the arithmetic genus of an algebraic variety. Trans. Amer. Math. Soc. **69** (1950).

ZARISKI, O., and P. SAMUEL: Commutative Algebra. New York-London-Melbourne-New Delhi-Toronto: Van Nostrand 1960.

ZARISKI, O., and O. F. G. SCHILLING: On the linearity of pencils of curves on an algebraic surface. Amer. J. Math. **60** (1938).

Index

Springer-Verlag
and the Environment

We at Springer-Verlag firmly believe that an international science publisher has a special obligation to the environment, and our corporate policies consistently reflect this conviction.

We also expect our business partners – paper mills, printers, packaging manufacturers, etc. – to commit themselves to using environmentally friendly materials and production processes.

The paper in this book is made from low- or no-chlorine pulp and is acid free, in conformance with international standards for paper permanency.

M. Aigner Combinatorial Theory ISBN 978-3-540-61787-7
A. L. Besse Einstein Manifolds ISBN 978-3-540-74120-6
N. P. Bhatia, G. P. Szegő Stability Theory of Dynamical Systems ISBN 978-3-540-42748-3
J. W. S. Cassels An Introduction to the Geometry of Numbers ISBN 978-3-540-61788-4
R. Courant, F. John Introduction to Calculus and Analysis I ISBN 978-3-540-65058-4
R. Courant, F. John Introduction to Calculus and Analysis II/1 ISBN 978-3-540-66569-4
R. Courant, F. John Introduction to Calculus and Analysis II/2 ISBN 978-3-540-66570-0
P. Dembowski Finite Geometries ISBN 978-3-540-61786-0
A. Dold Lectures on Algebraic Topology ISBN 978-3-540-58660-9
J. L. Doob Classical Potential Theory and Its Probabilistic Counterpart ISBN 978-3-540-41206-9
R. S. Ellis Entropy, Large Deviations, and Statistical Mechanics ISBN 978-3-540-29059-9
H. Federer Geometric Measure Theory ISBN 978-3-540-60656-7
S. Flügge Practical Quantum Mechanics ISBN 978-3-540-65035-5
L. D. Faddeev, L. A. Takhtajan Hamiltonian Methods in the Theory of Solitons
                                   ISBN 978-3-540-69843-2
I. I. Gikhman, A. V. Skorokhod The Theory of Stochastic Processes I ISBN 978-3-540-20284-4
I. I. Gikhman, A. V. Skorokhod The Theory of Stochastic Processes II ISBN 978-3-540-20285-1
I. I. Gikhman, A. V. Skorokhod The Theory of Stochastic Processes III ISBN 978-3-540-49940-4
D. Gilbarg, N. S. Trudinger Elliptic Partial Differential Equations of Second Order
                                ISBN 978-3-540-41160-4
H. Grauert, R. Remmert Theory of Stein Spaces ISBN 978-3-540-00373-1
H. Hasse Number Theory ISBN 978-3-540-42749-0
F. Hirzebruch Topological Methods in Algebraic Geometry ISBN 978-3-540-58663-0
L. Hörmander The Analysis of Linear Partial Differential Operators I – Distribution Theory
                  and Fourier Analysis ISBN 978-3-540-00662-6
L. Hörmander The Analysis of Linear Partial Differential Operators II – Differential
                  Operators with Constant Coefficients ISBN 978-3-540-22516-4
L. Hörmander The Analysis of Linear Partial Differential Operators III – Pseudo-
                  Differential Operators ISBN 978-3-540-49937-4
L. Hörmander The Analysis of Linear Partial Differential Operators IV – Fourier
                  Integral Operators ISBN 978-3-642-00117-8
K. Itô, H. P. McKean, Jr. Diffusion Processes and Their Sample Paths ISBN 978-3-540-60629-1
T. Kato Perturbation Theory for Linear Operators ISBN 978-3-540-58661-6
S. Kobayashi Transformation Groups in Differential Geometry ISBN 978-3-540-58659-3
K. Kodaira Complex Manifolds and Deformation of Complex Structures ISBN 978-3-540-22614-7
Th. M. Liggett Interacting Particle Systems ISBN 978-3-540-22617-8
J. Lindenstrauss, L. Tzafriri Classical Banach Spaces I and II ISBN 978-3-540-60628-4
R. C. Lyndon, P. E Schupp Combinatorial Group Theory ISBN 978-3-540-41158-1
S. Mac Lane Homology ISBN 978-3-540-58662-3
C. B. Morrey Jr. Multiple Integrals in the Calculus of Variations ISBN 978-3-540-69915-6
D. Mumford Algebraic Geometry I – Complex Projective Varieties ISBN 978-3-540-58657-9
O. T. O'Meara Introduction to Quadratic Forms ISBN 978-3-540-66564-9
G. Pólya, G. Szegő Problems and Theorems in Analysis I – Series. Integral Calculus.
                        Theory of Functions ISBN 978-3-540-63640-3
G. Pólya, G. Szegő Problems and Theorems in Analysis II – Theory of Functions. Zeros.
                        Polynomials. Determinants. Number Theory. Geometry
                        ISBN 978-3-540-63686-1
W. Rudin Function Theory in the Unit Ball of $\mathbb{C}^n$ ISBN 978-3-540-68272-1
S. Sakai C*-Algebras and W*-Algebras ISBN 978-3-540-63633-5
C. L. Siegel, J. K. Moser Lectures on Celestial Mechanics ISBN 978-3-540-58656-2
T. A. Springer Jordan Algebras and Algebraic Groups ISBN 978-3-540-63632-8
D. W. Stroock, S. R. S. Varadhan Multidimensional Diffusion Processes ISBN 978-3-540-28998-2
R. R. Switzer Algebraic Topology: Homology and Homotopy ISBN 978-3-540-42750-6
A. Weil Basic Number Theory ISBN 978-3-540-58655-5
A. Weil Elliptic Functions According to Eisenstein and Kronecker ISBN 978-3-540-65036-2
K. Yosida Functional Analysis ISBN 978-3-540-58654-8
O. Zariski Algebraic Surfaces ISBN 978-3-540-58658-6